Banach, Fréchet, Hilbert and Neumann Spaces

"Analysis is sometimes a means
of becoming disgusted piecemeal
with something that
as a whole was bearable."

PAUL VALERY
Bad Thoughts and Others

"For the Nation, Science and Glory."

NAPOLEON BONAPARTE
(Motto for the *Ecole Polytechnique)*

Analysis for PDEs Set

coordinated by
Jacques Blum

Volume 1

Banach, Fréchet, Hilbert and Neumann Spaces

Jacques Simon

iSTE

WILEY

First published 2017 in Great Britain and the United States by ISTE Ltd and John Wiley & Sons, Inc.

ISTE Ltd
27-37 St George's Road
London SW19 4EU
UK

www.iste.co.uk

John Wiley & Sons, Inc.
111 River Street
Hoboken, NJ 07030
USA

www.wiley.com

Library of Congress Control Number: 2017932960

British Library Cataloguing-in-Publication Data
A CIP record for this book is available from the British Library
ISBN 978-1-78630-009-6

Contents

Introduction

Purpose. This book is the first of a set of books dedicated to the mathematical tools for partial derivative equations existing in physics.

This first volume is dedicated to normed or semi-normed vector spaces, including Banach, Fréchet and Hilbert spaces, with new developments concerning *Neumann spaces* — which designate those in which every Cauchy sequence converges — and *extractable spaces* — those in which every bounded sequence has a weakly convergent subsequence.

The book presents the main properties of these spaces employed for the construction of distributions, Lebesgue and Sobolev spaces and traces, with real or Neumann space values, that are the subject of Volumes II [102], III [103] and IV [104], respectively, and in the resolution of partial differential equations, in Volume V [105].

To this end, differential calculus is extended to semi-normed spaces in the present Volume I.

Target audience. We have looked for simple methods requiring a minimal level of knowledge in order to make these tools accessible to the largest possible audience — doctoral students, PhD students, engineers — without restricting their generality and even by generalizing some results, so that this book is also intended for researchers. This has led us to a non-conventional approach that emphasizes semi-norms and sequential properties, whether regarding completeness, compactness or continuity.

Why semi-normed spaces? We do not restrict ourselves to normed spaces because essential spaces for PDEs are not normable, such as $L^p_{\text{loc}}(\Omega)$, $\mathcal{D}'(\Omega)$ or $H^m(\Omega)$-weak.

We are interested in vector spaces E endowed with a family of semi-norms, rather than with a locally convex topology which is equivalent, in order to *define the differentiability* (p. 275) therein by comparing the semi-norms of a variation in the data to the semi-norms of the variation in the mapping. This also allows us to *define* $L^p(\Omega; E)$ (Volume III), because a semi-norm can be raised to a power p, not a neighborhood!

In addition, the study is easier with semi-norms than with the topology, although less usual: it follows that of normed spaces, the main difference consisting of working on several semi-norms or norms instead of a single norm.

Neumann spaces. We particularly focus on the spaces that are sequentially complete — we call them *Neumann* spaces — because it is essential that E has this property to define $\mathcal{D}'(\Omega; E)$ and $L^p(\Omega; E)$: it is the one that guarantees them satisfactory properties and especially that $\int_\Omega f \in E$ when $f \in \mathcal{C}(\Omega; E)$ and that $\mathcal{C}(\Omega; E) \subset \mathcal{D}'(\Omega; E)$, as will be seen in Volume II. The study of these spaces is a step in that of spaces useful for evolution PDEs such as $L^2((0,T); L^2_{\text{loc}}(\Omega))$, $L^p((0,T); E\text{-weak})$ or $\mathcal{D}'((0,T); \mathcal{D}'(\Omega))$.

Despite the name *sequentially complete* being less simple, the concept itself is simpler than *complete* (see Definition 4.8, p. 55) and, above all, more general: for example, if E is a Hilbert space with infinite dimension, E-weak is sequentially complete but is not complete (see (4.11), p. 63).

Extractable and reflexive spaces. We exhaustively study spaces in which any bounded sequence has a weakly convergent subsequence — we call them *extractable* — because numerous results of existence of solutions for PDEs are based on this property.

We also study reflexivity in detail because it provides extractability results. We balance and simplify its definition by introducing a new notion, *pre-reflexivity* (Definition 17.1, p. 243).

Other sequential properties. We also insist on:
— *Sequential compactness* because it is the basis for extractability: it ensures that every sequence has a convergent subsequence, which compactness does not ensure (see (2.6), p. 27).
— *Sequential density* that provides better approximations than density.
— *Sequential continuity* because important mappings are sequentially continuous but not continuous, such as the duality bilinear form of a non-normable space (Theorem 13.22, p. 204) or the composition of linear mappings (Theorem 12.12, p. 171).

Moreover, we do not overlook topological properties because they are distinct from their sequential counterparts in non-metrizable spaces.

Weak topologies. The topology of E-weak is very simply defined by the semi-norms $\|e\|_{E\text{-weak};e'} = |\langle e', e\rangle|$ indexed by $e' \in E'$. Similarly, the topology of E'-$*$weak is defined by $\|e'\|_{E'\text{-}*\text{weak};e} = |\langle e', e\rangle|$ indexed by $e \in E$.

Differential calculus. We study differentiability in any separated semi-normed space E because it is an important tool for the study of distributions with values in such a space. It seemed that it was only achieved in a normed space or in particular cases such as *scalar differentiation* that corresponds to differentiation in E-weak.

Numerous conventional properties remain but not all. For example, we are not able to prove the second- or higher-order differentiability of a composite mapping when the intermediary space is not normed (see Theorem 21.6, p. 308) or the continuity of the differential of a composite of continuously differentiable mappings (see Theorem 19.18, p. 287).

Novelties. The extension of differential calculus to separated semi-normed spaces (Chapters 19 to 21) appears as new to us. The same happens with the characterizations of the dual and of the weak topology of an intersection of semi-normed spaces (Theorems 14.9, p. 215 and 15.11, p. 227), and with the extractability property of such an intersection (Theorem 18.14, p. 271). We can also include the introduction of Neumann spaces (Definition 4.10, p. 55), extractable spaces (Definition 18.1, p. 265), pre-reflexivity (Definition 17.1, p. 243) and compactant mappings (Definition 8.14, p. 125).

The pre-reflexivity of metrizable spaces (Theorem 17.8, p. 249) and, more generally, of infra-barreled spaces (Theorem 17.11, p. 250) is new in this form (which is equivalent to previous results, see Note 4, p. 248).

Prerequisites. The proofs in the body of text are not based on any external result, with the exception of countability and $\mathbb{R}$ properties whose statements are recalled in Chapter 1. Indeed, it seemed interesting to recall all of the necessary knowledge given the unorthodox course of this book: the study of semi-normed spaces without resorting to the general topology, distributions construction (in Volume II) before that of integrable functions (in Volume III), etc.

Comments. Unlike the body of the text, the comments written in small characters may make use of external or not yet established results. The first chapter, *1 Prerequisites*, also appears in small print because it could also be admitted or omitted, as well as the last section, *22.4 Derivatives of the power, logarithm and exponential functions*.

Reminders. This book is written such that it can be read in an out-of-order fashion by a non-expert: the proofs are detailed including arguments that may be trivial for an expert and the numbers of the theorems being used are systematically recalled. The author asks for the reader's indulgence regarding the heaviness that may result thereof.

History. The origin of the concepts and of the results is specified as much as possible, in footnotes, hoping that the unavoidable injustices will be forgiven and, above all, reported to the author (for re-editions!). In addition to the opportunity offered to state what is believed to be new, history shows that behind every theorem there is a man or woman, sometimes contemporary, sometimes ancient and wise, that our distant ancestors — Greeks included — reasoned as well as us and that simple ideas yet remain(ed) to be found.

Navigation through this book:

— The **table of contents**, at the beginning of the book, gives the list of the topics addressed.
— The **index**, p. 335, provides another thematic access.
— The **notation table**, p. xvii, clarifies their meaning when doubt remains.
— Assumptions are all stated inside the theorems themselves.

Acknowledgments. Enrique FERNÁNDEZ-CARA suggested to me a large number of pertinent improvements. Olivier BESSON and Fulbert MIGNOT have also contributed to substantial improvements. The manuscript and the proofs have been carefully re-read by Jérôme LEMOINE and Pierre DREYFUSS; they are thus entirely responsible for any error that may subsist. Jacques BLUM has the great merit of having snatched the manuscript from its torpor.

Donald KNUTH has graciously provided the scientific community with his remarkable TEX software, utilized to compose this book.

I am extremely pleased to thank them.

Jacques SIMON
Chapdes-Beaufort
March 2017

Familiarization with Semi-normed Spaces

A **semi-normed space** is a vector space E endowed with a family $\{\| \ \|_{E;\nu} : \nu \in \mathcal{N}_E\}$ of semi-norms:
— The set $\mathcal{N}_E$ indexing the semi-norms is, *a priori*, any.
— A **normed** space is a particular case where this family is reduced to a norm.
— Any locally convex topological vector space can be endowed with a family of semi-norms generating its topology (Neumann's theorem).
— We only consider **separated** spaces, namely in which, if $\|u\|_{E;\nu} = 0$ for all $\nu \in \mathcal{N}_E$, then $u = 0_E$.

Utilization:
— $u_n \to u$ in E means that $\|u_n - u\|_{E;\nu} \to 0$ for all $\nu \in \mathcal{N}_E$.
— U is **bounded** in E means that $\sup_{u \in U} \|u\|_{E;\nu} < \infty$ for all $\nu \in \mathcal{N}_E$.
— T is **continuous** from E into F at u means that, for any $\mu \in \mathcal{N}_F$ and $\epsilon > 0$, there is a finite set $N \subset \mathcal{N}_E$ and $\eta > 0$ such that $\sup_{\nu \in N} \|v - u\|_{E;\nu} \leq \eta$ yields $\|T(v) - T(u)\|_{F;\mu} \leq \epsilon$.

Examples — real-valued functional spaces:
— $L^p(\Omega)$ is endowed with the norm $\|f\|_{L^p(\Omega)} = (\int_\Omega |f|^p)^{1/p}$.
— $L^p_{\text{loc}}(\Omega)$ is endowed with the semi-norms $\|f\|_{L^p_{\text{loc}}(\Omega);\omega} = (\int_\omega |f|^p)^{1/p}$ indexed by the bounded open sets ω such that $\overline{\omega} \subset \Omega$.
— $\mathcal{D}(\Omega)$ is endowed with the semi-norms $\|\varphi\|_{\mathcal{D}(\Omega);q} = \sup_{x \in \Omega} \sup_{0 \leq |\beta| \leq q(x)} q(x) |\partial^\beta \varphi(x)|$ indexed by the functions $q \in \mathcal{C}^+(\Omega)$ (they generate the topology of inductive limit of the $\mathcal{D}_K(\Omega)$).
— $\mathcal{D}'(\Omega)$ is endowed (here) with the semi-norms $\|f\|_{\mathcal{D}'(\Omega);\varphi} = |\langle f, \varphi \rangle|$ indexed by $\varphi \in \mathcal{D}(\Omega)$ (they generate the weak topology).

Examples — vector-valued functional spaces:
— $L^p(\Omega; E)$ is endowed with the semi-norms $\|f\|_{L^p(\Omega;E);\nu} = (\int_\Omega \|f\|^p_{E;\nu})^{1/p}$ indexed by $\nu \in \mathcal{N}_E$.
— $\mathcal{D}'(\Omega; E)$ is endowed with the semi-norms $\|f\|_{\mathcal{D}'(\Omega;E);\varphi,\nu} = \|\langle f, \varphi \rangle\|_{E;\nu}$ indexed by $\varphi \in \mathcal{D}(\Omega)$ and $\nu \in \mathcal{N}_E$ (they generate the topology of pointwise convergence on $\mathcal{D}(\Omega)$).

Examples — weak spaces, dual spaces:
— E-weak is endowed with the semi-norms $\|e\|_{E\text{-weak};e'} = |\langle e', e \rangle|$ indexed by $e' \in E'$.
— E' is endowed with the semi-norms $\|e'\|_{E';B} = \sup_{e \in B} |\langle e', e \rangle|$ indexed by the bounded sets B of E.
— E'-weak is endowed with the semi-norms $\|e'\|_{E'\text{-weak};e''} = |\langle e'', e' \rangle|$ indexed by $e'' \in E''$.
— E'-$*$weak is endowed with the semi-norms $\|e'\|_{E'\text{-}*\text{weak};e} = |\langle e', e \rangle|$ indexed by $e \in E$.

Neumann spaces and others:

— A **Neumann** space is a separated sequentially complete semi-normed space.

— A **Fréchet** space is a sequentially complete metrizable semi-normed space.

— A **Banach** space is a sequentially complete normed space.

— A separated semi-normed space is **extractable** if every bounded sequence has a weakly convergent subsequence.

Advantages of using semi-norms rather than topology:

— The semi-norms allow the definition of the differentiability (by comparing the semi-norms of a variation in the data to the semi-norms of the variation in the mapping).

— They allow the definition of $L^p(\Omega; E)$ (by raising the semi-norms to the power p).

— Their use, which is easy, follows that of normed spaces, the main difference consisting of working on several semi-norms or norms instead of a single norm.

— Some definitions are simpler, for example that of a bounded set U: "$\sup_{v \in U} \|v\|_{E;\nu} < \infty$ for any semi-norm $\| \ \|_{E;\nu}$ of E" would be expressed in terms of topology, in the more abstract form "for any open set V containing 0_E, there is $t > 0$ such that $tU \subset V$".

Notations

SETS

$\mathbb{N}$	set of natural numbers: $\mathbb{N} = \{0, 1, 2, \ldots\}$	3
$\mathbb{N}^*$	set of non-zero natural numbers: $\mathbb{N}^* = \{1, 2, \ldots\}$	3
$\mathbb{Z}$	set of integers: $\mathbb{Z} = \{\ldots, -2, -1, 0, 1, 2, \ldots\}$	3
$\mathbb{Q}$	space of rational numbers	4
$\mathbb{R}$	space of real numbers	4
$\emptyset$	empty set	
$\subset$	algebraic inclusion	1
$\setminus$	set difference: $U \setminus V = \{u \in U : u \notin V\}$	1
$\times$	product: $U \times V = \{(u, v) : u \in U,\ v \in V\}$	2

SEPARATED SEMI-NORMED SPACES

E	separated semi-normed space	10
$\\| \ \\|_{E;\nu}$	semi-norm of E of index ν	10
$\mathcal{N}_E$	set indexing the semi-norms of E	10
$B_{E;\nu}(u, r)$	closed semi-ball: $B_{E;\nu}(u, r) = \{v \in E : \\|v - u\\|_{E;\nu} \leq r\}$	22
$\underset{\leftrightarrows}{\equiv}$	equality of the families of semi-norms	35
$\underset{\leftrightarrows}{=}$	topological equality	35
$\underset{\rightarrow}{\subset}$	topological inclusion	35
E-weak	space E endowed with pointwise convergence on E'	221
E-strong	space E endowed with uniform convergence on the bounded subsets of E'	251

SETS OF A SEMI-NORMED SPACE

$\mathring{U}$	interior of a set U	21
$\overline{U}$	closure of U	21
$\overline{U}^{\mathrm{seq}}$	sequential closure of U	21
$U^{\lim \mathrm{seq}}$	set of the limits of the sequences of U	21
∂U	boundary of U	21
$U + V$	set of sums: $U + V = \{u + v : u \in U,\ v \in V\}$	46
$U - V$	set of differences: $U - V = \{u - v : u \in U,\ v \in V\}$	46

DUAL OF A SEMI-NORMED SPACE

COMPOSITE OF SEMI-NORMED SPACES

SPACES OF MAPPINGS

INTERVALS AND SEGMENTS

POINTS AND SUBSETS OF $\mathbb{R}^d$

MAPPING DIFFERENTIAL AND FUNCTION DERIVATIVE

PARTICULAR NUMBERS AND FUNCTIONS

TYPOGRAPHY

▮	end-of-statement
□	end of proof or remark

Chapter 1

Prerequisites

The purpose of this preliminary chapter is to clarify terminology and notation and to recall elementary properties of countability and of real numbers. The heart of the matter begins in Chapter 2, p. 11.

1.1. Sets, mappings, orders

Set theory notations. We utilize the usual notations of set theory[1], namely $=$ for "*is equal to*", $\in$ for "*is an element of*", $\forall$ for "*for all*", $\exists$ for "*there exists*", $\Rightarrow$ for "*implies*", $\Leftarrow$ for "*results from*", $\Leftrightarrow$ for "*is equivalent to*". A barred symbol designates the negation of the property, for example $\neq$ means "*is not equal to*".

We denote $\{a, b, \ldots, z\}$ the set consisting of a, b, ... and z, $\{a : P(a)\}$ or $\{a\}_{P(a)}$ the set of the a having the property $P(a)$, and $\emptyset$ the empty set. We denote $\subset$ set inclusion, $\cup$ their union and $\cap$ their intersection. We denote $\supset$ for " *is a superset of*" and $U \setminus V = \{u \in U : u \notin V\}$.

We denote $(a, b, \ldots, z)$ the ordered set composed of a, b, ... and z. Therefore, $(a, b) \neq (b, a)$ if $a \neq b$. On the contrary, $\{a, b\} = \{b, a\}$ always holds.

---

1 **History of the set theory notation.** The symbol $=$ was introduced by Robert RECORDE in 1557 [89]. The symbol $\in$ was introduced by Bertrand RUSSEL in 1903 [95] (Giuseppe PEANO used ϵ in 1889 [84]). The symbols $\forall$ and $\exists$ were introduced by Giuseppe PEANO in the *Formulario Mathematico* in five volumes published from 1895 to 1905 [85]. The symbols $\Rightarrow$, $\Leftarrow$ and $\Leftrightarrow$ were introduced in 1954 by Nicolas BOURBAKI [16] (William ACKERMANN utilized $\leftrightarrow$ in 1936 [1]). The symbol $\neq$ was introduced (with a vertical bar) by Leonhard EULER.
The symbol $\emptyset$ was introduced in 1939 by Nicolas BOURBAKI [14]. The symbols $\cup$ and $\cap$ were introduced by Hermann GRASSMANN in 1844 [48] as operation symbols; they were used for the union and the intersection by Giuseppe PEANO [82] in 1888. The symbols $\subset$ and $\supset$ were introduced by Ernst SCHRÖDER in 1877.
The symbol $\times$ was introduced in 1631 by William OUGHTRED [81] (for the product of numbers). The use of the symbol $\prod$ for the product is due to Carl Friedrich GAUSS according to Florian CAJORI [22, vol. 2, p. 78] (GULLBERG attributes it to René DESCARTES).
The rather convenient usage of designating the data and the unknowns of an equation by letters was introduced by Francis VIÈTE [113] in 1646.

Bibliography. The list of cited works is in p. 325. The authors are listed in p. 331.

We denote $U_1 \times U_2 \times \cdots \times U_d = \{(u_1, u_2, \ldots, u_d) : u_i \in U_i \text{ for all } i\}$ the *product*[2] of the sets $U_1, \ldots, U_d$ and denote $U^d = \{(u_1, u_2, \ldots, u_d) : u_i \in U \text{ for every } i\}$. In particular, $U^2 = U \times U$.

Given a set I and for each $i \in I$, a set U_i, we denote $\bigcup_{i \in I} U_i$ the union of all the U_i, $\bigcap_{i \in I} U_i$ their intersection and $\prod_{i \in I} U_i = \{(u_i)_{i \in I} : u_i \in U_i, \ \forall i \in I\}$ their product. These quantities are related by **De Morgan's laws**[3]:

$$\bigcup_{i \in I} E \setminus U_i = E \setminus \bigcap_{i \in I} U_i, \tag{1.1}$$

$$\bigcap_{i \in I} E \setminus U_i = E \setminus \bigcup_{i \in I} U_i. \tag{1.2}$$

The axiomatic construction of the set theory is, for instance, achieved in [SCHWARTZ, 100, chap. I] or [BOURBAKI, 16].

Mappings. A **mapping** T from a set X into a set Y is the data, for each $u \in X$, of an element $T(u) \in Y$. It is said that T is defined on X and that $T(u)$ is the **image** by T of the point u. The **image**[4] by T of a subset U of X is the set

$$T(U) \stackrel{\text{def}}{=} \{T(u) : u \in U\}. \tag{1.3}$$

It is said that T is **injective** if $T(u) = T(v)$ yields $v = u$, that it is **surjective** if $T(X) = Y$, and that it is **bijective** if it is injective and surjective. It is also said that T is an **injection**, a **surjection** or a **bijection**.

The **preimage** by T of a subset W of Y is the set

$$T^{-1}(W) \stackrel{\text{def}}{=} \{u \in X : T(u) \in W\}.$$

If T is bijective, there exists a unique mapping T^{-1} from Y into X, known as **inverse** of T, such that, for all $u \in X$,

$$T^{-1}(T(u)) = u.$$

Then, T is the inverse mapping of T^{-1}, that is $T(T^{-1}(u)) = u$.

The **restriction** of T to a subset U of X is the mapping $T|_U$ defined on U by $T|_U(u) = T(u)$ for all $u \in U$. An **extension** of T (to a set $V \supset X$) is any mapping (defined on V) of which T is a restriction.

Ordering. An **order** on a set U is a relation $\preceq$ among some pairs of its elements such that, for all u, v and w in U: $u \preceq u$; $u \preceq v$ and $v \preceq$ imply $u = v$; $u \preceq v$ and $v \preceq w$ imply $u \preceq w$. It is then said that U is **ordered** by $\preceq$. We denote $u \prec v$ if $u \preceq v$ and $u \neq v$; and $u \succeq v$ if $v \preceq u$; and $u \succ v$ if $v \prec u$.

It is said that U is **totally ordered** by $\preceq$ if, in addition, for all u and v in U, there is $u \preceq v$ or $v \preceq u$.

If V is a subset of a set U ordered by $\preceq$, $V \subset U$ and $m \in U$, it is said that m is a **majorant** or an **upper bound** of V if every $v \in V$ verifies $v \preceq m$. It is said that m is the **supremum** or the **least upper bound** of V if m is a majorant of V and if any other majorant m' of V verifies $m \preceq m'$; if it exists, it is unique and denoted by $\sup V$. It is said that m is the **maximum** of V if m is an element of V and is a majorant of V; if it exists, it is unique and denoted by $\max V$.

We respectively call **minorant, lower bound, infimum** or **greatest lower bound**, and **minimum** the elements obtained by replacing $\preceq$ by $\succeq$ in these definitions. If they exist, the infimum is denoted $\inf V$ and the minimum is denoted $\min V$ and they are unique. A maximum is a supremum, and the latter is a majorant; a minimum is an infimum, and the latter is a minorant.

---

2 **History of the notion of set product.** The product (often called *Cartesian*) of any two sets was introduced by Georg CANTOR [24].

3 **History of De Morgan's laws.** The relations (1.1) and (1.2) were demonstrated for two sets by Augustus DE MORGAN in 1858 and Charles PEIRCE in 1867.

4 **History of the concept of image by a mapping.** The introduction of the image of a set by a mapping seems pertaining to Richard DEDEKIND [31] (= [32, t. 3, p. 348]).

1.2. Countability

We denote $\mathbb{N} = \{1, 2, \ldots\}$ the set of "natural" numbers (or positive integers)[5], $\mathbb{N}^* = \{0, 1, 2, \ldots\}$[6] the set of non-negative integers and $\mathbb{Z} = \{\ldots, -2, -1, 0, 1, 2, \ldots\}$ the set of all integers[7].

A **sequence** of a set U is the data, for every $n \in \mathbb{N}$, of an element $u_n \in U$. It is an ordered set that is denoted $(u_n)_{n\in\mathbb{N}}$. This is equivalent to considering that it is a mapping from $\mathbb{N}$ into U, namely $n \mapsto u_n$. A **subsequence** of $(u_n)_{n\in\mathbb{N}}$ is any sequence of the form $(u_{\sigma(n)})_{n\in\mathbb{N}}$, where $\sigma(n+1) > \sigma(n)$, that is where σ is a strictly increasing mapping from $\mathbb{N}$ into itself.

A set U is said **countable** if there is a bijection from U onto a subset of $\mathbb{N}$. It is said that U is **finite** if there is a bijection from U on an **integer interval**, that is a set of the form

$$[\![m, n]\!] \stackrel{\text{def}}{=} \{i \in \mathbb{Z} : m \le i \le n\}$$

with m and n in $\mathbb{Z}$. It is said that U is **infinite** if it is not finite (an infinite countable set is sometimes said to be *denumerable* [LANG, 66, p. 6]).

Let us give countability properties, see, for instance, [SCHWARTZ, 100, p. 104–108] or [SIMON, 107].

THEOREM 1.1.– (a) *Every finite or empty set is countable.*

(b) *Every subset of a countable set is countable.*

(c) *Every image of a countable set by a mapping is countable.*

(d) *Every finite product of countable sets is countable.*

(e) *Every countable union of countable sets is countable.*

(f) *The set of the finite subsets of a countable set is countable.*

(g) *The set of the finite sequences of a countable set is countable.* □

---

5 **History of the notation of natural numbers.** There are very old ways to denote and handle natural numbers that are similar to our decimal system, see [BOURBAKI, 20, p. 64], with the exception of the number "zero" introduced in 628 by the Indian mathematician BRAHMAGUPTA. "Zero" originates from the Sanskrit *sunya* which means *nothing*; in Arabic, it became *sifr*, latinized into *zephirum* by Leonardo FIBONACCI in the 13th Century; it permanently became "zero" in 1491 in a Florentine treaty.
The *Arabic numerals* 0, 1, 2 ... 9 were introduced in the Mediterranean countries by Muhammad Ibn MŪSĀ AL-KHUWĀRIZMĪ around 810.

History of the notation $\mathbb{N}$. The notation N (for *Naturale*, naturals in Italian) was introduced by Giuseppe PEANO in 1895 [85] (to refer to the set of strictly positive integers).

6 **French notation.** The French and Italian denote by $\mathbb{N}$ the set of non-negative integers (that we denote $\mathbb{N}^*$) and they denote by $\mathbb{N}^*$ our set $\mathbb{N}$!

7 **History of the notation $\mathbb{Z}$.** The $\mathbb{Z}$ notation (for *Zahlen*, numbers in German) seems to be attributed to Nicolas BOURBAKI.

1.3. Construction of $\mathbb{R}$

Motivation. The heart of the matter, which begins in Chapter 2, calling on $\mathbb{R}$ (for Definitions 2.1 of vector spaces and 2.2 of norms and semi-norms, etc.) it had seemed to us useful to recall a self-contained construction of the real numbers set and to establish its properties that we will use. Thus, here are the outlines of its construction by cuts (the construction by completion, frequently used, is based on more developed notions — Cauchy sequences, equivalent sequences — and is longer). For a detailed presentation, the reader is, for instance, referred to [ENDERTON, 39, p. 112–120] or [SIMON, 107].

The set $\mathbb{Q}$ of rational numbers. A **rational number**[8] is any subset of $\mathbb{Z} \times \mathbb{N}$ of the form

$$\frac{m}{n} \stackrel{\text{def}}{=} \{(m', n') \in \mathbb{Z} \times \mathbb{N} : mn' = m'n\} \tag{1.4}$$

where $m \in \mathbb{Z}$ and $n \in \mathbb{N}$. The set of all rational numbers is denoted by $\mathbb{Q}$. Its addition, its multiplication and its order are defined by

$$\frac{m}{n} + \frac{k}{\ell} \stackrel{\text{def}}{=} \frac{m\ell + nk}{n\ell}, \quad \frac{m}{n}\frac{k}{\ell} \stackrel{\text{def}}{=} \frac{mk}{n\ell}, \quad \left[\frac{m}{n} \leq \frac{k}{\ell} \stackrel{\text{def}}{\Leftrightarrow} m\ell \leq nk\right].$$

The set $\mathbb{R}$ of real numbers. A **real number**[9] is any **semi-cut** of $\mathbb{Q}$, that is any set x such that:

$$x \subset \mathbb{Q};\ x \neq \emptyset;\ x \neq \mathbb{Q};\ x \text{ has no maximum; } \alpha \in x \text{ and } \alpha' \in \mathbb{Q} \setminus x \text{ imply } \alpha < \alpha'. \tag{1.5}$$

The set of real numbers is denoted by $\mathbb{R}$.

Given $\alpha \in \mathbb{Q}$, a real number is defined by $\dot{\alpha} = \{\gamma \in \mathbb{Q} : \gamma < \alpha\}$. We identify $\alpha \in \mathbb{Q}$ to $\dot{\alpha}$ which gives $\mathbb{Q} \subset \mathbb{R}$. This is permissible because $\dot{\alpha} = \dot{\beta}$ gives $\alpha = \beta$. By identifying in addition any integer m to the rational number $m/1$, we get

$$\mathbb{N} \subset \mathbb{Z} \subset \mathbb{Q} \subset \mathbb{R}. \tag{1.6}$$

8 **History of the construction of $\mathbb{Q}$.** The "representation" of $\mathbb{Q}$ by classes of pairs of integers is attributed to Karl WEIERSTRASS in his unpublished lectures [BOURBAKI, 20, p. 37]. Previously, rational numbers were associated with the idea of breaking down a "quantity" into equal parts or with the ratio between two "quantities of the same nature".

History of the notation $\mathbb{Q}$. The $\mathbb{Q}$ notation (for *Quotients*) was introduced by Nicolas BOURBAKI in 1969 [18, chap. 1].

9 **History of the construction of $\mathbb{R}$.** Richard DEDEKIND [32, t. II, p. 315–334] defined $\mathbb{R}$ as the set of cuts of $\mathbb{Q}$ (a **cut** is the pair composed of a semi-cut and of the biggest open rational set included in its complementary).

Construction of $\mathbb{R}$ by completion. Almost simultaneously, Karl WEIERSTRASS [unpublished classes], Charles MÉRAY and Georg CANTOR [24, p. 92–97] defined, independently one from another, $\mathbb{R}$ as a *completion* of $\mathbb{Q}$ by various methods close to the HAUSDORFF's completion method of a metric space.

Eudoxus ratios. Previously, real numbers were related to the ratio between two "geometrical quantities" of the same nature (length, area, etc.), the ratio possibly being irrational as Pythagoreans had proven. DEDEKIND cuts are close to the definition of EUDOXUS of the inequality $a/a' < b/b'$ between two ratios (real, thus) by means of the implication $na > n'a' \Rightarrow nb > n'b'$ for arbitrary integers n and n' [BOURBAKI, 20, p. 187 and 194].

Operations on $\mathbb{R}$. A total **order** on $\mathbb{R}$ is defined by[10]

$$x \leq y \quad \stackrel{\text{def}}{\Leftrightarrow} \quad x \subset y.$$

We denote $x < y$ if $x \leq y$ and $x \neq y$. We denote $x \geq y$ if $y \leq x$ and $x > y$ if $y < x$. The **absolute value** of $x \in \mathbb{R}$ is defined by $|x| = x$ if $x \geq 0$ and $|x| = -x$ otherwise.

The **addition** is defined from $\mathbb{R} \times \mathbb{R}$ into $\mathbb{R}$ by

$$x + y \stackrel{\text{def}}{=} \{\alpha + \beta : \alpha \in x,\ \beta \in y\}.$$

Given $x \in \mathbb{R}$, there is a unique real number $-x$ such that $-x + x = 0$. We denote $x - y = x + (-y)$.

The **multiplication** is defined from $\mathbb{R} \times \mathbb{R}$ into $\mathbb{R}$, if x and y are positive, by

$$xy \stackrel{\text{def}}{=} \{\alpha\beta : \alpha \in x,\ \beta \in y,\ \alpha \geq 0,\ \beta \geq 0\} \cup \{\gamma \in \mathbb{Q} : \gamma < 0\}$$

and else by reverting to this case. The **quotient** x/y of $x \in \mathbb{R}$ by $y \in \mathbb{R}$, $y \neq 0$, is defined by

$$x/y \stackrel{\text{def}}{=} \{\alpha/\beta : \alpha \in x,\ \alpha \geq 0,\ \beta \in \mathbb{Q} \setminus y\} \cup \{\gamma \in \mathbb{Q} : \gamma < 0\}$$

if x and y are positive and else by reverting to this case.

These definitions easily yield the usual properties of the operations on $\mathbb{R}$. The operations on $\mathbb{Q}$ are recovered, and thus it is allowed to denote them similarly and the identification (1.6) is justified.

1.4. Properties of $\mathbb{R}$

Bounds in $\mathbb{R}$. Let us show that any non-empty majored subset of $\mathbb{R}$ has a least upper bound[11].

THEOREM 1.2.– (a) *Any non-empty majored subset of $\mathbb{R}$ has a least upper bound.*

(b) *Any non-empty minored subset of $\mathbb{R}$ has a greatest lower bound.*

(c) *Any non-empty finite subset of $\mathbb{R}$ has a minimum and a maximum.* □

Proof. If V is a non-empty majored subset of $\mathbb{R}$, $x = \bigcup_{v \in V} v$ is a semi-cut, that is a real number, which is the least upper bound of V. If V is a non-empty minored subset of $\mathbb{R}$, the set $-V = \{-v : v \in V\}$ is majored and $-\sup -V$ is the greatest lower bound of V.

Should a non-empty finite set have no minimum (resp. maximum), each of its n elements would be strictly minored (resp. majored) by another of its elements that would be impossible at most the n-th time since the order of $\mathbb{R}$ is total. □

---

10 **History of notations.** The symbols $<$ and $>$ appeared in a treatise by Thomas HARRIOT [52], published posthumously in 1631. The symbols $\leq$ and $\geq$ were used by John WALLIS in 1670.
The notation $|x|$ of the absolute value was introduced by Karl WEIERSTRASS.
The symbols $+$ and $-$ appeared in an arithmetic of Johannes WIDMANN in 1489 [117] (they were divulgated by Michael STIFEL in 1544 [109]). The use of the symbol $\sum$ for summation is attributed to Leonhard EULER [41] in 1755.

11 **History of Theorem 1.2.** Bernard BOLZANO proved in 1817 [13] that any non-empty minored subset of $\mathbb{R}$ has a greatest lower bound.

The order of $\overline{\mathbb{R}}$. For any subset of $\mathbb{R}$ to have a greatest lower bound and a least upper bound, two new elements are introduced denoted as $-\infty$ and $+\infty$[12] and the **extended real set** is defined by $\overline{\mathbb{R}} \stackrel{\text{def}}{=} \mathbb{R} \cup \{-\infty, +\infty\}$. Its order is defined by $x \leq y$ if: x and y are real and $x \leq y$; or $x = -\infty$; or $y = +\infty$. Every subset of $\overline{\mathbb{R}}$ indeed has a least upper bound, namely: its least upper bound in $\mathbb{R}$ if it is majored in $\mathbb{R}$ and non-empty; $+\infty$ if it is not majored in $\mathbb{R}$; $-\infty$ if it is empty. It similarly has a greatest lower bound.

For a subset V of $\mathbb{R}$, the expression "$\sup V = \infty$" is thus always meaningful in $\overline{\mathbb{R}}$ and it is equivalent to "V is majored in $\mathbb{R}$". We will more simply say "V is majored" ("majored" will never be used in the sense of $\overline{\mathbb{R}}$ since it is always true). The expression $\sup V = \infty$, for its part, is equivalent to "V is not majored".

A **real interval** is any set having one of the following forms[13]: $(a, b) = \{x \in \mathbb{R} : a < x < b\}$ (open interval), where $a \in \overline{\mathbb{R}}$ and $b \in \overline{\mathbb{R}}$; $[a, b] = \{x \in \mathbb{R} : a \leq x \leq b\}$ (closed interval), where $a \in \mathbb{R}$ and $b \in \mathbb{R}$; $(a, b] = \{x \in \mathbb{R} : a < x \leq b\}$ (left semi-open interval), where $a \in \overline{\mathbb{R}}$ and $b \in \mathbb{R}$; $[a, b) = \{x \in \mathbb{R} : a \leq x < b\}$ (right semi-open interval), where $a \in \mathbb{R}$ and $b \in \overline{\mathbb{R}}$.

Convergence in $\mathbb{R}$. A real sequence $(x_n)_{n\in\mathbb{N}}$ is said to **converge to a limit**[14] $x \in \mathbb{R}$, and it is denoted as $x_n \to x$ when $n \to \infty$, if:

$$\forall \epsilon > 0, \text{ there is } m \in \mathbb{N} \text{ such that: } \quad n \geq m \Rightarrow |x_n - x| \leq \epsilon. \tag{1.7}$$

A real sequence is **increasing** if $u_{n+1} \geq u_n$ and **decreasing** if $u_{n+1} \leq u_n$.

THEOREM 1.3.– *Every increasing and majored, or decreasing and minored, real sequence converges.* □

Proof. If an increasing sequence $(u_n)_{n\in\mathbb{N}}$ is majored, it has (Theorem 1.2 (a)) a least upper bound $b \in \mathbb{R}$. According to the definition of the bound, for every $\epsilon > 0$, there is $m \in \mathbb{N}$ such that $b - \epsilon \leq u_m \leq b$. For all $n \geq m$, it holds $b - \epsilon \leq u_m \leq u_n \leq b$, hence $|u_n - b| \leq \epsilon$. Thus, $u_n \to b$. If $(u_n)_{n\in\mathbb{N}}$ is decreasing and minored, then $(-u_n)_{n\in\mathbb{N}}$ is increasing and majored, and thus it converges and so does $(u_n)_{n\in\mathbb{N}}$. □

THEOREM 1.4.– *Let $x \in \mathbb{R}$ and $n \in \mathbb{N}$. If $0 < x < 1$, then $x^n \to 0$ when $n \to \infty$.* □

Proof. Since $0 < x < 1$, x^n is positive and decreases with n, and thus (Theorem 1.3) it has a limit ℓ. The equality $x^{n+1} = x^n x$ gives at the limit $\ell = \ell x$, from which $\ell(1 - x) = 0$ and thus $\ell = 0$. □

Inequalities usage convention. According to Laurent SCHWARTZ, **we reserve strict inequalities to situations in which large inequalities are not suitable**, and therefore to the case in which care must be taken. For example, in Definition (1.7) of convergence, the large inequality $n \geq m$ could be replaced by $n > m$, but the strict inequality $\epsilon > 0$ cannot be replaced by $\epsilon \geq 0$. □

Approximations in $\mathbb{R}$. Let us show that any real number is the limit of rational numbers, which is useful since $\mathbb{Q}$ is countable[15].

---

12 **History of the notion of infinite element.** The Baron Augustin CAUCHY *completed* $\mathbb{R}$ by adjoining the elements $+\infty$ and $-\infty$ to it.
The symbol ∞ was introduced by John WALLIS in 1655 [114] to designate infinity.

13 **French notation.** The French denote $]a, b[$ the open interval. They denote, like we do, $[a, b]$ the closed interval, and furthermore $]a, b]$ and $[a, b[$ the semi-open intervals.

14 **History of the notion of convergent sequence.** The Baron Augustin CAUCHY defined the convergence of real numbers, in 1821, in its *Cours d'analyse à l'École Royale Polytechnique* [25, p. 19] which paved the way for modern analysis. Niels ABEL contributed to the emergence of this concept.

15 **History of Theorem 1.6.** Richard DEDEKIND proved in 1873 [32] that $\mathbb{Q}$ is countable and that $\mathbb{R}$ is not.

THEOREM 1.5.– *Let $x \in \mathbb{R}$. For every $n \in \mathbb{N}$, there exist rational numbers x_n and x'_n such that:*

$$x_n \leq x \leq x'_n, \quad |x'_n - x_n| \leq \frac{1}{n}. \square$$

Proof. If $x \geq 0$, the set $\{i \in \mathbb{N}^* : i/n \leq x\}$ is majored (by nx) and non-empty (it contains 0), and therefore it has a maximum m; then $x_n = m/n$ and $x'_n = (m+1)/n$ have the required properties. If $x \leq 0$, $x_n = -(-x)'_n$ and $x'_n = -(-x)_n$ are suitable. $\square$

THEOREM 1.6.– *The set $\mathbb{Q}$ is countable.* $\square$

Proof. The set $\mathbb{Q}$ is the image of $\mathbb{Z} \times \mathbb{N}$ which is countable (Theorem 1.1 (d)) by the mapping g defined by $g(n, m) = n/m$, and therefore it is countable (Theorem 1.1 (c)). $\square$

Powers. The n-th **root** $x^{1/n}$ and, more generally, the **fractional power**[16] $x^{m/n}$ of a real positive number x is defined, for $n \in \mathbb{N}$ and $m \in \mathbb{Z}$, if m is positive by

$$x^{m/n} \stackrel{\text{def}}{=} \{\alpha \in \mathbb{Q} : \alpha^n < x^m, \ \alpha \geq 0\} \cup \{\alpha \in \mathbb{Q} : \alpha < 0\}$$

and else by $x^{m/n} = 1/x^{-m/n}$. The square root[17] of x is $\sqrt{x} = x^{1/2}$.

For a real exponent s, x **to the power of** s is defined by: if $x \geq 1$,

$$x^s \stackrel{\text{def}}{=} \sup\{x^\alpha : \alpha \in \mathbb{Q}, \ \alpha \leq s\} \tag{1.8}$$

(this upper bound exists since the set of these x^α is majored because they increase with α when $x > 1$) and $x^s = (1/x)^{-s}$ if $0 < x < 1$. By convention, $x^0 = 1$ for $x > 0$ and $0^s = 0$ for $s > 0$ (0^0 is not defined). Fractional powers are thus recovered (thanks to Theorem 1.5) and the usual rules below are successively recovered for integer, rational and real numbers:

$$x^{s+t} = x^s x^t, \quad x^{st} = (x^s)^t, \quad (xy)^s = x^s y^s. \tag{1.9}$$

Let us finally majorize the sum of a geometric series. Let $x \in \mathbb{R}$, $x \neq 1$ and two integers such that $0 \leq n \leq m$. Then:

$$0 < x < 1 \quad \Rightarrow \quad 0 < x^n + x^{n+1} + \cdots + x^m < \frac{x^n}{1-x}. \tag{1.10}$$

This follows from the equality $x^n + x^{n+1} + \cdots + x^m = (x^n - x^{m+1})/(1-x)$ which is verified by multiplying the two sides by $1 - x$.

---

16 **History of the computation of fractional powers.** Around 300 B.C., EUCLID gave calculation rules which are equivalent to $x^{m+n} = x^m x^n$ for integers m and n in its *Elements* [40, book IX, proposition 11]. Nicolay ORESME introduced the notion of positive fractional power of a number in the 14th Century and gave the rules of calculation of (1.9) [29] with notations close to ours. His treatise, too ahead of the mathematics of his time, disappeared into oblivion.
In the 15th Century, Nicolas CHUQUET introduced negative integer powers and again expressed Euclid's rule [27]. Michael STIFEL introduced negative fractional powers and recovered the calculation laws of ORESME in 1544 [109, p. 35 and 249–250].

17 **History of the root notation.** Christoff RUDOLFF introduced in 1526, in a treatise on arithmetic, the $\surd x$ notation to designate the square root of x.
René DESCARTES used the notation $\sqrt{x}$ in 1637 in the appendix *Géometrie* of the *Discours de la Méthode* [33, p. 299].

PART 1

SEMI-NORMED SPACES

Chapter 2

Semi-normed Spaces

2.1. Definition of semi-normed spaces

Let us define vector spaces[1].

DEFINITION 2.1.– *A **vector space** is a set E endowed with an **addition** denoted by $+$ from $E \times E$ into E, a **scalar multiplication** from $\mathbb{R} \times E$ into E and an element $0_E \in E$ such that: for all $u \in E$, $v \in E$, $w \in E$, $s \in \mathbb{R}$ and $t \in \mathbb{R}$,*

$$(u+v)+w = u+(v+w), \quad u+v = v+u, \quad u+0_E = u,$$

$$t(u+v) = tu+tv, \quad (s+t)u = su+tu, \quad s(tu) = (st)u, \quad 1u = u,$$

there is an element $-u \in E$ such that $-u+u = 0_E$. ■

Terminology. Definition 2.1 is, in fact, that of a **vector space over** $\mathbb{R}$. We deliberately omit "over $\mathbb{R}$" because we will only consider this case. □

We denote by $u-v \stackrel{\text{def}}{=} u+(-v)$ and often 0 instead of 0_E if there is no ambiguity.

Let us define norms and semi-norms[2].

DEFINITION 2.2.– (a) *A **semi-norm** is a mapping p from a vector space E into $\mathbb{R}$ such that, for all $u \in E$, $v \in E$ and $t \in \mathbb{R}$,*

$$p(u) \geq 0, \quad p(tu) = |t|\, p(u), \quad p(u+v) \leq p(u)+p(v).$$

---

1 **History of the notion of vector space.** Giuseppe PEANO [83, chap. IX] gave in 1888 the general definition of a vector space (not necessarily with finite dimension).
The first use of geometric language to describe another space than $\mathbb{R}^d$, in this case ℓ^2, was made in 1907, by Erhard SCHMIDT [98] and Maurice FRÉCHET [44].

2 **History of the notion of norm.** The first norm was introduced in 1907, on ℓ^2, by Erhard SCHMIDT [98] and Maurice FRÉCHET [44] (with the current notation $\| \;\|$ and the *subadditivity*).

History of the concept of semi-norm. John VON NEUMANN introduced semi-norms in 1935 [78].

(b) *A **norm** is a semi-norm p such that $p(u) > 0$ for all $u \neq 0_E$.* ▮

Let us define normed spaces and separated semi-normed spaces[3].

DEFINITION 2.3.– (a) *A **normed space** is a vector space endowed with a norm. The norm of a normed space E is denoted by $\| \ \|_E$.*

(b) *A **semi-normed space** is a vector space endowed with a non-empty family of semi-norms. The family of semi-norms of a semi-normed space E is denoted by*

$$\{\| \ \|_{E;\nu} : \nu \in \mathcal{N}_E\}.$$

(c) *A semi-normed space E is said to be **separated** if, for all $u \in E$,*

$$[\,\forall \nu \in \mathcal{N}_E,\ \|u\|_{E;\nu} = 0\,] \quad \Rightarrow \quad u = 0_E.$$ ▮

The set $\mathcal{N}_E$ indexing the semi-norms is, *a priori*, any.

A semi-normed space E, as any vector space, is never empty since it contains at least one element, namely 0_E.

Caution. Definition 2.3 (b) is general but not universal. For Laurent SCHWARTZ [99, p. 240], a semi-normed space is a vector space endowed with a *filtering* family of semi-norms, which is the same because (Theorem 3.15) any family is equivalent to a filtering family. For Robert EDWARDS [38, p. 80], a semi-normed space is a vector space endowed with a *single* semi-norm, which drastically changes. □

Let us give an obvious property.

THEOREM 2.4.– *Every normed space is a separated semi-normed space.* ▮

Proof. This results from Definition 2.2 (b) of a norm and Definition 2.3. □

We will only make use of separated spaces because this ensures the uniqueness of the limit of any convergent sequence (see Theorem 2.7 (a) and property (2.2), p. 16).

Examples of separated semi-normed spaces:

— The space $\mathbb{R}$ endowed with $\|x\|_{\mathbb{R}} = |x|$ is normed and thus separated semi-normed.
— The space $\mathcal{C}_b(\mathbb{R})$ of continuous and bounded functions endowed with $\|f\|_{\mathcal{C}_b(\mathbb{R})} = \sup_{x \in \mathbb{R}} |f(x)|$ is

3 **History of the notion of normed space.** The axioms of a (sequentially complete) normed space were introduced, in the example of $\mathcal{C}(I)$, I being a compact interval, by Frigyes RIESZ in 1918 in a famous essay [93]. The general definition of a normed space was given by Stefan BANACH in 1922 in his thesis [7].

History of the concept of semi-normed space. John VON NEUMANN introduced semi-norms in 1935 [78] (with an unnecessary countability condition). He also showed [78, thm. 26, p. 19] that they coincide with the locally convex topological vector spaces that Andrey KOLMOGOROV had introduced in 1934 [63, p. 29].

normed (this norm defines the topology of uniform convergence) and thus separated semi-normed.
— The space $\mathcal{C}(\mathbb{R})$ of continuous functions endowed with the $\|f\|_{\mathcal{C}(\mathbb{R});n} = \sup_{|x|\leq n} |f(x)|$ indexed by $n \in \mathbb{N}$ is separated semi-normed (these semi-norms define the topology of uniform convergence on compact sets).
— The space $\mathcal{C}(\mathbb{R})$ endowed with the $\|f\|_{\mathcal{C}(\mathbb{R})\text{-pt},x} = |f(x)|$ indexed by $x \in \mathbb{R}$ is separated semi-normed (these semi-norms define the topology of pointwise convergence). □

Some separated semi-normed spaces used in the following volumes for the study of PDEs:
— The space of distributions $\mathcal{D}'(\Omega)$ endowed with the $\|f\|_{\mathcal{D}'(\Omega);\varphi} = |\langle f, \varphi\rangle|$ indexed by $\varphi \in \mathcal{D}(\Omega)$ (here Ω is an open subset of $\mathbb{R}^d$).
— The "local" Lebesgue space $L^p_{\text{loc}}(\Omega)$ endowed with the $\|f\|_{L^p_{\text{loc}}(\Omega);\omega} = (\int_\omega |f|^p)^{1/p}$ indexed by the bounded open sets ω such that $\overline{\omega} \subset \Omega$.
— A Hilbert space E endowed with the $|||u|||_{E\text{-weak};v} = |(u,v)_E|$ indexed by $v \in E$ (in practice, E will be $L^2(\Omega)$ or $H^1_0(\Omega)$). □

Space endowed with several norms. A vector space endowed with several norms is a separated semi-normed space but is **not a normed space** (on the other hand, it can be *normable*).

For example, the space $\mathcal{D}(\Omega)$ of indefinitely differentiable functions with compact support in Ω endowed with the norms (which generate the topology of inductive limit of the $\mathcal{D}_K(\Omega)$) indexed by $q \in \mathcal{C}^+(\Omega)$,

$$\|\varphi\|_{\mathcal{D}(\Omega);q} \stackrel{\text{def}}{=} \sup_{x\in\Omega} \sup_{0\leq|\beta|\leq q(x)} q(x)|\partial^\beta \varphi(x)| \tag{2.1}$$

is separated semi-normed, but it is neither normed nor normable. □

Motivation for the study of separated semi-normed spaces. We study in detail the separated semi-normed spaces because they play a crucial role in the study of PDEs. Although the most frequently used spaces are normed, such as L^p and $W^{m,p}$, or metrizable, such as $W^{m,p}_{\text{loc}}$, we cannot be restricted to those cases because other essential separated semi-normed spaces, such as $\mathcal{D}'$ or $W^{m,p}$-weak, are neither normable nor metrizable. □

Comparison to locally convex spaces. *Semi-normed space* is equivalent to a *locally convex topological vector space* according to **von Neumann's theorem** [NEUMANN, 78, thm. 26, p. 19]. □

Advantages of the construction using semi-norms. The construction with semi-norms has three advantages (the first is decisive) compared to the construction with a locally convex topology:

1) *Effectiveness.* The use of semi-norms allows **the definition of the differentiation in a general separated locally convex space** (Definition 19.1) by comparing the semi-norms of a variation in the data to the semi-norms of the variation in the mapping. It also allows **the definition of L^p spaces with values in such a space**, see Volume III, because it is possible to raise to a power p a semi-norm (but not an open set!).

This is the reason why, when only resorting to topologies, metrics or norms, the differentiation and the L^p construction are solely done for values in a normed space or in certain particular cases, for example *in the scalar sense* which corresponds to the weak topology.

2) *Simplicity.* The construction follows that of normed spaces, the main difference consisting of working on several (semi-)norms instead of a single norm. We now give two examples:
— A set U is bounded if, see Definition 2.9, "$\sup_{v\in U} \|v\|_{E;\nu} < \infty$ *for every semi-norm* $\| \ \|_{E;\nu}$ *of* E". With the topology only, the definition takes the more abstract form "for any open set V containing 0_E, there is $t > 0$ such that $tU \subset V$".
— The semi-norms defined by (2.1) generate the topology of inductive limit of the $\mathcal{D}_K(\Omega)$ that is complex enough in its introducing to be excluded from the majority of books dedicated to PDEs. In terms of semi-norms, $T \in \mathcal{D}'(\Omega)$ is expressed as "there exist $q \in \mathcal{C}^+(\Omega)$ and $c \in \mathbb{R}$ such that $|T(\varphi)| \leq c\,\|\varphi\|_{\mathcal{D}(\Omega);q}$

for any $\varphi \in \mathcal{D}(\Omega)$" instead of "for any open $\mathcal{O}$ of $\mathbb{R}$, the set $T^{-1}(\mathcal{O})$ is open in $\mathcal{D}(\Omega)$ endowed with the topology of inductive limit of the $\mathcal{D}_K(\Omega)$".

3) *Numerical values.* Convergence is measured by the (real) value of the semi-norms, see Definition 2.6, as well as the fact of being bounded (we have just seen it) which is suitable for numerical computing. □

Comparison to construction using a distance in the case of a metrizable space. Construction with semi-norms is also advantageous, for a metrizable space E, compared to the construction using a distance d because this one is not enough for building $L^p(\Omega; E)$. Moreover, the condition $\sup_{u \in U} d(u, 0) < \infty$ does not characterize the bounded sets U if E is not normable, see (4.1) p. 49. □

Examples of non-separated semi-normed spaces:
— The space $\mathbb{R}^2$ endowed with the semi-norm $\|(x_1, x_2)\| = |x_1|$ is not separated.
— A vector space endowed with the zero semi-norm is not separated (this semi-norm defines the coarse topology). □

We now give a few calculation rules and show that any semi-norm is *contracting*.

THEOREM 2.5.– *Let E be a vector space, $u \in E$, $v \in E$, $t \in \mathbb{R}$ and p a semi-norm on E. Then:*

(a) $$0_E \text{ and } -u \text{ are unique.}$$

(b) $$0u = 0_E, \quad t0_E = 0_E, \quad p(0_E) = 0, \quad (-1)u = -u.$$

(c) $$|p(v) - p(u)| \le p(v - u). \ \square$$

Proof. (a) If $u + v = u$, then $v = 0_E + v = -u + u + v = -u + u = 0_E$, and thus there is a single element 0_E such that $u + 0_E = u$.

If $v + u = 0_E$, then $v = 0_E + v = -u + u + v = -u + 0_E = -u$ thus, u being given, there is a single element $-u$ such that $-u + u = 0_E$.

(b) We have $0u = 0_E + 0u = -0u + 0u + 0u = -0u + (0+0)u = -0u + 0u = 0_E$.

And $t0_E = 0_E + t0_E = -t0_E + t0_E + t0_E = -t0_E + t(0_E + 0_E) = -t0_E + t0_E = 0_E$.

And also, $p(0_E) = p(00_E) = 0p(0_E) = 0$.

Finally, $(-1)u + u = (-1)u + 1u = (-1 + 1)u = 0u = 0_E$, therefore $(-1)u = -u$ since the latter is unique.

(c) We have $p(v) = p(u + (v - u)) \le p(u) + p(v - u)$ hence the stated inequality if $p(v) \ge p(u)$; otherwise, we exchange u and v in this calculation. □

2.2. Convergent sequences

Let us define convergent sequences and Cauchy sequences[4] of a separated semi-normed space.

DEFINITION 2.6.– *Let E be a separated semi-normed space, $\{\| \ \|_{E;\nu} : \nu \in \mathcal{N}_E\}$ its family of semi-norms and $(u_n)_{n\in\mathbb{N}} \subset E$.*

(a) *$(x_n)_{n\in\mathbb{N}}$ **converges to a limit** $u \in E$, which is denoted by $u_n \to u$ when $n \to \infty$, if, for all $\nu \in \mathcal{N}_E$,*

$$\|u_n - u\|_{E;\nu} \to 0 \quad \textit{when } n \to \infty.$$

That is if, for all $\nu \in \mathcal{N}_E$ and $\epsilon > 0$, there exists $m \in \mathbb{N}$ such that $n \geq m$ yields $\|u_n - u\|_{E;\nu} \leq \epsilon$.

(b) *$(u_n)_{n\in\mathbb{N}}$ is a **Cauchy sequence** if, for all $\nu \in \mathcal{N}_E$,*

$$\sup_{n'\geq n} \|u_{n'} - u_n\|_{E;\nu} \to 0 \quad \textit{when } n \to \infty.$$

That is if, for all $\nu \in \mathcal{N}_E$ and $\epsilon > 0$, there exists $m \in \mathbb{N}$ such that $n' \geq n \geq m$ yields $\|u_{n'} - u_n\|_{E;\nu} \leq \epsilon$.

(c) *$(u_n)_{n\in\mathbb{N}}$ is **bounded** if, for all $\nu \in \mathcal{N}_E$,*

$$\sup_{n\in\mathbb{N}} \|u_n\|_{E;\nu} < \infty.$$

That is if, for all $\nu \in \mathcal{N}_E$, there exists $b \in \mathbb{R}$ such that $\|u_n\|_{E;\nu} \leq b$ for all $n \in \mathbb{N}$. ■

Consistency. When $E = \mathbb{R}$, we recover the definition of a convergent real sequence, see (1.7), p. 6. □

When there is no ambiguity, we simply denote "$u_n \to u$" omitting "when $n \to \infty$".

Let us give some properties of convergent sequences.

4 **History of the notion of convergent sequence.** The Baron Augustin CAUCHY gave Definition 2.6 (a) of convergence, for $\mathbb{R}$, in 1821 [25, p. 19]. Niels ABEL contributed to the emergence of this concept.
History of the concept of Cauchy sequence. Augustin CAUCHY introduced the convergence criterion of Definition 2.6 (b), for real series, in 1821 [25, p. 115-116], by admitting it (that is by implicitly considering $\mathbb{R}$ as the completion of $\mathbb{Q}$). Bernard BOLZANO had stated this criterion in 1817 in [13] and tried without success to justify it because of the lack of a consistent definition of $\mathbb{R}$.

THEOREM 2.7.– *For any sequence of a separated semi-normed space:*

(a) *Convergent* $\Rightarrow$ *bounded, Cauchy, has a unique limit.*

(b) *Cauchy* $\Rightarrow$ *bounded.* ∎

Proof. Let E be a separated semi-normed space and $\{\| \ \|_{E;\nu} : \nu \in \mathcal{N}_E\}$ its family of semi-norms.

(b) Let $(u_n)_{n\in\mathbb{N}}$ be a Cauchy sequence of E. Let $\nu \in \mathcal{N}_E$ and m be such that $n' \geq n \geq m$ leads to $\|u_{n'} - u_n\|_{E;\nu} \leq 1$. For all $n' \geq m$, we get $\|u_{n'}\|_{E;\nu} \leq \|u_m\|_{E;\nu} + 1$. Therefore, for all $n \in \mathbb{N}$,

$$\|u_n\|_{E;\nu} \leq \sup\{\|u_0\|_{E;\nu}, \ldots, \|u_{m-1}\|_{E;\nu}, \|u_m\|_{E;\nu} + 1\}.$$

The right-hand side being finite, $(u_n)_{n\in\mathbb{N}}$ is bounded.

(a) Let $(u_n)_{n\in\mathbb{N}}$ be a convergent sequence of E, with limit u.

Cauchy property. Let $\nu \in \mathcal{N}_E$, $\epsilon > 0$ and $m \in \mathbb{N}$ be such that $n \geq m$ leads to $\|u_n - u\|_{E;\nu} \leq \epsilon/2$. For all $n' \geq n \geq m$,

$$\|u_{n'} - u_n\|_{E;\nu} \leq \|u_{n'} - u\|_{E;\nu} + \|u - u_n\|_{E;\nu} \leq \epsilon.$$

Therefore, $(u_n)_{n\in\mathbb{N}}$ is a Cauchy sequence. Furthermore, it is bounded by (b).

Uniqueness of the limit. Assume that $(u_n)_{n\in\mathbb{N}}$ have another limit u'. For all $\nu \in \mathcal{N}_E$,

$$\|u' - u\|_{E;\nu} \leq \|u' - u_n\|_{E;\nu} + \|u_n - u\|_{E;\nu}.$$

The right-hand side tends to 0 when $n \to \infty$, thus $\|u' - u\|_{E;\nu} = 0$. The separation (Definition 2.3 (c)) then gives $u' = u$. □

Convergent sequences of a non-separated semi-normed space. As many other definitions, Definition 2.6 of a convergent sequence could be extended to not necessarily separated spaces. We do not do it because we would lose the uniqueness of the limit. In effect:

$$\begin{cases} \textit{In a semi-normed space that is not separated,} \\ \textit{every convergent sequence has infinitely many limits.} \end{cases} \tag{2.2}$$

Proof of (2.2). If $u_n \to u$ in a space E semi-normed by a family $\{\| \ \|_{E;\nu} : \nu \in \mathcal{N}_E\}$, then $u_n \to u + v$ for all $v \in E_0$ where $E_0 = \{v \in E : \|v\|_{E;\nu} = 0, \ \forall \nu \in \mathcal{N}_E\}$. If E is not separated, E_0 contains at least one non-zero v and therefore it contains an infinity (of which tv for all $t \in \mathbb{R}$). □

For example, in the non-separated space $\mathbb{R}^2$ endowed with the semi-norm $\|(x_1, x_2)\| = |x_1|$, the null sequence converges to $(0, 0)$ and also to $(0, 1)$ (but it does not converge to $(1, 0)$). □

Let us give properties of subsequences.

THEOREM 2.8.– *Let E be a separated semi-normed space, $(u_n)_{n\in\mathbb{N}} \subset E$, $u \in E$ and $(u_{\sigma(n)})_{n\in\mathbb{N}}$ a subsequence of $(u_n)_{n\in\mathbb{N}}$. Then:*

(a) $$u_n \to u \quad \Rightarrow \quad u_{\sigma(n)} \to u.$$

(b) $$(u_n)_{n\in\mathbb{N}} \text{ is a Cauchy sequence and } u_{\sigma(n)} \to u \quad \Rightarrow \quad u_n \to u.$$ ▮

Proof. We denote by $\{\| \ \|_{E;\nu} : \nu \in \mathcal{N}_E\}$ the family of semi-norms of E.

(a) Assume that $u_n \to u$. For all $\nu \in \mathcal{N}_E$ and $\epsilon > 0$, there exists $n_0 \in \mathbb{N}$ such that, for all $n \geq n_0$, we have $\|u_n - u\|_{E;\nu} \leq \epsilon$. Then, $\sigma(n) \geq n \geq n_0$ thus $\|u_{\sigma(n)} - u\|_{E;\nu} \leq \epsilon$. Which proves that $u_{\sigma(n)} \to u$.

(b) Assume that $(u_n)_{n\in\mathbb{N}}$ is a Cauchy sequence and $u_{\sigma(n)} \to u$. For every $\nu \in \mathcal{N}_E$ and $\epsilon > 0$, there exists $n_0 \in \mathbb{N}$ such that, for every $n \geq n_0$ and $k \geq n_0$, we have $\|u_n - u_k\|_{E;\nu} \leq \epsilon/2$. In addition, there is $n_1 \in \mathbb{N}$ such that, for all $n \geq n_1$, we have $\|u_{\sigma(n)} - u\|_{E;\nu} \leq \epsilon/2$. For all $n \geq \sup\{n_0, n_1\}$, we get $\sigma(n) \geq n_0$ and thus

$$\|u_n - u\|_{E;\nu} = \|(u_n - u_{\sigma(n)}) + (u_{\sigma(n)} - u)\|_{E;\nu} \leq \epsilon.$$

This proves that $u_n \to u$. □

2.3. Bounded, open and closed sets

Let us define bounded, open and closed sets[5] of a separated semi-normed space.

DEFINITION 2.9.– *Let E be a separated semi-normed space, $\{\| \ \|_{E;\nu} : \nu \in \mathcal{N}_E\}$ its family of semi-norms and $U \subset E$.*

(a) *U is **bounded** if, for every $\nu \in \mathcal{N}_E$,*

$$\sup_{u\in U} \|u\|_{E;\nu} < \infty.$$

That is if, for every $\nu \in \mathcal{N}_E$, there exists $b \in \mathbb{R}$ such that $\|u\|_{E;\nu} \leq b$ for all $u \in U$.

5 **History of the notion of sequentially closed set.** Sequentially closed sets were introduced, for $\mathbb{R}$ and then for $\mathbb{R}^d$, by George CANTOR, see his works [24].

History of the concept of bounded set. Bounded sets in a semi-normed space were introduced in 1935 by John VON NEUMANN [78]. Andrey KOLMOGOROV had introduced them in 1934 [63] for a topological vector space.

(b) *U is **open** if, for every $u \in U$, there exist a finite set $N \subset \mathcal{N}_E$ and $\eta > 0$ such that:*

$$v \in E, \; \sup_{\nu \in N} \|v - u\|_{E;\nu} \leq \eta \quad \Rightarrow \quad v \in U.$$

(c) *U is **closed** if $E \setminus U$ is open.*

(d) *U is **sequentially closed** if every sequence of U that converges in E has its limit in U.* ▮

Consistency. We find again the notion of bounded introduced for sequences in Definition 2.6 (c). □

Let us compare closed and sequentially closed.

THEOREM 2.10.– *For any subset of a separated semi-normed space:*

$$\textit{Closed} \quad \Rightarrow \quad \textit{sequentially closed.}$$ ▮

Proof. Let U be a closed subset of a separated semi-normed space E, endowed with semi-norms $\{\| \ \|_{E;\nu} : \nu \in \mathcal{N}_E\}$. If U were not sequentially closed, there would be $u \in E \setminus U$ and $(u_n)_{n \in \mathbb{N}} \subset U$ such that $u_n \to u$. Since $E \setminus U$ is open, there would exist a finite set $N \subset \mathcal{N}_E$ and $\eta > 0$ such that $\sup_{\nu \in N} \|v - u\|_{E;\nu} \leq \eta$ leads to $v \in E \setminus U$. For n large enough, we would have $\sup_{\nu \in N} \|u_n - u\|_{E;\nu} \leq \eta$ and thus $u_n \notin U$ which would contradict the assumption on u_n. Therefore, U is sequentially closed □.

Converse. *Closed* coincides with *sequentially closed* in every metrizable space (Theorem 4.6 (a)), but not in every separated semi-normed space. For example, in the space $\mathcal{F}$ of functions from $\mathbb{R}$ into itself endowed with the semi-norms $\|f\|_x = |f(x)|$ indexed by $x \in \mathbb{R}$ (that is endowed with the pointwise convergence), a set U that is sequentially closed but not closed is defined by

$$U = \{f \in \mathcal{F} : \{x : f(x) \neq 0\} \textit{ is countable}\}. \tag{2.3}$$

□

Let us show that open sets are stable under union and finite intersection.

THEOREM 2.11.– *In every separated semi-normed space:*

(a) *Any (finite or infinite) union of open sets is open.*

(b) *Any finite intersection of open sets is open.* ▮

Proof. Let E be a separated semi-normed space and $\{\| \ \|_{E;\nu} : \nu \in \mathcal{N}_E\}$ its semi-norms.

(a) Let u be a point of a union $\bigcup_{i\in\mathcal{I}} U_i$ of open subsets of E. Therefore, u is an element of one of them, say U_k, and thus there is a finite set $N_k \subset \mathcal{N}_E$ and $\eta_k > 0$ such that $\sup_{\nu\in N_k} \|v-u\|_{E;\nu} \le \eta_k$ leads to $v \in U_k$, and thus to $v \in \bigcup_{i\in\mathcal{I}} U_i$. As a consequence, $\bigcup_{i\in\mathcal{I}} U_i$ is open.

(b) Let u be a point of a finite intersection $\bigcap_{i\in I} U_i$ of open subsets of E. For each $i \in I$, there exist a finite set $N_i \subset \mathcal{N}_E$ and $\eta_i > 0$ such that $\sup_{\nu\in N_i} \|v-u\|_{E;\nu} \le \eta_i$ yields $v \in U_i$. The set $N = \bigcup_{i\in I} N_i$ is finite, and thus $\min_{i\in I} \eta_i$ does exist (Theorem 1.2 (c)) and is > 0. In addition, $\sup_{\nu\in N} \|v-u\|_{E;\nu} \le \min_{i\in I} \eta_i$ leads to $v \in U_i$ for every $i \in I$ and therefore to $v \in \bigcap_{i\in I} U_i$. Consequently, $\bigcap_{i\in I} U_i$ is open. □

Convention of usage of sup **and** max. According to Laurent SCHWARTZ, **we reserve** *max* **and** *min* **to the cases in which it is helpful that the bound be reached**, and thereby in the case where caution must be taken, since *sup* and *inf* are no longer suitable.

For instance, in the proof of Theorem 2.11 (b), we write $\min_{i\in I} \eta_i$ since we use the fact that it is attained to get it > 0. On the opposite, in Definition 2.9 (b) of an open set, we write $\sup_{\nu\in N} \|v-u\|_{E;\nu}$ although this bound also is attained because it is useless. □

Let us show that closed, sequentially closed and bounded sets are stable under intersection and finite union.

THEOREM 2.12.– *In every separated semi-normed space:*

(a) *Any (finite or infinite) intersection of closed sets is closed.*

(b) *Any (finite or infinite) intersection of sequentially closed sets is sequentially closed.*

(c) *Any (finite or infinite) intersection of bounded sets is bounded.*

(d) *Any finite union of closed sets is closed.*

(e) *Any finite union of sequentially closed sets is sequentially closed.*

(f) *Any finite union of bounded sets is bounded.* ▮

Proof. (a) Let $\bigcap_{i\in\mathcal{I}} U_i$ be an intersection of closed subsets of a separated semi-normed space E. It is closed because its complement is $\bigcup_{i\in\mathcal{I}} E \setminus U_i$ (De Morgan's relation (1.1), p. 2) which is open (Theorem 2.11 (a) because the $E \setminus U_i$ are open).

(b) This follows from Definition 2.9 (d) of a sequentially closed set.

(d) Let $\bigcup_{i \in I} U_i$ be a finite union of closed sets. It is closed because its complement is $\bigcap_{i \in I} E \setminus U_i$ (De Morgan's relation (1.2)) which is open (Theorem 2.11 (b)).

(e) Let now $\bigcup_{i \in I} U_i$ be a finite union of sequentially closed sets. If a sequence of $\bigcup_{i \in I} U_i$ converges to an element u, at least one of the U_i contains a subsequence; the latter necessarily converges to u (Theorem 2.8 (a)) which is thus an element of this U_i and *a fortiori* of their union. It is therefore sequentially closed.

(c) *and* (f). These are consequences of Definition 2.9 (a) of a bounded set. □

It should be observed that the whole space and the empty set are open and closed.

THEOREM 2.13.– *In every separated semi-normed space E:*

$$E \text{ and } \emptyset \text{ are open, closed and sequentially closed.} \blacksquare$$

Proof. The sets E and $\emptyset$ satisfy Definition 2.9 (b) of an open set.

Since they are the complement of one another, they are thereby closed according to Definition 2.9 (c). Being closed, they are sequentially closed (Theorem 2.10). □

Topology of a separated semi-normed space. We call **topology** of a separated semi-normed space the set of its open sets. □

Open sets of a topological space. We now give a broader definition of open sets.

DEFINITION 2.14.– *A* ***topological space*** *is any set E endowed with a* ***topology****, that is a family of subsets of E called* ***open sets*** *having the following properties:*

(a) *E and $\emptyset$ are open sets.*

(b) *Any (finite or infinite) union of open sets is open.*

(c) *Any finite intersection of open sets is open.* ■

Consistency. The open sets of a separated semi-normed space verify these properties according to Theorems 2.13 and 2.11. Such a space endowed with its open sets is therefore a topological space. □

Other example of topological space. A subset U of a separated semi-normed space endowed with the induced topology, that is with the open sets defined by (8.3), p. 117, is a topological space. □

Topology insufficiency, necessity for a uniform structure. Much of the concepts studied in this book (convergent sequences, continuous mappings, etc.) may be extended to topological spaces, but others (Cauchy sequences, uniformly continuous mappings, sequentially complete spaces), essential for the study of PDEs, require a **uniform space**[6], see [SCHWARTZ, 99, chap. XVIII, § 3, p. 237] or [BOURBAKI, 19, chap. II, § 1, def. 1].

We do not study general topological spaces because, as seen above, they are not sufficient. And we do not study general uniform spaces because separated semi-normed spaces, which are enough in our situation, are a very simple particular case thereof; actually, the vector structure, and more specifically the subtraction, very simply gives the expected properties of a uniform structure. □

2.4. Interior, closure, balls and semi-balls

Let us define the interior, closure and boundary of a subset of a separated semi-normed space.

DEFINITION 2.15.– *Let U be a subset of a separated semi-normed space:*

(a) *The **interior** of U is the largest open set included in U. It is denoted by $\mathring{U}$.*

(b) *The **closure** of U is the smallest closed set containing U. It is denoted by $\overline{U}$.*

(c) *The **sequential closure** of U is the smallest sequentially closed set containing U. It is denoted by $\overline{U}^{\mathrm{seq}}$.*

(d) *We denote by $U^{\mathrm{lim\ seq}}$ the set of the limits of the sequences of U that converge.*

(e) *The **boundary** of U is the set $\overline{U} \setminus \mathring{U}$. It is denoted by ∂U.* ▮

Justification.– (a) There is indeed a larger open set included in U, namely the union of open sets included in U. It is actually open due to Theorem 2.11 (a).

(b) There is indeed a smaller closed set containing U, namely the intersection of the closed sets containing U. It is actually closed due to Theorem 2.12 (a).

(c) *Closed* may be replaced by *sequentially closed* in the justification of (b). □

Terminology. It is also referred to as **adherence** instead of closure. □

Notation. The notation U^{lim} would be lighter than $U^{\mathrm{lim\ seq}}$ but it designates the set of limits of the generalized sequences of U, a set that may be distinct from $U^{\mathrm{lim\ seq}}$ (but which coincides with $\overline{U}$). □

Let us compare the closures and the set of limits.

THEOREM 2.16.– *For any subset U of aseparated semi-normed space:*

$$U \subset U^{\mathrm{lim\ seq}} \subset \overline{U}^{\mathrm{seq}} \subset \overline{U}. \ ▮$$

6 **History of the notion of uniform space.** Uniform spaces were introduced in 1937 by André WEIL [116].

Proof. The first inclusion is the result of the fact that every $u \in E$ is the limit of the constant sequence $(u, u, u, \ldots)$. The second is the result of the fact that a sequentially closed set contains the limits of its convergent sequences (Definition 2.9 (d)). The third results from the fact that every closed set is sequentially closed (Theorem 2.10). □

Closure of the set of limits. There exist separated semi-normed spaces in which some sets U satisfy:

$$U^{\lim\,\mathrm{seq}} \textbf{ is not } \textit{sequentially closed}. \tag{2.4}$$

Such a set U is shown in [NEUMANN, 77, p. 380].

On the other hand, in a metrizable space, $U^{\lim\,\mathrm{seq}}$ is always sequentially closed according to Theorem 4.6 (f). □

Let us characterize the interior and closure and connect their complements.

THEOREM 2.17.– *Let E be a separated semi-normed space, $\{\| \ \|_{E;\nu} : \nu \in \mathcal{N}_E\}$ its family of semi-norms and $U \subset E$. Then:*

(a)
$$\mathring{U} = \left\{ \begin{array}{l} u \in U : \exists \textit{ a finite } N \subset \mathcal{N}_E, \exists \eta > 0 \textit{ such that} \\ \qquad v \in E, \ \sup_{\nu \in N} \|v - u\|_{E;\nu} \leq \eta \ \Rightarrow \ v \in U \end{array} \right\}.$$

(b)
$$\overline{U} = \left\{ \begin{array}{l} u \in E : \forall \textit{ finite } N \subset \mathcal{N}_E, \forall \eta > 0, \exists v \in U \\ \qquad \textit{such that } \sup_{\nu \in N} \|v - u\|_{E;\nu} \leq \eta \end{array} \right\}.$$

(c)
$$E \setminus \overline{U} = (E \setminus U)^{\circ}. \ \blacksquare$$

Proof. (a) The set in the right-hand side is open (Definition 2.9 (b)) and it contains any point of any open set included in U. This is thus the greatest open set included in U.

(c) This results from the fact that a closed set is (Definition 2.9 (c)) the complement of an open set.

(b) This is resulting from (a) and (c). □

We now define balls and semi-balls.

DEFINITION 2.18.– *Le E be a separated semi-normed space.*

(a) *We call* **semi-ball** *or* **closed semi-ball** *of E any set of the form*

$$B_{E;\nu}(u, r) \stackrel{\mathrm{def}}{=} \{v \in E : \|v - u\|_{E;\nu} \leq r\}$$

where $u \in E$, $r > 0$ and $\| \ \|_{E;\nu}$ is a semi-norm of E.

(b) *We call* ***open semi-ball*** *any set of the form*

$$\mathring{B}_{E;\nu}(u,r) \stackrel{\text{def}}{=} \{v \in E : \|v-u\|_{E;\nu} < r\}.$$

(c) *In a normed space, a* ***ball*** *designates a semi-ball. We denote it by $B_E(u,r)$ or $\mathring{B}_E(u,r)$ depending on whether it is closed or open.* ▮

Justification.– (a) The semi-ball $B_{E;\nu}(u,r)$ is indeed closed because its complement $V = E \setminus B_{E;\nu}(u,r)$ is open (Definition 2.9 (b)) since, if z is an element of V, the same holds for every $v \in E$ such that $\|v-z\|_{E;\nu} \le \eta$ where $\eta = (\|z\|_{E;\nu} - r)/2$.

(b) The semi-ball $\mathring{B}_{E;\nu}(u,r)$ is open because, when z is one of its elements, the same happens for every $v \in E$ such that $\|v-z\|_{E;\nu} \le \eta$ where $\eta = (r - \|z\|_{E;\nu})/2$. The notation $\mathring{B}_{E;\nu}(u,r)$ is justified because it is the interior of $B_{E;\nu}(u,r)$ according to Theorem 2.17 (a). □

Convention. For us, a ball or a semi-ball is therefore closed by default. This choice is consistent with the convention of L. SCHWARTZ for using large inequalities by default, see p. 6. This is not general, many authors designate an open ball by $B_E(u,r)$. □

2.5. Density, separability

Let us define the density of a set in another.

DEFINITION 2.19.– *Let V and U be subsets of a separated semi-normed space.*

(a) *V is* ***dense*** *in U if $V \subset U \subset \overline{V}$.*

(b) *V is* ***sequentially dense*** *in U if $V \subset U \subset V^{\text{lim seq}}$.*

That is, if $V \subset U$ and every $u \in U$ is the limit of a sequence of V. ▮

Let us compare these two notions of density.

THEOREM 2.20.– *For any pair of subsets of aseparated semi-normed space:*

$$\textit{Sequentially dense} \quad \Rightarrow \quad \textit{dense}. \ ▮$$

Proof. This follows from $U^{\text{lim seq}} \subset \overline{U}$ (Theorem 2.16). □

Converse. *Dense* coincides with *sequentially dense* in every metrizable space (Theorem 4.6 (e)), but not in every separated semi-normed space. For example, the set U defined by (2.3), p. 18, is dense but not sequentially dense in its closure $\overline{U}$. □

Density in the closure. According to Definition 2.19, U is dense in its closure $\overline{U}$ and is sequentially dense in the set $U^{\text{lim seq}}$ of its limits. On the other hand, there are separated semi-normed spaces in which there are sets U such that:

$$U^{\text{lim seq}} \textbf{ is not } \textit{sequentially dense in } \overline{U}^{\text{seq}}. \tag{2.5}$$

This is the case of any set U verifying (2.4) p. 22. This never happens in a metrizable space because then $U^{\text{lim seq}} = \overline{U}^{\text{seq}} = \overline{U}$ (Theorem 4.6 (f)). □

Let us define separable sets.

DEFINITION 2.21.– *Let U be a subset of a separated semi-normed space.*

(a) *U is* ***separable*** *if it contains a countable and dense subset.*

That is, if there exists a countable set V such that $V \subset U \subset \overline{V}$.

(b) *U is* ***sequentially separable*** *if it contains a countable and sequentially dense subset.*

That is, if there exists a countable set $V \subset U$ such that every $u \in U$ is the limit of a sequence of V. ■

Let us compare these two notions of separability.

THEOREM 2.22.– *For any subset of a separated semi-normed space:*

$$\textit{Sequentially separable} \quad \Rightarrow \quad \textit{separable}. \;■$$

Proof. This follows from the fact that sequential density results in density (Theorem 2.20). □

QUESTION: Converse of Theorem 2.22. Does *separable* coincide with *sequentially separable*? We will see that this is the case in every metrizable space (Theorem 4.6 (d)), but the author of these lines does not know whether this is, or not, true for all separated semi-normed spaces. □

CAUTION: Separable has nothing to do with separated (Definition 2.3). This terminology, although unfortunate (*approximable* would be preferable than *separable*), is universally used. □

Motivation for the study of the sequential properties of density. *Sequential* properties guarantee much better approximations than *topological* properties. In this way, the sequential density of V in U ensures the existence, for any given u in U, of a sequence $(v_n)_{n \in \mathbb{N}}$ of V such that v_n be "close" to u as soon as n is large enough (with density only, the approximations set is no longer necessarily countable). Similarly, the sequential separability of U ensures the existence of approximations belonging to a sequence independent of u.

Furthermore, the sequential separability of a Fréchet space (which coincides with its separability) allows the extraction of a $*$weakly convergent subsequence of any bounded sequence of its dual (Banach–Bourbaki theorem 13.16). □

2.6. Compact sets

Let us define notions of compactness related to covering properties[7]. A **cover** of a set U is a family of sets whose union contains U. It is said **open** if its elements are open, and **finite** if it has a finite number of elements. A **subcover** is a subfamily that constitutes a cover of U.

DEFINITION 2.23.– *Let U be a subset of a separated semi-normed space E.*

(a) *U is **compact** if every cover of U by open sets has a finite subcover.*

(b) *U is **relatively compact** if $\overline{U}$ is compact.*

(c) *U is **precompact** if, for every semi-norm $\| \ \|_{E;\nu}$ of E and $\epsilon > 0$, there exists a finite number of $u \in E$ such that U is covered by the semi-balls $B_{E;\nu}(u, \epsilon)$, that is by the sets $\{v \in E : \|v - u\|_{E;\nu} \leq \epsilon\}$.* ∎

Let us now define notions of compactness related to the convergence of subsequences[8].

DEFINITION 2.24.– *Let U be a subset of a separated semi-normed space E.*

(a) *U is **sequentially compact** if every sequence U has a convergent subsequence whose limit is an element of U.*

(b) *U is **relatively sequentially compact** if every sequence of U has a convergent subsequence (in E).* ∎

CAUTION. Some authors [LANG, 66, p. 31] call *sequentially compact* a set in which every sequence has an accumulation point, which is weaker than (b). □

Motivation for the study of compactness. Compactness plays a crucial role in Heine's theorem 8.10 (uniform continuity on compact sets) and in the Schauder–Tychonoff fixed point theorem which are very useful for solving PDEs, see Volume V. □

---

7 **History of the notion of compact set.** David VAN DANTZIG gave Definition 2.23 (a) of a compact set (in a topological space) in 1932 [112, p. 595]. It was then referred to as "Borel–Lebesgue property" because they had established it for the closed segments of $\mathbb{R}$, see note 7 p. 59.
Pavel ALEKSANDROV and Pavel URYSOHN had introduced in 1924 [3, p. 260] the closely related notion of "bicompact" set (namely of a set that is included in a finite sum of domains of the set as soon as it is included in any sum).

History of the notion of the precompact set. This concept was introduced by John VON NEUMANN in 1935 [78, Definition II, p. 1] in the given general form here, under the name of *totally bounded* set.

8 **History of the notion of sequentially compact set.** Sequential compactness, formerly called *Bolzano–Weierstrass principle*, was designed by Bernard BOLZANO, and also by Karl WEIERSTRASS in his teachings, unpublished but which had considerable influence.

Motivation for the study of sequential compactness. The extraction of a convergent subsequence of a sequence of *approximated solutions*, that is the relative sequential compactness of the sequence, is also an important tool for solving PDEs that we will intensively use in Volume V, in particular by means of the notion of *extractable space* studied below (Chapter 18). □

Compactness versus sequential compactness. The relationship between these two notions is not obvious in the light of their definitions; in general, each of both does not imply the other, see (2.6) and (2.7), p. 27, and their study is motivated, as we have just seen, by very different uses. We are studying them in parallel to bring forward their differences and their similarities.

These notions can be connected by using filters: a set is compact if and only if any Cauchy filter is convergent thereto [SCHWARTZ, 99, corollary, p. 91]; therefore we could say that it is "compact for filters" as opposed to "compact for sequences". In a metrizable space, these notions are equivalent (Theorem 4.6 (b)), which makes that some authors then refer to *compactness* even though they make use of *sequential compactness*. □

Let us compare these notions of compactness[9].

THEOREM 2.25.– *For any subset of a separated semi-normed space:*

(a) *Compact* $\Rightarrow$ *relatively compact* $\Rightarrow$ *precompact* $\Rightarrow$ *bounded.*

(b) *Sequentially compact* $\Rightarrow$ *relatively sequentially compact* $\Rightarrow$ *precompact.*

∎

Proof. Let U be a subsequence of a separated semi-normed space E, endowed with semi-norms $\{\| \ \|_{E;\nu} : \nu \in \mathcal{N}_E\}$.

(a) *Compact* $\Rightarrow$ *relatively compact.* Assume that U is compact. For a moment, admit that it be closed. Then, $\overline{U}$ coincides with U and thereby it is compact. This proves that U is relatively compact. The following property (that will be developed in Theorem 2.26) remains to be established.

Compact $\Rightarrow$ *closed.* Assume that U is compact. Let $u \in E \setminus U$. For each $w \in U$, there exist $\nu \in \mathcal{N}_E$ and $a > 0$ such that $\|u - w\|_{E;\nu} = a$ and an open set is defined by $\mathcal{O}_w = \{v \in E : \|v - w\|_{E;\nu} < a/2\}$. The $\mathcal{O}_w$ cover U, therefore a finite number of them cover it, corresponding to w_i, ν_i and a_i where $i \in \mathcal{I}$. Then

$$\sup_{i \in \mathcal{I}} \|v - u\|_{E;\nu_i} \leq \min_{i \in \mathcal{I}} \frac{a_i}{2} \quad \Rightarrow \quad v \notin \bigcup_{i \in \mathcal{I}} \mathcal{O}_{w_i} \quad \Rightarrow \quad v \in E \setminus U.$$

This shows that $E \setminus U$ is open, and thus that U is closed.

(a) *Relatively compact* $\Rightarrow$ *precompact.* Assume that U is relatively compact and let $\nu \in \mathcal{N}_E$ and $\epsilon > 0$. When u spans U, the open sets $\mathcal{O}_u = \{v \in E : \|v - u\|_{E;\nu} < \epsilon\}$

9 **History of Theorem 2.25.** The fact that every precompact set be bounded was established by John VON NEUMANN in 1935 [78, Theorem 10, p. 9].

cover $\overline{U}$, which is compact. There exists therefore a finite subcover, which proves that U is precompact.

Precompact $\Rightarrow$ *bounded.* Assume that U is precompact. Let $\nu \in \mathcal{N}_E$ and let a finite number of elements $u_i \in E$ be such that the semi-balls $B_{E;\nu}(u_i, 1)$ cover U. Then $\sup_{u \in U} \|u\|_{E;\nu} \leq \sup_i \|u_i\|_{E;\nu} + 1$, which is finite. Therefore, U is bounded.

(b) *Sequentially compact* $\Rightarrow$ *relatively sequentially compact.* This results from Definition 2.24 of these two notions.

Relatively sequentially compact $\Rightarrow$ *precompact.* Assume that U is relatively sequentially compact. If it were not precompact, there would exist $\nu \in \mathcal{N}_E$, $\epsilon > 0$ and a sequence $(u_n)_{n \in \mathbb{N}}$ of U such that $\|u_n - u_m\|_{E;\nu} \geq \epsilon$ as soon as $m \neq n$. None of its subsequences would converge, which would contradict the assumption. Thereby, U is precompact. □

CAUTION. In a metrizable space, *compact* coincides with *sequentially compact* (Theorem 4.6 (b)), but in general, each of both does not imply the other:

$$\textit{Compact} \;\not\Rightarrow\; \textit{sequentially compact}. \tag{2.6}$$

$$\textit{Sequentially compact} \;\not\Rightarrow\; \textit{compact}. \tag{2.7}$$

Proof of (2.6). In the space $\mathcal{F}$ of real functions on $[0, 1]$ endowed with pointwise convergence, the set $U = \{f \in \mathcal{F} : |f(x)| \leq 1, \forall x\}$ is compact (Theorem 11.9) but is not sequentially compact because a sequence $(f_n)_{n \in \mathbb{N}}$ of U having no convergent subsequence is defined by: for every $x \in \mathbb{R}$ and $k \in \mathbb{N}$,

$$f_n(x) = c_k \text{ if } 2^{-n}k \leq x < 2^{-n}(k+1),$$

where $c_k = 1$ if k is even and $c_k = 0$ otherwise (thus f_n "oscillates" increasingly more as n increases). □

Proof of (2.7). Let $\mathcal{E}$ be the space of real functions f on $[0, 1]$ such that $\{x : f(x) \neq 0\}$ is countable endowed with pointwise convergence. The set $U = \{f \in \mathcal{E} : |f(x)| \leq 1, \forall x\}$ is sequentially compact but not compact in $\mathcal{E}$. Being not immediate, the proof will be given in a more comprehensive book on semi-normed spaces "to be published". □

Let us show that compact sets are relatively compact closed sets and characterize the other notions of compactness.

THEOREM 2.26.– *For any subset of a separated semi-normed space:*

(a) *Compact* $\Leftrightarrow$ *closed and relatively compact.*

(b) *Sequentially compact* $\Leftrightarrow$
$\Leftrightarrow$ *sequentially closed and relatively sequentially compact.*

(c) *Relatively compact* $\Leftrightarrow$ *included in a relatively compact set*
$\Leftrightarrow$ *included in a compact set.*

(d) *Precompact* $\Leftrightarrow$ *included in a precompact set.* ∎

Proof. Let U and V be subsets of the same semi-normed space.

(a) Every compact set is relatively compact and closed, as it has been established in the proof of Theorem 2.25 (a).

Conversely, if U is closed it is equal to $\overline{U}$ (according to Definition 2.15 (b)) and the latter is compact if U is relatively compact (Definition 2.23 (b)).

(b) This is obtained from Definitions 2.24 (a) and (b) of (relatively) sequentially compact sets and 2.9 (d) of sequentially closed sets.

(c) If U is relatively compact, it is included in the relatively compact set U itself.

And, if U is included in a relatively compact set V, it is included in the compact set $\overline{V}$.

Conversely, if U is included in a compact set V, the point is to prove that it is relatively compact, in other words that $\overline{U}$ is compact. Now, V is closed according to (a), thus $\overline{U} \subset \overline{V} = V$. Consider a cover of $\overline{U}$ by open sets. By adding thereto the open set $E \setminus \overline{U}$, a cover of V is obtained. A finite cover of V and thus of $\overline{U}$ can be extracted therefrom. By removing $E \setminus \overline{U}$ (if it is part of it), we still obtain a cover of $\overline{U}$; it is a finite subcover of the original cover, thereby $\overline{U}$ is indeed compact.

(d) Definition 2.23 (c) shows that any subset of a precompact set is precompact. The converse is obvious. □

Let us show that compact sets are characterized by the following **finite intersection property**.

THEOREM 2.27.– *A subset U of a separated semi-normed space is compact if and only if:*

$$\left\{\begin{array}{l}\textit{Any family } \{\mathcal{F}_i\}_{i\in\mathcal{I}} \textit{ of closed subsets of } E \textit{ such that } \bigcap_{i\in J}\mathcal{F}_i \cap U \neq \emptyset\\ \textit{for any finite subset } J \textit{ of } \mathcal{I} \textit{ verifies:}\\ \qquad\qquad \bigcap_{i\in\mathcal{I}}\mathcal{F}_i \cap U \neq \emptyset.\end{array}\right.$$ ∎

Proof. The stated property is equivalent to:

$$\left\{\begin{array}{l}\text{For any family } \{\mathcal{F}_i\}_{i\in\mathcal{I}} \text{ of closed subsets of } E \text{ such that } \bigcap_{i\in\mathcal{I}}\mathcal{F}_i \cap U = \emptyset,\\ \text{there exists a finite set } J \subset \mathcal{I} \text{ such that } \bigcap_{i\in J}\mathcal{F}_i \cap U = \emptyset.\end{array}\right.$$

Now, if U satisfies this property, for any open cover $\{\mathcal{O}_i\}_{i\in\mathcal{I}}$ of U, the closed sets $\mathcal{F}_i = E \setminus \mathcal{O}_i$ verify $\bigcap_{i\in\mathcal{I}} \mathcal{F}_i \cap U = \emptyset$; there is thus a finite set $J \subset \mathcal{I}$ such that $\bigcap_{i\in J} \mathcal{F}_i \cap U = \emptyset$; then $\{\mathcal{O}_i\}_{i\in J}$ is a finite subcover of U, which is thereby compact.

Conversely, if U is compact, for any family of closed set $\{\mathcal{F}_i\}_{i\in\mathcal{I}}$ which satisfies $\bigcap_{i\in\mathcal{I}} \mathcal{F}_i \cap U = \emptyset$, the open sets $\mathcal{O}_i = E \setminus \mathcal{F}_i$ cover U therefore there is a finite subcover $\{\mathcal{O}_i\}_{i\in J}$; then $\bigcap_{i\in J} \mathcal{F}_i \cap U = \emptyset$. □

Let us show that, in Definition 2.23 (c) of precompactness, the semi-balls can be centered on the set itself.

THEOREM 2.28.– *Let U be a precompact subset of a separated semi-normed space E, $\| \ \|_{E;\nu}$ a semi-norm of E and $\epsilon > 0$. Then, there exists a finite number of elements $u \in U$ such that semi-balls $B_{E;\nu}(u,\epsilon)$ cover U.* ▮

Proof. Definition 2.23 provides a cover of U by means of a finite number of semi-balls $B_{E;\nu}(w,\epsilon/2)$. For each w for which $B_{E;\nu}(w,\epsilon/2) \cap U$ is not empty, we choose one of its points, that is denoted by u. Then, the $B_{E;\nu}(u,\epsilon)$ cover U. □

Topological definition of precompactness. Definition 2.23 (c) of the precompactness of a set U by semi-norms seems weaker than the topological definition, namely that, for any non-empty open set $\mathcal{O}$ of E, there is a finite number of elements $u \in U$ such that sets $\mathcal{O} + u$ cover U. In fact, these two definitions are equivalent according to Theorem 2.29 below since any open set contains (Definition 2.9 (b)) a finite intersection of semi-balls.

THEOREM 2.29.– *Let E be a separated semi-normed space, $\{\| \ \|_{E;\nu} : \nu \in \mathcal{N}_E\}$ its family of semi-norms and $U \subset E$. If*

$$U \text{ is precompact}$$

then:

$$\begin{cases} \text{for any finite set } N \subset \mathcal{N}_E \text{ and any } \epsilon > 0, \text{ there is a finite number} \\ \text{of elements } u \in U \text{ such that the sets } \bigcap_{\nu\in N} B_{E;\nu}(u,\epsilon) \text{ cover } U. \end{cases} \square$$

Proof. Theorem 2.28 gives this property when N is reduced to a single element. We will show the general case by induction on the number of its elements. Thus, let N be a finite subset of $\mathcal{N}_E$ verifying this property and let $\mu \in \mathcal{N}_E$.

Then, there is a finite number of elements $u \in U$ such that the sets $X_u = \bigcap_{\nu\in N} B_{E;\nu}(u,\epsilon/2)$ cover U. Each subset $X_u \cap U$ of U, being precompact (Theorem 2.26 (d)), contains (Theorem 2.28) a finite set W_u of points w such that the $B_{E;\mu}(w,\epsilon)$ cover it. Then, $B_{E;\nu}(w,\epsilon) \supset B_{E;\nu}(u,\epsilon/2)$ and therefore $\bigcap_{\nu\in N} B_{E;\nu}(w,\epsilon) \supset X_u$ and the $Y_w = \bigcap_{\nu\in N\cup\{\mu\}} B_{E;\nu}(w,\epsilon)$ cover X_u when w spans W_u. Thus, the Y_w cover U when w spans the union of the W_u, which is finite. This proves the stated property for $N \cup \{\mu\}$. The latter thus holds for the general case. □

Separability of compact sets. The various types of compact sets are separable:

$$\begin{cases} \text{Compact, or sequentially compact,} \\ \text{or precompact, or relatively compact} \\ \text{or relatively sequentially compact} \end{cases} \Rightarrow \begin{cases} \text{separable and} \\ \text{sequentially separable.} \end{cases} \tag{2.8}$$

Proof. Definition 2.23 (c) of a precompact set implies that it is sequentially separable (Definition 2.21 (b)) which leads to the other results according to Theorems 2.22 and 2.25. □

2.7. Connected and convex sets

Let us define connected sets and connected components.

DEFINITION 2.30.– *Let U be a subset of a separated semi-normed space.*

(a) *U is **connected** if it cannot be covered by two open sets whose intersections with it are disjoint and not empty.*

That is if, for all open sets $\mathcal{O}_1$ and $\mathcal{O}_2$,

$$U \subset \mathcal{O}_1 \cup \mathcal{O}_2, \; U \cap \mathcal{O}_1 \cap \mathcal{O}_2 = \emptyset \quad \Rightarrow \quad U \cap \mathcal{O}_1 = \emptyset \text{ or } U \cap \mathcal{O}_2 = \emptyset.$$

(b) *The **connected component** of U generated by a point $u \in U$ is the greatest connected set included in U that contains u.* ▮

Justification of (b). There exists such a set, namely the union $\mathcal{U}$ of all connected sets V included in U containing u. It has the required properties because:
— $\mathcal{U}$ is connected. Indeed, let $\mathcal{O}_1$ and $\mathcal{O}_2$ be some open sets such that $\mathcal{U} \subset \mathcal{O}_1 \cup \mathcal{O}_2$ and $\mathcal{U} \cap \mathcal{O}_1 \cap \mathcal{O}_2 = \emptyset$. If $\mathcal{U} \cap \mathcal{O}_1 \neq \emptyset$, one of the V intersects $\mathcal{O}_1$, and therefore does not intersect $\mathcal{O}_2$, and thus is included in $\mathcal{O}_1$. From which $u \in \mathcal{O}_1$. Then, each V is included in $\mathcal{O}_1$. Hence, $\mathcal{U}$ is included in $\mathcal{O}_1$. Thus, $\mathcal{U} \cap \mathcal{O}_2 = \emptyset$, which proves that $\mathcal{U}$ is connected.
— $\mathcal{U}$ contains u because $\{u\}$ is one of the V. Finally, $\mathcal{U}$ is included in U. □

Let us show that any real interval (defined p. 6) is connected.

THEOREM 2.31.– *Any real interval is connected.* ▮

Proof. Let I be an interval of $\mathbb{R}$. If it were not connected, there would exist two open sets $\mathcal{O}_1$ and $\mathcal{O}_2$ of $\mathbb{R}$ such that

$$I \subset \mathcal{O}_1 \cup \mathcal{O}_2, \quad I \cap \mathcal{O}_1 \cap \mathcal{O}_2 = \emptyset \tag{2.9}$$

and points $x_1 \in I \cap \mathcal{O}_1$ and $x_2 \in I \cap \mathcal{O}_2$. By exchanging $\mathcal{O}_1$ and $\mathcal{O}_2$ if necessary, we could assume $x_1 < x_2$. Let

$$Z = \{z \in I : [x_1, z] \subset \mathcal{O}_1\}.$$

This set would be majored (by x_2) and non-empty (it would contain x_1), and therefore it would have (Theorem 1.2 (a)) a least upper bound $\zeta \in \mathbb{R}$. It would verify

$$[x_1, \zeta) \subset \mathcal{O}_1 \tag{2.10}$$

since, if $x_1 \leq z < \zeta$, there would exist $z' \in Z$ such that $z' \geq z$ and thus $[x_1, z'] \subset \mathcal{O}_1$.

Now, $\zeta \notin \mathcal{O}_1$ otherwise ($\mathcal{O}_1$ being open) there would exist $\eta > 0$ such that $[\zeta - \eta, \zeta + \eta] \subset \mathcal{O}_1$, hence (with (2.10)) $[x_1, \zeta + \eta] \subset \mathcal{O}_1$ and therefore $\zeta + \eta \in Z$, so that ζ would not be the least upper bound of Z.

In addition, $\zeta \notin \mathcal{O}_2$ otherwise ($\mathcal{O}_2$ being open) there would exist $\eta > 0$ such that $[\zeta - \eta, \zeta + \eta] \subset \mathcal{O}_2$, which would contradict (2.10). Thus, $\zeta \notin \mathcal{O}_1 \cup \mathcal{O}_2$ which would contradict the first property of (2.9). Therefore, I is connected. □

Let us now define convex sets.

DEFINITION 2.32.– *A subset U of a vector space is **convex** if, for all $u \in U$, $v \in U$ and $t \in [0,1]$,*

$$tu + (1-t)v \in U. \blacksquare$$

We denote by

$$[u, v] = \{tu + (1-t)v : 0 \le t \le 1\}$$

the (line) **segment** connecting two points u and v of a vector space. This notation is consistent because, if $E = \mathbb{R}$, we find again the closed interval $[u, v]$. Then:

$$U \text{ is convex} \quad \Leftrightarrow \quad [u, v] \subset U \text{ for all } u \text{ and } v \text{ in } U.$$

Let us show that any convex set is connected.

THEOREM 2.33.– *In a separated semi-normed space, any convex set is connected.* ▮

Proof. Let E be a separated semi-normed space, $\{\| \ \|_{E;\nu} : \nu \in \mathcal{N}_E\}$ its family of semi-norms and U a convex subset of E. If U were not connected, there would be two open sets $\mathcal{O}_1$ and $\mathcal{O}_2$ of E such that $U \subset \mathcal{O}_1 \cup \mathcal{O}_2$, $U \cap \mathcal{O}_1 \cap \mathcal{O}_2 = \emptyset$ and points $u_1 \in U \cap \mathcal{O}_1$ and $u_2 \in U \cap \mathcal{O}_2$. Denoting by, for $i = 1$ or 2,

$$\mathcal{T}_i = \{t \in \mathbb{R} : u_1 + t(u_2 - u_1) \in \mathcal{O}_i\},$$

we would get $0 \in \mathcal{T}_1$, $1 \in \mathcal{T}_2$ and

$$[0,1] \subset \mathcal{T}_1 \cup \mathcal{T}_2, \quad [0,1] \cap \mathcal{T}_1 \cap \mathcal{T}_2 = \emptyset$$

because $\{u_1 + t(u_2 - u_1) : 0 \le t \le 1\} \subset U$ according to the convexity assumption. The $\mathcal{T}_i$ being open as we will verify hereafter, the interval $[0,1]$ would not be connected, which would contradict Theorem 2.31. Thereby, U is connected.

Would they exist, the $\mathcal{T}_i$ would be indeed open because, if $t \in \mathcal{T}_i$, then the point $u_t = u_1 + t(u_2 - u_1)$ would belong to $\mathcal{O}_i$, and thus there would exist (Definition 2.9 (b) since $\mathcal{O}_i$ is open) $\eta > 0$ and a finite set $N \subset \mathcal{N}_E$ such that $\sup_{\nu \in N} \|u - u_t\|_{E;\nu} \le \eta$ would yield $u \in \mathcal{O}_i$. Then, $|s - t| \le \eta / \sup_{\nu \in N} \|u_2 - u_1\|_{E;\nu}$ would yield $u_s \in \mathcal{O}_i$, and therefrom $s \in \mathcal{T}_i$. □

Let us give properties of connected components.

THEOREM 2.34.– *In any separated semi-normed space:*

(a) *Any set is the union of its connected components.*

(b) *The connected components of a set are pairwise disjoint (or equal).*

(c) *The connected components of an open set are open.* ▮

Proof. Let U be a subset of a separated semi-normed space E. We denote by $\mathcal{U}_u$ its connected component generated by (one of its points) u.

(a) Each $u \in U$ is an element of $\mathcal{U}_u$. Conversely, all the $\mathcal{U}_u$ are included in U.

(b) Whenever $\mathcal{U}_u$ and $\mathcal{U}_v$ share one point w in common, they are both equal to $\mathcal{U}_w$. Indeed, $\mathcal{U}_u$ is a connected set included in U containing w, and therefore $\mathcal{U}_u \subset \mathcal{U}_w$. Subsequently, $u \in \mathcal{U}_w$ from which, similarly, $\mathcal{U}_w \subset \mathcal{U}_u$. Thus, $\mathcal{U}_u = \mathcal{U}_w$. Similarly, $\mathcal{U}_v = \mathcal{U}_w$.

(c) Let $\{\| \ \|_{E;\nu} : \nu \in \mathcal{N}_E\}$ be the family of semi-norms of E. Assume that U is open and let $u \in U$. There exists (Definition 2.9 (b)) a finite set $N \subset \mathcal{N}_E$ such that the set $B = \{v \in E : \sup_{\nu \in N} \|v - u\|_{E;\nu} \leq \eta\}$ be included in U. Being convex, B is connected (Theorem 2.33) and thus (Definition 2.30 (b)) $B \subset \mathcal{U}_u$. Therefore, $\mathcal{U}_u$ is open. □

Let us show that the closure of a convex set is convex.

THEOREM 2.35.– *In a separated semi-normed space, the closure of every convex set is convex.* ▮

Proof. Let E be a separated semi-normed space, $\{\| \ \|_{E;\nu} : \nu \in \mathcal{N}_E\}$ its family of semi-norms and $U \subset E$.

Let $u \in \overline{U}$, $v \in \overline{U}$, $t \in [0, 1]$ and $w = tu + (1-t)v$. For any finite set $N \subset \mathcal{N}_E$ and every $\eta > 0$, there exist (Theorem 2.17 (b)) $u' \in U$ and $v' \in U$ such that $\sup_{\nu \in N} \|u' - u\|_{E;\nu} \leq \eta$ and $\sup_{\nu \in N} \|v' - v\|_{E;\nu} \leq \eta$. Then, $w' = tu' + (1-t)v'$ verifies

$$\|w' - w\|_{E;\nu} \leq t \|u' - u\|_{E;\nu} + (1-t) \|v' - v\|_{E;\nu} \leq \eta.$$

If U is convex, then $w' \in U$ from which $w \in \overline{U}$ according to the characterization of the closure of Theorem 2.17 (b). Thus, $\overline{U}$ is convex. □

Chapter 3

Comparison of Semi-normed Spaces

3.1. Equivalent families of semi-norms

Let us define the equivalence of two families of semi-norms.

DEFINITION 3.1.– *Let $\{\| \ \|_{1;\nu} : \nu \in \mathcal{N}_1\}$ and $\{\| \ \|_{2;\mu} : \mu \in \mathcal{N}_2\}$ be two families of semi-norms on the same vector space E.*

(a) *The first family* ***dominates*** *the second if, for all $\mu \in \mathcal{N}_2$, there exist a finite set $N_1 \subset \mathcal{N}_1$ and $c_1 \in \mathbb{R}$ such that, for every $u \in E$,*

$$\|u\|_{2;\mu} \leq c_1 \sup_{\nu \in N_1} \|u\|_{1;\nu}.$$

(b) *Both families are* ***equivalent*** *if each dominates the other. It is also said that they* ***generate the same topology***. ▮

Terminology. It is said that families of semi-norms *generate the same topology* when they are equivalent because the equivalence of the families of semi-norms results in the equality of the families of open sets (Theorem 3.4 (b)), that is of topologies (Definition 2.14), and conversely as it can be easily verified. □

Observe that, if a vector space is separated by a family of semi-norms, it is also by any equivalent or dominating family.

THEOREM 3.2.– *Let $\mathcal{F}_1$ and $\mathcal{F}_2$ be two families of semi-norms on the same vector space E. Then:*

$$\begin{cases} \mathcal{F}_1 \text{ dominates or is equivalent to } \mathcal{F}_2 \text{ and} \\ \mathcal{F}_2 \text{ makes } E \text{ separated} \end{cases} \Rightarrow \quad \mathcal{F}_1 \text{ makes } E \text{ separated.}$$ ▮

Proof. Let $u \in E$. If $\|u\|_{1;\nu} = 0$ for all $\nu \in \mathcal{N}_1$, the inequality of Definition 3.1 (a) shows that $\|u\|_{2;\mu} = 0$ for all $\mu \in \mathcal{N}_2$. If the second family separates E (Definition 2.3 (c)), it follows therefrom that $u = 0_E$, thus the first family separates it as well. □

Observe that the equivalence of families of semi-norms is *transitive*.

THEOREM 3.3.– *Let $\mathcal{F}_1$, $\mathcal{F}_2$ and $\mathcal{F}_3$ be three families of semi-norms on the same vector space E. Then:*

(a) *$\mathcal{F}_1$ dominates $\mathcal{F}_2$ and $\mathcal{F}_2$ dominates $\mathcal{F}_3$ $\Rightarrow$ $\mathcal{F}_1$ dominates $\mathcal{F}_3$.*

(b) *$\mathcal{F}_1$ is equivalent to $\mathcal{F}_2$ and $\mathcal{F}_2$ is equivalent to $\mathcal{F}_3$ $\Rightarrow$ $\mathcal{F}_1$ is equivalent to $\mathcal{F}_3$.* ▮

Proof. If $\mathcal{F}_1$ dominates $\mathcal{F}_2$ which dominates $\mathcal{F}_3$, Definition 3.1 (a) shows that $\mathcal{F}_1$ dominates $\mathcal{F}_3$, hence (a). And vice versa, from which (b). □

Let us show that the already defined concepts (open, bounded and compact sets, etc.) are preserved when the family of semi-norms is replaced by an equivalent family.

THEOREM 3.4.– *If the family of semi-norms of a separated semi-normed space E is replaced by an equivalent family, it preserves:*

(a) *Convergent sequences, their limits, Cauchy sequences and bounded sequences of E.*

(b) *Bounded, open, closed, sequentially closed, separable, sequentially separable, compact, sequentially compact, relatively compact, relatively sequentially compact, precompact and connected subsets of E.*

(c) *Dense or sequentially dense subsets of a set of E.*

(d) *The interior, the closure, the sequential closure and the boundary of the subsets of E.* ▮

Proof. These are consequences of the properties of the topological inclusions given in Theorem 3.8. □

3.2. Topological equalities and inclusions

Let us define vector spaces.

DEFINITION 3.5.– *A **vector subspace** of a vector space F is any subset E of F that is a vector space for the addition and the multiplication of E.*

That is any subset E of F such that

$$E \neq \emptyset \quad \textit{and} \quad u+v \in E, \; tu \in E, \;\; \forall u \in E, \; v \in E, \; t \in \mathbb{R}. \tag{3.1}$$

■

Justification of (3.1).– Properties (3.1) are included in Definition 2.1 of a vector space. Conversely, if (3.1) is satisfied, the other properties of Definition 2.1 are verified with $0_E = 0_F$; especially, $0_F \in E$ because there exists $w \in E$, thus $0w = 0_F \in E$, see Theorem 2.5 (b), and $-u \in E$ if $u \in E$ because $-u = (-1)u$. □

Let us now define equalities and topological inclusions (of not necessarily separated semi-normed spaces).

DEFINITION 3.6.– *Let E and F be two semi-normed spaces.*

(a) *We denote*

$$E \underset{\leftrightarrow}{\equiv\!\equiv} F$$

if $E = F$ and if their additions, their multiplications and their families of semi-norms coincide. That is, if they have the same vector space structure and the same semi-norms.

(b) *The space E is **topologically equal** to F and we denote*

$$E \underset{\leftrightarrow}{\equiv} F$$

if $E = F$, if their additions and their multiplication coincide and if their families of semi-norms are equivalent.

(c) *The space E is **topologically included** in F and we denote*

$$E \underset{\to}{\subseteq} F$$

if E is a vector subspace of F and if the family of semi-norms of E dominates the family of restrictions to E of semi-norms of F.

That is to say, if $E \subset F$, if their additions and their multiplications coincide on E and if, for any semi-norm $\| \; \|_{F;\mu}$ of F, there is a finite set $N \subset \mathcal{N}_E$ (set indexing the semi-norms of E) and $c \in \mathbb{R}$ such that, for all $u \in E$,

$$\|u\|_{F;\mu} \leq c \sup_{\nu \in N} \|u\|_{E;\nu}. \;\blacksquare$$

Justification of the topological naming. *Topological* equality is equivalent to the equality of families of open sets, namely the equality of topologies (see the observation "*Terminology*", p. 33).

Topological inclusion $E \underset{\to}{\subseteq} F$ designates, for its part, the fact that any subset of E that is open in F be open in E (Theorem 3.8 (c)). If $E = F$, this means that the topology of F is included in that of E (it is said that it is *coarser*). Otherwise, this only means that the "topology induced on E by F" (see p. 117) is coarser than the topology of E. □

Let us observe that every semi-normed space topologically included in a separated space is separated.

THEOREM 3.7.– *Let E and F be two semi-normed spaces such that*

$$E \underset{\leftrightarrow}{=} F \quad \text{or} \quad E \underset{\to}{\subseteq} F.$$

Then:

$$F \text{ is separated} \quad \Rightarrow \quad E \text{ is separated.} ∎$$

Proof. Let $\{\| \ \|_{E;\nu} : \nu \in \mathcal{N}_E\}$ and $\{\| \ \|_{F;\mu} : \mu \in \mathcal{N}_F\}$ be the families of semi-norms of E and F. Let $u \in E$ be such that $\|u\|_{E;\nu} = 0$ for every $\nu \in \mathcal{N}_E$.

For all $\mu \in \mathcal{N}_F$ there exist (Definition 3.6 (b) or (c)) a finite set $N \subset \mathcal{N}_E$ and $c \in \mathbb{R}$ such that $\|u\|_{F;\mu} \leq c \sup_{\nu \in N} \|u\|_{E;\nu} = 0$. If F is separated (Definition 2.3 (c)), then $u = 0_F$ and therefore $u = 0_E$ because E and F have the same zero element (since their additions coincide on E). Thereby, E is separated. □

Let us give properties of topological inclusions.

THEOREM 3.8.– *Let E and F be two separated semi-normed spaces such that*

$$E \underset{\to}{\subseteq} F.$$

Let $(u_n)_{n\in\mathbb{N}} \subset E$, $u \in E$, $U \subset E$, $V \subset E$ and $W \subset F$. Then:

(a) *If $(u_n)_{n\in\mathbb{N}}$ has one of the following properties in E, it has it in F:*

convergent, convergent to u, Cauchy sequence, bounded.

(b) *If U has one of the following properties in E, it has it in F:*

bounded, compact, relatively compact, precompact, sequentially compact, relatively sequentially compact, dense in V, sequentially dense in V, separable, sequentially separable, connected.

(c) *If U has one of the following properties in F, it has it in E:*

open, closed, sequentially closed.

(d) $$\mathring{U}^E \supset \mathring{U}^F,$$

$$\overline{U}^E \subset \overline{U}^F, \ \overline{U}^{\text{seq} E} \subset \overline{U}^{\text{seq} F}, \ U^{\lim \text{seq} E} \subset U^{\lim \text{seq} F}, \ \partial_E U \subset \partial_F U.$$

(e) *If W has one of the following properties in F, $W \cap E$ has it in E:*

open, closed, sequentially closed. ■

Proof. Let $\{\| \ \|_{E;\nu} : \nu \in \mathcal{N}_E\}$ and $\{\| \ \|_{F;\mu} : \mu \in \mathcal{N}_F\}$ be the families of semi-norms of E and F. Since $E \underset{\to}{\subsetneq} F$, for every $\mu \in \mathcal{N}_F$, there exist (Definition 3.6 (c)) a finite set $N_\mu \subset \mathcal{N}_E$ and $c_\mu > 0$ such that, for every $v \in E$,

$$\|v\|_{F;\mu} \leq c_\mu \sup_{\nu \in N_\mu} \|v\|_{E;\nu}. \tag{3.2}$$

(a) *Sequence.* If the sequence $(u_n)_{n\in\mathbb{N}}$ is convergent to u, a Cauchy sequence, or bounded in E, it is so in F according to the inequality (3.2) and Definitions 2.6 (a), (b) and (c) of these notions.

(e) *Open.* If W is open in F, there exist (Definition 2.9 (b)) a finite set $M \subset \mathcal{N}_F$ and $\epsilon > 0$ such that every $w \in F$ such that $\sup_{\mu\in M} \|w - u\|_{F;\mu} \leq \epsilon$ is an element of W. Then, $N = \bigcup_{\mu\in M} N_\mu$ is finite, $\eta = \epsilon / \sup_{\mu\in M} c_\mu$ is > 0 and, with (3.2), $v \in E$ and $\sup_{\nu\in N} \|v - u\|_{E;\nu} \leq \epsilon$ lead to $v \in W \cap E$. The latter is therefore indeed open.

Closed. If W is closed in F, then $F \setminus W$ is open therein, thus, as it has just been seen, $(F \setminus W) \cap E$ is open in E. Its complement $E \setminus ((F \setminus W) \cap E)$, that is $W \cap E$, is therefore closed in E.

Sequentially closed. Let $(u_n)_{n\in\mathbb{N}} \subset U$ be such that $u_n \to u$ in E. Then, $u_n \to u$ in F according to (a). If U is sequentially closed in F, then $u \in U$ and as a result U is in fact sequentially closed in E.

(c) These properties are the result of (e) since $U \cap E = U$.

(d) *Interior.* The set $\mathring{U}^F$, being included in E (since it is included in U) and open in F, is open in E according to (c). It is therefore included in $\mathring{U}^E$ since the latter is (Definition 2.15 (a)) the largest open subset of E included in U.

Closure. The set $\overline{U}^F \cap E$ is closed in E according to (e), since $\overline{U}^F$ is closed in F, and it contains U, and thus $\overline{U}^E$ since the latter is (Definition 2.15 (b)) the smallest closed set E containing U. Therefore, $\overline{U}^F \supset \overline{U}^F \cap E \supset \overline{U}^E$.

Sequential closure. The set $\overline{U}^{\text{seq}\,F} \cap E$ is sequentially closed in E according to (e), since $\overline{U}^{\text{seq}\,F}$ is sequentially closed in F, and it contains U, and thus $\overline{U}^{\text{seq}\,E}$ because it is (Definition 2.15 (c)) the smallest sequentially closed set containing U. Consequently, $\overline{U}^{\text{seq}\,F} \supset \overline{U}^{\text{seq}\,F} \cap E \supset \overline{U}^{\text{seq}\,E}$.

Limits. The sequences of U that converge in E converging in F according to (a), the set $U^{\lim\,\text{seq}\,E}$ of their limits (Definition 2.15 (d)) is included in $U^{\lim\,\text{seq}\,F}$.

Boundaries. By Definition 2.15 (e), $\partial_E U = \overline{U}^E \setminus \mathring{U}^E$ that is included in the set $\overline{U}^F \setminus \mathring{U}^F = \partial_F U$ since, as it has been seen above, $\overline{U}^E \subset \overline{U}^F$ and $\mathring{U}^E \supset \mathring{U}^F$.

(b) *Bounded.* If U is bounded in E, that is if $\sup_{v \in U} \|v\|_{E;\nu} < \infty$ for every $\nu \in \mathcal{N}_E$, then $\sup_{v \in U} \|v\|_{F;\mu} < \infty$ for every $\mu \in \mathcal{N}_F$ according to (3.2) and thus U is indeed bounded in F.

Compact. Let $\{\mathcal{O}_i\}_{i \in \mathcal{I}}$ be a family of open subsets of F covering U. The sets $\mathcal{O}_i \cap E$ cover U and are open in E following (e), so, if U is compact in E, there exists a finite subcover $\{\mathcal{O}_i \cap E\}_{i \in J}$. Then $\{\mathcal{O}_i\}_{i \in J}$ is a finite cover of U, which is therefore indeed compact in F.

Relatively compact. If U is relatively compact in E, it is included in a compact set (Theorem 2.26 (c)) of E, and thus of F just as it has been seen. As a result, it is (Theorem 2.26 (c), again) indeed relatively compact in F.

Sequentially compact. If U is sequentially compact (Definition 2.24 (a)) in E, every sequence of U has a subsequence that converges therein to a limit which is an element of U. Since this subsequence converges in F according to (a), U is indeed sequentially compact in F.

Relatively sequentially compact. If U is relatively sequentially compact (Definition 2.24 (a)) in E, every sequence of U has a subsequence that converges therein. This subsequence converging in F according to (a), U is indeed sequentially compact in F.

Precompact. Assume that U is precompact in E and let $\mu \in \mathcal{N}_F$, N_μ satisfy (3.2) and $\epsilon > 0$. For each $\nu \in N_\mu$, there exists (Definition 2.23 (c)) a finite set $\mathcal{Z}_\nu \subset E$ such that the semi-balls $B_{E;\nu}(z_\nu, \epsilon/2c) = \{v \in E : \|v - z_\nu\|_{E;\nu} \leq \epsilon/2c\}$ cover U when z_ν spans $\mathcal{Z}_\nu$. The sets $A(z) = \bigcap_{\nu \in N_\mu} B_{E;\nu}(z_\nu, \epsilon/2c)$ therefore cover U when $z = (z_\nu)_{\nu \in N_\mu}$ spans the finite set $\mathcal{Z} = \prod_{\nu \in N_\mu} \mathcal{Z}_\nu$.

For each $z \in \mathcal{Z}$, if $A(z)$ is not empty, one of its points is chosen that is denoted by z^*. Any other point v of $A(z)$ verifies $\|v - z^*\|_{E;\nu} \leq \|v - z_\nu\|_{E;\nu} + \|z^* - z_\nu\|_{E;\nu} \leq \epsilon/c$ for each $\nu \in N_\mu$, and thus $\|v - z^*\|_{F;\mu} \leq \epsilon$. In other words,

$$A(z) \subset B_{F;\mu}(z^*, \epsilon).$$

The semi-balls $B_{F;\mu}(z^*, \epsilon)$ therefore cover U when z spans $\mathcal{Z}$, which proves that U is precompact in F.

Dense. If, in E, U is dense in V, that is if $U \subset V \subset \overline{U}^E$, then we get $U \subset V \subset \overline{U}^F$ according to (d), so the density is actually achieved in F.

Sequentially dense. The procedure is just as it has been done with the density, with (Definition 2.19 (b)) $U^{\text{lim seq}}$ instead of $\overline{U}$.

Separable. If in E, U is separable, that is (Definition 2.21 (a)) if there exists a dense countable set in U, it is also achieved in F according to the density property established above.

Sequentially separable. One proceeds just as it has been carried out with separable, with (Definition 2.21 (b)) the sequential density instead of the density.

Connected. If U is connected in E, it is so in F, otherwise (Definition 2.30) it could be covered by two open sets $\mathcal{O}_1$ and $\mathcal{O}_2$ of F having disjoint and non-empty intersections with U. Then, $\mathcal{O}_1 \cap E$ and $\mathcal{O}_2 \cap E$ would be open subsets of E according to (e) and their intersections with U would be disjoint and non-empty, which would contradict the connectedness of U. □

3.3. Topological subspaces

Let us define topological subspaces.

DEFINITION 3.9.– *Let E and F be two semi-normed spaces.*

The space E is a ***topological subspace*** *of F if it is a vector subspace of F and if it is endowed with the restrictions of the semi-norms of F or, more generally, of a family that is equivalent to the family of these restrictions.* ▮

Let us observe that any topological subspace of a separated space is separated.

THEOREM 3.10.– *Let E and F be two semi-normed spaces such that*

$$E \text{ is a topological subspace of } F.$$

Then:

(a) $$E \subsetneqq F.$$

(b) $$F \text{ is separated} \quad \Rightarrow \quad E \text{ is separated.}$$ ▮

Proof. (a) This inclusion follows from Definitions 3.9 of a topological subspace and 3.6 (c) of a topological inclusion.

(b) This is the result of Theorem 3.7, according to (a). □

We next give properties of topological subspaces.

THEOREM 3.11.– *Let E and F be two separated semi-normed spaces such that*

$$E \text{ is a topological subspace of } F.$$

Let $(u_n)_{n\in\mathbb{N}} \subset E$, $u \in E$, $U \subset E$ and $V \subset E$. Then:

(a) $(u_n)_{n\in\mathbb{N}}$ *has one of the following properties in* E *if and only if it has it in* F:

$$\text{convergent to } u, \text{ Cauchy sequence, bounded.}$$

(b) U *has one of the following properties in* E *if and only if it has it in* F:

$$\begin{cases} \text{bounded, compact, sequentially compact,} \\ \text{precompact, dense in } V, \text{ separable.} \end{cases}$$

(c) U *is open in* E $\Leftrightarrow$ *there is an open set* $\mathcal{O}$ *of* F *such that* $U = \mathcal{O} \cap E$.

(d) U *is closed in* E $\Leftrightarrow$ *there is a closed set* $\mathcal{F}$ *of* F *such that* $U = \mathcal{F} \cap E$.

(e) $$\overline{U}^E = \overline{U}^F \cap E. \blacksquare$$

Let us complete these results with properties of sequentially closed subspaces.

THEOREM 3.12.– *Let* E *and* F *be separated semi-normed spaces such that*

$$E \text{ is a sequentially closed (or closed) topological subspace of } F.$$

Let $(u_n)_{n\in\mathbb{N}} \subset E$ *and* $U \subset E$. *Then:*

(a) $(u_n)_{n\in\mathbb{N}}$ *converges in* E *if and only if it converges in* F.

(b) U *has one of the following properties in* E *if and only if it has it in* F:

$$\begin{cases} \text{sequentially closed, relatively sequentially compact,} \\ \text{sequentially dense in } V, \text{ sequentially separable.} \end{cases}$$

(c) $$\overline{U}^{\text{seq}\,E} = \overline{U}^{\text{seq}\,F}, \quad U^{\lim\text{seq}\,E} = U^{\lim\text{seq}\,F}. \blacksquare$$

The above results are now completed by properties of closed subspaces.

THEOREM 3.13.– *Let* E *and* F *be separated semi-normed spaces such that*

$$E \text{ is a closed topological subspace of } F.$$

Let $U \subset E$. *Then:*

(a) U *has one of the following properties in* E, *if and only if it has it in* F:

$$\text{closed, relatively compact.}$$

(b) $$\overline{U}^E = \overline{U}^F. \blacksquare$$

Proof of Theorem 3.11. Let $\{\| \ \|_{F;\nu} : \nu \in \mathcal{N}_F\}$ be the family of semi-norms of F and $\{||| \ |||_{E;\nu} : \nu \in \mathcal{N}_F\}$ be the family of its restrictions to E, that is, for all $v \in E$ and $\nu \in \mathcal{N}_F$,

$$|||v|||_{E;\nu} = \|v\|_{F;\nu}. \tag{3.3}$$

Let us endow E with this family of semi-norms which preserves (Theorem 3.4) the notions being studied (convergent suites, open sets, etc.) since it is equivalent (Definition 3.9) to its own family.

(a) *Sequence.* The sequence $(u_n)_{n\in\mathbb{N}}$ is a Cauchy sequence, or convergent to u, or bounded in E if and only if it is so in F according to the equality (3.3) and Definitions 2.6 (a), (b) and (c) of these notions.

(c) *Open set.* Assume that U is an open set in E and let $v \in U$. There exists (Definition 2.9 (b)) a finite set $N \subset \mathcal{N}_F$ and $\eta > 0$ such that every $z \in E$ such that $\sup_{\nu \in N} \|z - v\|_{E;\nu} \le \eta$ is an element of U, that is $\bigcap_{\nu\in N} B_{E;\nu}(v;\eta) \subset U$. Let

$$\mathcal{O}_{v,N,\eta} = \bigcap_{\nu \in N} \mathring{B}_{F;\nu}(v;\eta).$$

Each $\mathcal{O}_{v,N,\eta}$ is open in F (as any finite intersection of open sets, see Theorem 2.11 (b)) and thus also their union $\mathcal{O}$ (Theorem 2.11 (a)). In addition, $\mathcal{O} \cap E \subset U$ because each $\mathcal{O}_{v,N,\eta} \cap E$ is included in U (because $\mathring{B}_{F;\nu}(v;\eta) \cap E = \mathring{B}_{E;\nu}(v;\eta)$); in addition, $\mathcal{O} \cap E \supset U$ since v spans U, and thereby $\mathcal{O} \cap E = U$.

Conversely, if $U = \mathcal{O} \cap E$ where $\mathcal{O}$ is an open subset of F, then U is open in E according to Theorem 3.8 (e).

(d) *Closed.* By definition, U being closed in E is equivalent to $E \setminus U$ being open in E namely, just as it has been seen, $E \setminus U = \mathcal{O} \cap E$ where $\mathcal{O}$ is open in F, and thus it is equivalent to $U = \mathcal{F} \cap E$ where $\mathcal{F} = F \setminus \mathcal{O}$ is closed in F.

(e) *Closure.* According to (d), there exists a closed set $\mathcal{F}$ of F such that $\overline{U}^E = \mathcal{F} \cap E$. Then $\mathcal{F}$ contains U, therefore it contains $\overline{U}^F$ (since this is the smallest closed set containing U) and consequently $\overline{U}^E \supset \overline{U}^F \cap E$.

This inclusion is an equality because, conversely, $\overline{U}^E \subset \overline{U}^F$ according to Theorem 3.8 (d) and thus $\overline{U}^E \subset \overline{U}^F \cap E$.

(b) *Bounded.* By definition, U is bounded in F if and only if, for every $\nu \in \mathcal{N}_E$, there is $\sup_{u\in U} \|u\|_{F;\nu} < \infty$, that is $\sup_{u\in U} |||u|||_{E;\nu} < \infty$ according to (3.3), and thus if and only if U is bounded in E since the semi-norms of E are also indexed by $\mathcal{N}_F$.

Compact. Assume that U is compact in F and let $\{\mathcal{A}_i\}_{i\in\mathcal{I}}$ be open subsets of E covering U. For each $i \in \mathcal{I}$, there exists, as it has just been seen, an open set $\mathcal{O}_i$ of F

such that $\mathcal{A}_i = \mathcal{O}_i \cap E$. The $\mathcal{O}_i$ cover U so there exists (Definition 2.23 (a)) a finite subcover $\{\mathcal{O}_i\}_{i \in J}$. The $\{\mathcal{A}_i\}_{i \in J}$ then form a finite cover of U, which is therefore indeed compact in E.

Conversely, if U is compact in E, it is also in F according to Theorem 3.8 (b).

Sequentially compact. If U is sequentially compact (Definition 2.24 (a)) in E, every sequence of U has a subsequence that converges therein to a limit which is an element of U; this subsequence converging to E according to (a), U is indeed sequentially compact in E. The innverse was established in Theorem 3.8 (b).

Precompact. Assume that U is precompact in F and let $\nu \in \mathcal{N}_F$ and $\epsilon > 0$. There is then (Theorem 2.28) a finite number of points $z \in U$ such that semi-balls $B_{F;\nu}(z, \epsilon)$ cover U. Then, $B_{E;\nu}(z, \epsilon) = B_{F;\nu}(z, \epsilon) \cap E$, and therefore these semi-balls of E cover $U \cap E$, that is U which, consequently, is indeed precompact in E.

Conversely, if U is precompact in E it is also in F because, if some $B_{E;\nu}(z, \epsilon)$ cover it, the corresponding $B_{F;\nu}(z, \epsilon)$ cover it *a fortiori*.

Dense. If U is dense in V in the space F, that is if $U \subset V \subset \overline{U}^F$, since $V \subset E$ we have $U \subset V \subset \overline{U}^F \cap E = \overline{U}^E$ according to (e), therefore the density is also achieved in the space E. The innverse has been established in Theorem 3.8 (b).

Separable. The set U is separable in E, which means that there exists a dense countable subset of U in the space E, if and only if this is achieved in the space F according to the previous property. □

Proof of Theorem 3.12. Here, E is sequentially closed in F.

(a) *Convergence.* If $(u_n)_{n \in \mathbb{N}} \subset E$ converges in F, its limit is an element of E since the latter is sequentially closed in F. Therefore, $(u_n)_{n \in \mathbb{N}}$ converges to the above-mentioned limit in E following Theorem 3.11 (b). And vice versa.

(b) *(beginning) Sequentially closed.* Let $(u_n)_{n \in \mathbb{N}} \subset U$ be such that $u_n \to u$ in F. Then, $u_n \to u$ in E as it has been observed. If U is sequentially closed in E, then $u \in U$ and thus U is sequentially closed in F. The inverse has been established in Theorem 3.8 (b).

Relatively sequentially compact. If the set U is relatively sequentially compact (Definition 2.24 (b)) in F, every sequence of U has a subsequence that converges therein. Since this subsequence converges in E as previously noted, U is therein relatively sequentially compact. The inverse has been established in Theorem 3.8 (b).

Sequential closure. The set $\overline{U}^{\text{seq}\,E}$, being sequentially closed in E, is also in F according to (b). Since it contains U, it contains $\overline{U}^{\text{seq}\,F}$ which is (Definition 2.15 (c)) the smallest sequentially closed set containing U. Conversely, $\overline{U}^{\text{seq}\,E} \subset \overline{U}^{\text{seq}\,F}$ according to Theorem 3.8 (d), therefore these sets are equal.

Limits. A sequence converges in E if and only if it converges in F following (a) and its limit is unique (Theorem 2.7 (a)), therefore the sets of limits $U^{\lim \text{seq}\,E}$ and $U^{\lim \text{seq}\,F}$ (Definition 2.15 (d)) coincide.

(b) *(continuation) Sequentially dense.* The set U is sequentially dense in V in the space E, that is (Definition 2.19 (b)) $U \subset V \subset U^{\lim \text{seq}\,E}$ if and only if this is achieved in the space F since $U^{\lim \text{seq}\,E} = U^{\lim \text{seq}\,F}$ according to (c).

Sequentially separable. The set U is sequentially separable in E, that is, there is a sequentially dense countable set in U in the space E, if and only if this is achieved in the space F according to the previous property. □

Proof of Theorem 3.13. Here, E is closed in F.

(a) *Closed.* If U is closed in E, according to Theorem 3.11 (d) there exists a closed set $\mathcal{F}$ of F such that $U = \mathcal{F} \cap E$; the intersection of closed sets being closed (Theorem 2.12 (a)), U is therefore closed in F. The inverse has been established in Theorem 3.8 (b).

(b) *Closure.* According to Theorem 3.11 (e), $\overline{U}^E = \overline{U}^F \cap E$. Now, $\overline{U}^F \subset \overline{E}^F$, and $\overline{E}^F = E$ since E is closed in F, thus $\overline{U}^F \cap E = \overline{U}^F$. Therefore, $\overline{U}^E = \overline{U}^F$.

(a) *Relatively compact.* Assume that U is relatively compact in F, that is (Definition 2.23 (b)) $\overline{U}^F$ is compact in F. Then, $\overline{U}^F$ is compact in E according to Theorem 3.11 (b) and it is equal to $\overline{U}^E$ following (b). Consequently, $\overline{U}^E$ is compact in E, and then U is therein relatively compact.

Conversely, if U is relatively compact in E, it is also in F according to Theorem 3.8 (b). □

3.4. Filtering families of semi-norms

Let us define the filtering families of semi-norms.

DEFINITION 3.14.– *A family $\{\| \ \|_\nu : \nu \in \mathcal{N}\}$ of semi-norms on a vector space E is* **filtering** *if, for any finite subset N of $\mathcal{N}$, there exists $\mu \in \mathcal{N}$ such that, for all $u \in E$,*

$$\sup_{\nu \in N} \|u\|_\nu \leq \|u\|_\mu. \blacksquare$$

Let us show that any separated semi-normed space can be endowed with a filtering family of semi-norms.

THEOREM 3.15.– *Any family of semi-norms on the same vector space is equivalent to a filtering family.* □

Proof. Let $\mathcal{F} = \{\| \ \|_\nu : \nu \in \mathcal{N}\}$ be a family of semi-norms on a vector space E. For any finite subset N of $\mathcal{N}$, a semi-norm is defined on E, as we are going to verify in Theorem 3.16 (a) below, by $||| \ |||_N = \sup_{\nu \in N} \| \ \|_\nu$. Their family is filtering (Definition 3.14) and is equivalent (Definition 3.1) to $\mathcal{F}$. □

We must now verify that the upper envelope of semi-norms is a semi-norm. We are also going to verify that this is the case of the sum of semi-norms.

THEOREM 3.16.– *Let $p_1, p_2 \dots, p_n$ be semi-norms on the same vector space E.*

(a) *Two semi-norms are defined on E, the* ***sum*** $\sum_{1 \le i \le n} p_i$ *of p_i and their* ***upper envelope*** $\sup_{1 \le i \le n} p_i$, *by: for all $u \in E$,*

$$\Big(\sum_{1 \le i \le n} p_i\Big)(u) \stackrel{\text{def}}{=} \sum_{1 \le i \le n} p_i(u), \qquad \Big(\sup_{1 \le i \le n} p_i\Big)(u) \stackrel{\text{def}}{=} \sup_{1 \le i \le n} p_i(u).$$

(b) *If one of the p_i is a norm, then $\sum_{1 \le i \le n} p_i$ and $\sup_{1 \le i \le n} p_i$ are norms.* ∎

Proof. (a) *Sum.* The properties of Definition 2.2 (a) of a semi-norm are satisfied because $\sum_{1 \le i \le n} p_i(u) \ge 0$ and

$$\sum_{1 \le i \le n} p_i(tu) = \sum_{1 \le i \le n} |t|\, p_i(u) = |t| \sum_{1 \le i \le n} p_i(u),$$

$$\sum_{1 \le i \le n} p_i(u+v) \le \sum_{1 \le i \le n} (p_i(u) + p_i(v)) = \sum_{1 \le i \le n} p_i(u) + \sum_{1 \le i \le n} p_i(v).$$

Upper envelope. The above properties are verified with $\sup_{1 \le i \le n}$ instead of $\sum_{1 \le i \le n}$ observing, for subadditivity, that

$$\sup_{1 \le i \le n} (a_i + b_i) \le \sup_{1 \le i \le n} (a_i + \sup_{1 \le j \le n} b_j) \le \sup_{1 \le i \le n} a_i + \sup_{1 \le j \le n} b_j.$$

(b) If there exists $k \in [\![1, n]\!]$ such that $u \ne 0_E$ yields $p_k(u) > 0$, then we get $\sum_{1 \le i \le n} p_i(u) > 0$ and $\sup_{1 \le i \le n} p_i(u) > 0$ for such u. □

Let us show that the characterizations of an open set, of the interior, of the closure and of the topological inclusion are simplified with a filtering family: the upper envelope of a finite number of semi-norms is therein replaced by a single semi-norm.

THEOREM 3.17.– *Let E be a separated semi-normed space such that:*

$$\textit{Its family } \{\| \ \|_{E;\nu} : \nu \in \mathcal{N}_E\} \textit{ of semi-norms is filtering.}$$

Let $U \subset E$ and F be another separated semi-normed space. Then:

$$\text{(a)} \qquad U \text{ is open} \quad \Leftrightarrow \quad \begin{cases} \forall u \in U, \text{ there exist } \nu \in \mathcal{N}_E \text{ and } \eta > 0 \text{ such that} \\ \qquad v \in E,\ \|v-u\|_{E;\nu} \leq \eta \quad \Rightarrow \quad v \in U \\ \text{that is to say, such that } B_{E;\nu}(u,\eta) \subset U. \end{cases}$$

$$\text{(b)} \qquad \overset{\circ}{U} = \big\{u \in U : \exists \nu \in \mathcal{N}_E, \exists \eta > 0 \text{ such that } \|v-u\|_{E;\nu} \leq \eta \Rightarrow v \in U\big\}.$$

$$\text{(c)} \qquad \overline{U} = \big\{u \in E : \forall \nu \in \mathcal{N}_E, \forall \eta > 0, \exists v \in U \text{ such that } \|v-u\|_{E;\nu} \leq \eta\big\}.$$

$$\text{(d)} \qquad E \underset{\rightarrow}{\subseteq} F \quad \Leftrightarrow \quad \begin{cases} E \subset F \text{ and, for every semi-norm } \| \ \|_{F;\mu} \text{ of } F, \\ \text{there exist } \nu \in \mathcal{N}_E \text{ and } c \in \mathbb{R} \text{ such that} \\ \qquad \| \ \|_{F;\mu} \leq c \| \ \|_{E;\nu}. \end{cases}$$ ∎

Proof. (a) If U is open, for every $u \in U$ there exists (Definition 2.9 (b)) a finite set $N \subset \mathcal{N}_E$ such that $\sup_{\lambda \in N} \|v-u\|_{E;\lambda} \leq \eta$ results in $v \in U$. By Definition 3.14 of a filtering family, there is $\nu \in \mathcal{N}_E$ such that $\sup_{\lambda \in N} \| \ \|_{E;\lambda} \leq \| \ \|_{E;\nu}$. Then, $\|v-u\|_{E;\nu} \leq \eta$ indeed leads to $v \in U$.

Conversely, this property satisfies Definition 2.9 (b) of an open set with $N = \{\nu\}$.

(b) *and* (c). These characterizations of $\overset{\circ}{U}$ and $\overline{U}$ may be deduced from those of Theorem 2.17 (a) and (b) in a similar manner to what we just did for an open set. □

Usefulness of filtering families. The use of filtering families is essential to define the sum of semi-normed spaces, see the remark "*Need for the filtration assumption*", p. 85.

On the other hand, it simplifies, among other things, the characterization of open and closed sets (Theorem 3.17 vs. Theorem 2.17), that of continuous mappings (Theorem 7.3 *vs*. Definition 7.1) and some proofs, by substituting a single semi-norm to the upper envelope of a finite number of semi-norms. □

Spaces endowed with filtering families. The "natural" family of some spaces is filtering. For example, the dual E' is endowed with the family of semi-norms $\|e'\|_{E';B}$ indexed by the bounded sets B of E which is filtering (Theorem 13.6 (b)).

Other common spaces are endowed with filtering families, such as $\mathcal{D}(\Omega)$, see (2.1), p. 13, and $L^p_{\text{loc}}(\Omega)$. □

Normed spaces. A single norm constitute, on its own, a filtering family of semi-norms. More generally, any finite family of semi-norms is filtering. □

Usefulness of non-filtering families. The "natural" family of some spaces is not filtering. For example, E-weak is endowed with the semi-norms $\|e\|_{E\text{-weak};e'} = |\langle e', e\rangle|$ indexed by $e' \in E'$ (Definition 15.1). If the aim was to consider filtering families only, then E-weak should be endowed with the semi-norms

$|||e|||_{E\text{-weak};N} = \sup_{e' \in N} |\langle e', e\rangle|$ indexed by the finite subsets N of E'. The maximum over finite subsets would of course disappear from the definition of an open set but it would reappear here.

Similarly, E'-∗weak (Definition 13.4 (b)), $\mathcal{L}(E;F)$-pt (Definition 12.1 (b)) and $\mathcal{D}'(\Omega)$ endowed with the topology of pointwise convergence on $\mathcal{D}(\Omega)$ (p. 13) are, among others, endowed with non-filtering families. □

3.5. Sums of sets

If U and V are sets and x is an element of the same vector space, we denote:

$$U + V \overset{\text{def}}{=} \{u + v : u \in U,\, v \in V\}, \quad U + x \overset{\text{def}}{=} \{u + x : u \in U\}, \tag{3.4}$$
$$-U \overset{\text{def}}{=} \{-u : u \in U\}, \quad U - V \overset{\text{def}}{=} U + (-V), \quad U - x \overset{\text{def}}{=} U + (-x).$$

Let us next show that the sum of closures is included in the closure of the sum.

THEOREM 3.18.– *Let U and V be subsets of a separated semi-normed space. Then:*

$$\overline{U} + \overline{V} \subset \overline{U + V}. \ ▮$$

Proof. Let E be the separated semi-normed space being considered. We endow it (Theorem 3.15) with a filtering family of semi-norms, which does not change the closure (Theorem 3.4 (d)).

Let $u \in \overline{U}$ and $v \in \overline{V}$. According to the characterization of the closure of Theorem 3.17 (c), for every $\nu \in \mathcal{N}_E$ and $\eta > 0$, there exist $u' \in U$ and $v' \in V$ such that $\|u' - u\|_{E;\nu} \le \eta/2$ and $\|v' - v\|_{E;\nu} \le \eta/2$. Then $\|(u' + v') - (u + v)\|_{E;\nu} \le \eta$ and $u' + v' \in U + V$, and thus $u + v$ is indeed contained in $\overline{U + V}$. □

Let us give a separation property of disjoint sets.

THEOREM 3.19.– *Let U and K be subsets of a separated semi-normed space E such that*

U is closed, K is compact, $U \cap K = \emptyset$.

Then, there is a finite intersection $B = \bigcap_{\nu \in N} B_{E;\nu}(0_E, r)$ of semi-balls such that

$$(U + B) \cap K = \emptyset. \ ▮$$

Proof. Let $\{\| \ \|_{E;\nu} : \nu \in \mathcal{N}_E\}$ be the family of semi-norms of E and let $U^c = E \backslash U$. Every $v \in K$ is an element of the open set U^c, therefore (Definition 2.9 (c)) there exists a finite set $N_v \subset \mathcal{N}_E$ and $\eta_v > 0$ such that, for all $w \in E$,

$$\sup_{\nu \in N_v} \|w - v\|_{E;\nu} \le \eta_\nu \quad \Rightarrow \quad w \in U^c. \tag{3.5}$$

Let

$$\mathcal{O}_v = \{w \in E : \sup_{\nu \in N_v} \|w - v\|_{E;\nu} < \eta_v/2\}.$$

The open sets $\mathcal{O}_v$ cover the compact K so there is a finite set $\mathcal{V} \subset K$ such that

$$\{\mathcal{O}_v : v \in \mathcal{V}\} \text{ covers } K. \tag{3.6}$$

The stated property is satisfied with $N = \bigcup_{v \in \mathcal{V}} N_v$ and $r = \min_{v \in \mathcal{V}} \eta_v$ (this minimum does exist and is > 0 thanks to Theorem 1.2 (c)). In effect, let $\mathcal{O} = \bigcup_{v \in \mathcal{V}} \mathcal{O}_v$. On the one hand, $K \subset \mathcal{O}$ according to (3.6), and on the other hand, $\mathcal{O}_v + B \subset U^c$ according to (3.5) from which $\mathcal{O} + B \subset U^c$. As a result, $K + B \subset U^c$ which is equivalent to $(K + B) \cap U = \emptyset$ and thus to $K \cap (U + B) = \emptyset$. □

We now show that the sum of a closed set and a compact set is closed.

THEOREM 3.20.– *In any separated semi-normed space, the sum $U + K$ of a closed set U and a compact set K is closed.* ▮

Proof. Let $z \notin U + K$. Then $(-U + z) \cap K = \emptyset$.

Now, $-U+z$ is closed because its complement $(-U+z)^c = (-U)^c+z = -U^c+z$ is open since U^c is open (as a point of fact, if V, here U^c, satisfies Definition 2.9 (b) of an open set, $-V$ and $-V + z$ also satisfy it with the same parameters N and η).

Theorem 3.19 therefore shows that there is a finite intersection B of semi-balls such that $(-U + z + B) \cap K = \emptyset$. Then, $(z + B) \cap (U + K) = \emptyset$, that is, with the notations of the above-mentioned theorem, that $\sup_{\nu \in N} \|v - z\|_{E;\nu} \leq r$ results in $v \notin U + K$. Which proves that $(U + K)^c$ is open and thus that $U + K$ is closed. □

Chapter 4

Banach, Fréchet and Neumann Spaces

4.1. Metrizable spaces

Let us define metrizability[1].

DEFINITION 4.1.– (a) *We say that a semi-normed space is* ***metrizable*** *if it is separated and if its family of semi-norms is equivalent to a countable family of semi-norms.*

(b) *We say that a semi-normed space is* ***normable*** *if its family of semi-norms is equivalent to (a family reduced to) a norm.* ∎

Justification of the "metrizable" name. We refer here to *metrizable* space because a distance or *metric* d is associated to any countable family, or that which leads to the same thing, to any sequence $(\| \ \|_k)_{k\in\mathbb{N}}$ of semi-norms, which generates the same topology by means of

$$d(u,v) = \sum_{k\in\mathbb{N}} 2^{-k} \frac{\|u-v\|_k}{1+\|u-v\|_k}.$$

Definition 4.1 is, in fact, that of a **separated countably semi-normable**. We prefer to loosely talk about *metrizable* spaces because this equivalent notion is more familiar. A **metrizable** space is, more accurately, a space "which is topologically equal to a metric space". □

Superiority of a sequence of semi-norms over a distance. Semi-norms make it possible to characterize bounded sets U with $\sup_{u\in U} \|u\|_k < \infty$ for all $k \in \mathbb{N}$ (Definition 2.9 (a)). On the other hand, for a distance d on a *metrizable but not normable* semi-normed space E, $\sup_{u\in U} d(u, 0_E) < \infty$ does not characterize bounded sets. Worse, for $c > 0$:

$$\text{No "ball" } \{u \in E : d(u,z) \le c\} \text{ is bounded.} \tag{4.1}$$

1 **History of the notion of metrizable space.** Maurice FRÉCHET gave the general definition of metric space in 1906, in [44].

As a matter of fact, otherwise $\{u : d(u, z) < c\}$ would be a non-empty open bounded set, and the existence of such a set is equivalent to normability according to **Kolmogorov's theorem**[2]. □

Examples of metrizable spaces:
— The space $\mathcal{C}_b(\mathbb{R})$ of continuous bounded functions endowed with the norm $\|f\|_{\mathcal{C}_b(\mathbb{R})} = \sup_{x\in\mathbb{R}} |f(x)|$ (that is of uniform convergence) is metrizable, like every normed space.
— The space $\mathcal{C}(\mathbb{R})$ of continuous functions endowed with semi-norms $\|f\|_{\mathcal{C}(\mathbb{R});n} = \sup_{|x|\le n} |f(x)|$ indexed by $n \in \mathbb{N}$ (that is of the uniform convergence on compact sets) is metrizable. In fact, it is separated and its family of semi-norms is countable.
— The space $\mathcal{C}(\mathbb{R})$ endowed with the above semi-norms indexed by $n \in \mathbb{R}$ is also metrizable. Indeed, it is separated and its family of semi-norms is equivalent to the countable family related to $n \in \mathbb{N}$. □

Example of non-metrizable space:
— The space $\mathcal{C}(\mathbb{R})$ endowed with semi-norms $\|f\|_{\mathcal{C}(\mathbb{R})\text{-pt};x} = |f(x)|$ indexed by $x \in \mathbb{R}$ (that is with pointwise convergence) is not metrizable. Indeed, the set defined by (2.3) p. 18 is sequentially closed but not closed, which contradicts metrizability, see Theorem 4.6 (a). □

Motivation for the study of metrizable spaces. These spaces present many properties of interest, in particular topological properties (closed, compact continuous, etc.) coincide therein with their sequential equivalents, see Theorems 4.6 and 9.1, and a large number of important spaces for functional analysis are metrizable semi-normed. This is the case of normed spaces, such as Lebesgue spaces L^p or Sobolev spaces $W^{m,p}$, but also of non-normable spaces, such as the spaces $\mathcal{C}$ of continuous functions, or $\mathcal{C}^m$ of continuously differentiable functions or the spaces $W^{m,p}_{\text{loc}}$. □

Let us give an obvious property.

THEOREM 4.2.– *For every vector space:*

$$\textit{Normed} \quad \Rightarrow \quad \textit{normable} \quad \Rightarrow \quad \textit{metrizable.} \blacksquare$$

Proof. The family of semi-norms of a normed space being reduced to a norm, this results from Definitions 4.1 for normable and metrizable. □

Let us observe that metrizability and normability are preserved by topological equivalences.

THEOREM 4.3.– (a) *Every metrizable semi-normed space remains so when its family of semi-norms is replaced by an equivalent family.*

(b) *Every normable semi-normed space remains so when its family of semi-norms is replaced by an equivalent family.* ▮

Proof. (a) If a semi-normed space E is metrizable (Definition 4.1 (a)), it is separated and its family $\mathcal{F}$ of semi-norms is equivalent to a countable family $\mathcal{F}_1$. Any family equivalent to $\mathcal{F}$ causes E to become separated (Theorem 3.2) and is equivalent to $\mathcal{F}_1$ (Theorem 3.3 (b)), and therefore makes it metrizable.

2 **History of Kolmogorov's theorem.** Andrey KOLMOGOROV showed in 1934 [63, p. 33] that a topological vector space is normable if and only if there is a bounded convex neighborhood of the origin, which here is equivalent to the existence of an open bounded set.

(b) If a semi-normed space E is normable, its family $\mathcal{F}$ of semi-norms is equivalent to a norm, and any family equivalent to $\mathcal{F}$ is also equivalent to it (Theorem 3.3 (b)). □

Let us show that all metrizable spaces can be endowed with an increasing sequence of semi-norms, which simplifies some proofs.

THEOREM 4.4.– *The family of semi-norms of any metrizable semi-normed space is equivalent to an increasing sequence of semi-norms.* ▮

Proof. Let E be a metrizable semi-normed space (Definition 4.1). Its family $\mathcal{F}$ of semi-norms is countable, and therefore is indexable by a subset N of $\mathbb{N}$, that is $\mathcal{F} = \{\| \, \|_n : n \in N\}$. Since $\mathcal{F}$ is not empty (Definition 2.3 (a)), N is not empty and therefore has a minimum m. For all $k \in \mathbb{N}$, a semi-norm on E is defined (Theorem 3.16 (a)) by $||| \, |||_k = \sup_{n \in N,\, n \leq m+k} \| \, \|_n$. The sequence $(||| \, |||_k)_{k \in \mathbb{N}}$ is increasing with k and equivalent (Definition 3.1) to $\mathcal{F}$. □

Other increasing sequence. The semi-norms $|||| \, ||||_k = \sum_{n \in N,\, n \leq m+k} \| \, \|_n$ constitute another increasing sequence equivalent to $\mathcal{F}$. □

Observe that any subspace of a metrizable space is metrizable.

THEOREM 4.5.– *Every topological subspace of a metrizable semi-normed space is metrizable.* ▮

Proof. Let E be a topological subspace of a semi-normed space F. If F is metrizable, following Definition 4.1, its family $\mathcal{F}$ of semi-norms is equivalent to a countable family $\mathcal{F}_1$. According to Definition 3.9 of a topological subspace, the family of E is equivalent to $\{q|_E : q \in \mathcal{F}\}$ and thus, following Definition 3.1 of equivalent families, to $\{q|_E : q \in \mathcal{F}_1\}$ which is countable. In addition, E is separated (Theorem 3.10 (b), since F is separated), and it is consequently metrizable. □

4.2. Properties of sets in metrizable spaces

Let us show that, in a metrizable space, sequential properties coincide with topological properties.

THEOREM 4.6.– *For any subset U of a metrizable semi-normed space:*

(a) *Closed* $\Leftrightarrow$ *sequentially closed.*

(b) *Compact* $\Leftrightarrow$ *sequentially compact.*

(c) *Relatively compact* $\Leftrightarrow$ *relatively sequentially compact.*

(d) *Separable* $\Leftrightarrow$ *sequentially separable.*

(e) *Dense in* U $\Leftrightarrow$ *sequentially dense in* U.

(f)
$$\overline{U} = \overline{U}^{\text{seq}} = U^{\text{lim seq}}. \blacksquare$$

Proof. Let E be a metrizable semi-normed space and $(\| \ \|_n)_{n \in \mathbb{N}}$ be an increasing sequence of semi-norms defining its topology (Theorem 4.4), which does not modify (Theorem 3.4) the properties to be examined (closed, sequentially closed and others). Let $U \subset E$.

(a) *Closed* $\Rightarrow$ *sequentially closed.* This is always true (Theorem 2.10).

Sequentially closed $\Rightarrow$ *closed.* Assume that U is sequentially closed. If it were not closed, $E \setminus U$ would not be open and therefore there would exist (Theorem 3.17 (a)) $u \in E \setminus U$ and, for all $n \in \mathbb{N}$, $u_n \in U$ such that $\|u_n - u\|_n \leq 1/n$. For all $k \in \mathbb{N}$ and $n \geq k$, then we would have $\|u_n - u\|_k \leq \|u_n - u\|_n \leq 1/n$. Consequently, $u_n \to u$ that would contradict the sequential closure of U. This proves that U is closed.

(b) *Compact* $\Rightarrow$ *sequentially compact.* Assume that U is compact. Let $(u_n)_{n \in \mathbb{N}}$ be a sequence of U and denote $V_n = \{u_m : m \geq n\}$ and $\overline{V_n}$ its closure. Any finite intersection of closed sets $\overline{V_n}$ is one of them (because they decrease with n), namely $\overline{V_i}$, and therefore has a common point with U (that is u_i). According to the finite intersection property (Theorem 2.27), there is thus

$$u \in \bigcap_{n \in \mathbb{N}} \overline{V_n} \cap U.$$

Following the characterization of the closure of Theorem 3.17 (c), there exists therefore $m_n \geq n$ such that $\|u_{m_n} - u\|_n \leq 1/n$. This m_n can be chosen as being strictly increasing with n (using $u \in \overline{V_{m_n+1}}$) which provides a subsequence $(u_{m_n})_{n \in \mathbb{N}}$ of the initial sequence that converges to u. This proves that U is sequentially compact.

Sequentially compact $\Rightarrow$ *compact.* Assume that U is sequentially compact. Let $\mathcal{R}$ be an open cover of U. For each $\ell \in \mathbb{N}$ and $u \in E$, denote

$$B_\ell(u) = \left\{ v \in E : \|v - u\|_\ell \leq \frac{1}{\ell} \right\}.$$

As a first step, let us show that there exists $k \in \mathbb{N}$ such that,

$$\forall u \in U, \quad B_k(u) \text{ is included in one of the open sets of } \mathcal{R}. \tag{4.2}$$

Otherwise, for all $n \in \mathbb{N}$, there would exist $u_n \in U$ such that no open subset of $\mathcal{R}$ contains $B_n(u_n)$. We could (Definition 2.24 (a)) extract a subsequence $(u_{\sigma(n)})_{n \in \mathbb{N}}$

converging to a limit $u \in U$. This would be included in an open set $\mathcal{O}$ of $\mathcal{R}$, and therefore (Theorem 3.17 (a)) there would exist $m \in \mathbb{N}$ and $\epsilon > 0$ such that

$$\{v \in E : \|v - u\|_m \leq \epsilon\} \subset \mathcal{O}.$$

By choosing $n \geq \sup\{m, 2/\epsilon\}$ such that $\|u_{\sigma(n)} - u\|_m \leq \epsilon/2$, there would be, for all $v \in E$,

$$\|v - u\|_m \leq \|v - u_{\sigma(n)}\|_m + \frac{\epsilon}{2} \leq \|v - u_{\sigma(n)}\|_{\sigma(n)} + \frac{\epsilon}{2}$$

since $m \leq n \leq \sigma(n)$ and $\| \ \|_m$ grows with m. Then, $\|v - u_{\sigma(n)}\|_{\sigma(n)} \leq 1/\sigma(n)$ would lead to $\|v - u\|_m \leq \epsilon$, from which $v \in \mathcal{O}$. Hence, $B_{\sigma(n)}(u_{\sigma(n)})$ would be included in $\mathcal{O}$ that would contradict the choice of the u_n. This proves (4.2).

For the second step, let us show that there exists a finite set $W \subset U$ such that

$$\bigcup_{w \in W} B_k(w) \text{ covers } U. \tag{4.3}$$

Otherwise, for all $n \in \mathbb{N}$, there would exist $w_n \in U$ which would not belong to $B_k(w_0) \cup \ldots \cup B_k(w_{n-1})$. For all $m \neq n$, it would follow that $\|w_m - w_n\|_k \geq 1/k$ therefore this sequence would not contain any convergent subsequence, which would contradict the sequential compactness of U. This proves (4.3).

For each $w \in W$, according to (4.2), there is an open set $\mathcal{O}_w$ of $\mathcal{R}$ which contains $B_k(w)$. According to (4.3), $\{\mathcal{O}_w\}_{w \in W}$ is a finite subcover of $\mathcal{R}$, which proves that U is compact.

(a) *Relatively compact $\Rightarrow$ relatively sequentially compact.* Assume that U is relatively compact. Then, (Definition 2.23 (b)) $\overline{U}$ is compact and therefore sequentially compact according to (b). Any sequence of U, belonging to $\overline{U}$, thus contains a convergent subsequence, which proves that U is relatively sequentially compact.

Relatively sequentially compact $\Rightarrow$ relatively compact. Assume that U is relatively sequentially compact. First, let us show that

$$\overline{U} \text{ is sequentially compact.} \tag{4.4}$$

Let $(u_n)_{n \in \mathbb{N}}$ be a Cauchy sequence of $\overline{U}$. For all $n \in \mathbb{N}$, the characterization of closure given in Theorem 3.17 (c) shows that there is $v_n \in U$ such that $\|v_n - u_n\|_n \leq 1/n$. By assumption on U, $(v_n)_{n \in \mathbb{N}}$ contains a converging subsequence $(v_{\sigma(n)})_{n \in \mathbb{N}}$. The subsequence $(u_{\sigma(n)})_{n \in \mathbb{N}}$ then also converges. Its limit is an element of $\overline{U}$ because it is closed and therefore (Theorem 2.10) sequentially closed. This proves (4.4).

Being sequentially compact, $\overline{U}$ is compact according to (b), which proves that U is relatively compact.

(f) *Equality* $\overline{U} = \overline{U}^{\text{seq}} = U^{\lim\,\text{seq}}$. There is always $U^{\lim\,\text{seq}} \subset \overline{U}^{\text{seq}} \subset \overline{U}$ (Theorem 2.16). Conversely, if $u \in \overline{U}$, for all $n \in \mathbb{N}$, there would exist (Theorem 3.17 (c)) $u_n \in U$ such that $\|u_n - u\|_n \leq 1/n$. Then, $u_n \to u$ so $u \in U^{\lim\,\text{seq}}$ (Definition 2.15 (d)).

(e) *Dense* $\Leftrightarrow$ *sequentially dense.* This results from (f), see Definition 2.19.

(d) *Separable* $\Leftrightarrow$ *sequentially separable.* This results from (e), see Definition 2.21. □

Let us show that, in a metrizable space, any subset of a separable set is separable.

THEOREM 4.7.– (a) *In any metrizable semi-normed space, every subset of a separable set is separable.*

(b) *In* (a), *separable can be replaced by sequentially separable.* ▮

Proof. (a) Let E be a metrizable semi-normed space and $(\|\ \|_n)_{n\in\mathbb{N}}$ be an increasing sequence of semi-norms defining its topology (Theorem 4.4). Let U be a separable subset of E and $(x_m)_{m\in\mathbb{N}}$ be a dense sequence in U. Finally, let $V \subset U$. Denote

$$X_{m,n} = \{v \in V : \|v - x_m\|_n \leq 1/n\}.$$

If this set is not empty, let us choose therein an element $a_{m,n}$. The set A of these $a_{m,n}$ is a countable subset of V (because it is the image of $\mathbb{N}^2$ which is countable by the mapping $(n,m) \mapsto a_{m,n}$, see Theorem 1.1 (d) and (c)). It is therefore enough to show that it is dense in V, that is

$$V \subset \overline{A}. \tag{4.5}$$

Now, let $v \in V$. Since $(x_m)_{m\in\mathbb{N}}$ is dense in U, the characterization of closure given in Theorem 3.17 (c) shows that, for any $n \in \mathbb{N}$, there exists $\sigma(n) \in \mathbb{N}$ such that $\|v - x_{\sigma(n)}\|_n \leq 1/n$. Then, $X_{\sigma(n),n}$ contains v, and so, being not empty, it contains the element $a_{\sigma(n),n}$ chosen above. According to the definition of $X_{m,n}$, this means that $\|a_{\sigma(n),n} - x_{\sigma(n)}\|_n \leq 1/n$. Therefore,

$$\|v - a_{\sigma(n),n}\|_n \leq 2/n.$$

Given $k \in \mathbb{N}$ and $\epsilon > 0$, if $n \geq \sup\{k, 2/\epsilon\}$, then $\|v - a_{\sigma(n),n}\|_k \leq \epsilon$ since $\|\ \|_n$ grows with n. From which, again by Theorem 3.17 (c), $v \in \overline{A}$ which proves (4.5).

(b) Since E is metrizable, separable therein is equivalent to sequentially separable (Theorem 4.6 (d)). □

The case of a non-metrizable space. Theorem 4.7 does not extend to all semi-normed spaces. In other words, a subset of a separable subset of a separated semi-normed space is not necessarily separable. For example, in the space $\mathcal{F}$ of functions from $\mathbb{R}$ into itself endowed with the semi-norms $\|f\|_x = |f(x)|$ indexed by $x \in \mathbb{R}$ (i.e. of pointwise convergence), let

$$U = \{1_{[a,b]} : a \in \mathbb{R},\ b \in \mathbb{R},\ a \leq b\}, \quad V = \{1_{[a,a]} : a \in \mathbb{R}\}$$

where $1_{[a,b]}$ is the function equal to 1 in $[a,b]$ and 0 elsewhere. As a result, U is separable (and sequentially separable) but V is neither one. □

4.3. Banach, Fréchet and Neumann spaces

Let us define sequential completeness[3].

DEFINITION 4.8.– *A separated semi-normed space is* ***sequentially complete*** *if all its Cauchy sequences converge.* ▮

It should be observed that sequential completeness is maintained by topological equivalence.

THEOREM 4.9.– *Any sequentially complete separated semi-normed space remains so if its family of semi-norms is replaced by an equivalent family.* □

Proof. The space remains separated following Theorem 3.2, and its Cauchy sequences and its convergent sequences are preserved according to Theorem 3.4 (a). □

Let us define Neumann, Fréchet and Banach spaces[4].

DEFINITION 4.10.– (a) *A* ***Neumann space*** *is any sequentially complete separated semi-normed space.*

(b) *A* ***Fréchet space*** *is any sequentially complete metrizable semi-normed space.*

(c) *A* ***Banach space*** *is any sequentially complete normed space.* ▮

Terminology. Neumann space is a **new name** given in the honor of John von Neumann, see Note 4, therefore readers must recall its definition whenever they use it. Seldom used so far, sequentially complete spaces were not named. □

Let us compare these spaces.

THEOREM 4.11.– *For every vector space:*

$$\textit{Banach space} \quad \Rightarrow \quad \textit{Fréchet space} \quad \Rightarrow \quad \textit{Neumann space.}$$ ▮

Proof. The first implication results from the fact that any normed space is metrizable (Theorem 4.2), the second follows from Definition 4.10 of these notions. □

---

3 **History of the notion of sequential completeness.** Maurice FRÉCHET defines sequential completeness of a metric space in 1906, in [44].

4 **History of the notion of Banach space.** Stefan BANACH and also Hans HAHN defined sequentially complete normed vector spaces in 1920-1922.
The axioms defining them had been introduced in the example of $\mathcal{C}(I)$, I being a compact interval, by Frigyes RIESZ in 1918 [93].
History of the notion of Fréchet space. Maurice FRÉCHET defined sequentially complete metric spaces in 1906, in [44].
History of the notion of Neumann space. John VON NEUMANN introduced sequentially complete separated semi-normed spaces in 1935, in [78].

Let us show that any sequentially closed subspace of a sequentially complete space is sequentially complete.

THEOREM 4.12.– *Let F be a separated semi-normed space and*

$$E \text{ a closed or sequentially closed topological subspace of } F.$$

Then:

(a) F *is a Neumann space* $\Rightarrow$ E *is a Neumann space.*

(b) F *is a Fréchet space* $\Rightarrow$ E *is a Fréchet space.*

(c) F *is a Banach space* $\Rightarrow$ $\begin{cases} \text{The restriction to } E \text{ of the norm} \\ \text{of } F \text{ generates the topology of } E \\ \text{and turns it into a Banach space.} \end{cases}$ ▮

Proof. It suffices to assume that E is sequentially closed since this is the case if it is closed (Theorem 2.10).

(a) Any Cauchy sequence of E is a Cauchy sequence in F (Theorem 3.11 (a)) therefore, if F is sequentially complete, it converges therein. It then converges in its sequentially closed topological subspace E, (Theorem 3.12 (a)) that is thus sequentially complete.

(b) This results from (a) because, if in addition F is metrizable, E is also based on Theorem 4.5.

(c) This results from (a) because, if in addition F is normed, according to Definition 3.9 of a topological subspace, the restriction of the norm of F generates the topology of E and therefore turns it into a normed space. □

Let us show that, conversely, any sequentially complete subspace is sequentially closed.

THEOREM 4.13.– *In separated semi-normed spaces, every sequentially complete topological subspace is sequentially closed.* ▮

Proof. Let E be a topological subspace of a separated semi-normed space F. If a sequence of E converges in F to a limit ℓ, it is a Cauchy sequence (Theorem 2.7 (a)) in F and thus in E (Theorem 3.11 (a)). If E is sequentially complete, it has a limit therein; the latter is also the limit in F (Theorem 3.11 (a)), which is unique (Theorem 2.7 (a)), therefore this is ℓ. Hence, $\ell \in E$. Consequently, E is sequentially closed in F. □

Motivation for the study of Neumann spaces. The sequential completeness of E is crucial for the construction, which we will perform in Volume II, of distributions space $\mathcal{D}'(\Omega; E)$ and, in Volume III, that of its subspaces $L^p(\Omega; E)$, $W^{m,p}(\Omega; E)$ and others. More precisely, it ensures them satisfactory properties and specifically that $\int_\Omega f \in E$ when $f \in \mathcal{C}(\Omega; E)$ and thus that $\mathcal{C}(\Omega; E) \subset \mathcal{D}'(\Omega; E)$.

Sequential completeness is also used for solving PDEs (and other types of equations) by constructing a sequence of approximations satisfying the Cauchy criterion. □

Some Banach, Fréchet or Neumann spaces. As mentioned, sequential completeness being very fruitful, for the record let us indicate a few cases where it is satisfied (these spaces are defined in the following volumes; here, Ω is an open subset of $\mathbb{R}^d$ and $m \in \mathbb{N}$):
— The following spaces are Banach spaces: $\mathcal{C}_b(\Omega)$ and $\mathcal{C}_b^m(\Omega)$, see Volume II, as well as $L^p(\Omega)$, $W^{m,p}(\Omega)$, $W^{-m,p}(\Omega)$, $W_0^{m,p}(\Omega)$, $\dot{W}_0^{m,p}(\Omega)$ and $\dot{W}^{-m,p}(\Omega)$, see Volume III.
— In addition to those above, the following spaces are Fréchet spaces: $\mathcal{C}(\Omega)$, $\mathcal{C}^m(\Omega)$, $\mathcal{C}^\infty(\Omega)$, $\mathcal{C}_b^\infty(\Omega)$, see Volume II, as well as $L^p_{\text{loc}}(\Omega)$, $W^{m,p}_{\text{loc}}(\Omega)$ and $W^{-m,p}_{\text{loc}}(\Omega)$, see Volume III.
— In addition to those above, the following spaces are Neumann spaces: $\mathcal{K}(\Omega)$, $\mathcal{K}^m(\Omega)$, $\mathcal{K}^\infty(\Omega)$, $\mathcal{D}(\Omega)$ and $\mathcal{D}'(\Omega)$, see Volume II. And also $\mathcal{F}(\Omega)$ (Theorem 11.3, below).

Sequential completeness properties of these spaces for their weak topologies as well as their dual spaces are the result of properties (17.3) to (17.6), p. 248.

The product, the intersection, the direct sum and, under certain conditions, the sum of Neumann spaces are also Neumann spaces (Theorems 6.7, 6.19, 6.34 and 6.27). □

Example of non-sequentially complete spaces:
— The space $\mathcal{C}(\mathbb{R})$-pt is not sequentially complete, see (11.1), p. 154. □

4.4. Compacts sets in Fréchet spaces

Let us give a **Fréchet's theorem**[5] that relates compactness to precompactness in a Fréchet space.

THEOREM 4.14.– *For any subset of a Fréchet space:*

(a) *Compact* $\Leftrightarrow$ *closed precompact.*

(b) *Relatively compact* $\Leftrightarrow$ *precompact.*

(c) *In* (a) *and* (b), *it is possible to replace closed, compact and relatively compact with their sequential counterparts.* ∎

Proof. Let E be a Fréchet space and $U \subset E$.

(c) Since (Definition 4.10 (b)) E is metrizable, closed, compact and relatively compact sets therein are equivalent to their sequential counterparts (Theorem 4.6 (a), (b) and (c)).

5 **History of Fréchet's theorem 4.14.** Maurice FRÉCHET proved this result in 1906-1910 in [43] and [45].

(b) According to (c), it is enough to show that, for U:

$$\textit{Relatively sequentially compact} \quad \Leftrightarrow \quad \textit{precompact}.$$

The direct part being true in all separated semi-normed space (Theorem 2.25 (b)), it remains to prove the converse. Let us thus assume that U is precompact and let $(\| \ \|_k)_{k\in\mathbb{N}}$ be an increasing sequence (Theorem 4.4) of semi-norms defining the topology of E and $(u_n)_{n\in\mathbb{N}}$ be a sequence of U.

According to the precompactness, for all $k \in \mathbb{N}$, U can be covered by a finite number of semi-balls $\{v \in E : \|v - w\|_k \leq 1/k\}$. At least one of them contains a subsequence $(u_{k,n})_{n\in\mathbb{N}}$ of $(u_n)_{n\in\mathbb{N}}$. The latter verifies for all n and n' in $\mathbb{N}$,

$$\|u_{k,n} - u_{k,n'}\|_k \leq \frac{2}{k}.$$

Through successive extractions, $(u_{k+1,n})_{n\in\mathbb{N}}$ can be chosen to be a subsequence of $(u_{k,n})_{n\in\mathbb{N}}$. The diagonal sequence[6] $(u_{n,n})_{n\in\mathbb{N}}$ verifies $\|u_{n,n} - u_{n',n'}\|_k \leq 2/k$ when $n' \geq n \geq k$. Let $\epsilon > 0$ and $m \geq \sup\{k, 2/\epsilon\}$; then, for all $n' \geq n \geq m$,

$$\|u_{n,n} - u_{n',n'}\|_k \leq \|u_{n,n} - u_{n',n'}\|_m \leq \frac{2}{m} \leq \epsilon.$$

Therefore, $(u_{n,n})_{n\in\mathbb{N}}$ is a Cauchy sequence. Since E is sequentially complete, it converges which proves that U is relatively sequentially compact.

(a) Compact is equivalent to relatively compact closed (Theorem 2.26 (a)) and thus, here, according to (c), to precompact closed. □

4.5. Properties of $\mathbb{R}$

The space $\mathbb{R}$ is endowed with its addition and its product defined in § 1.3. They turn it into a vector space. It is endowed with the norm defined by

$$\| \ \|_{\mathbb{R}} = | \ |.$$

Let us show that $\mathbb{R}$ is sequentially complete.

THEOREM 4.15.– *The space $\mathbb{R}$ is a Banach space.* ▮

6 **History of the extraction of a diagonal subsequence.** The diagonal process for extracting a subsequence was introduced in 1891 by Georg CANTOR [23], a text recovered from his works [24].

Proof. The point is (Definition 4.10 (c)) to verify that $\mathbb{R}$ is sequentially complete. Let $(u_n)_{n\in\mathbb{N}}$ be one of its Cauchy sequence. It is bounded (Theorem 2.7 (b)) therefore (Theorem 1.2 (b)), for all $n \in \mathbb{N}$, a real number v_n is defined by

$$v_n = \inf\{u_i : i \geq n\}.$$

The sequence $(v_n)_{n\in\mathbb{N}}$ is increasing and majored thus (Theorem 1.3) it has a limit $\ell \in \mathbb{R}$. Let $\epsilon > 0$. There exists $m_1 \in \mathbb{N}$ such that, for all $n \geq m_1$,

$$|v_n - \ell| \leq \epsilon.$$

Since $(u_n)_{n\in\mathbb{N}}$ is a Cauchy sequence, there exists $m_2 \in \mathbb{N}$ such that, for every $i \geq n \geq m_2$, there is $|u_i - u_n| \leq \epsilon$. Then, $u_n - \epsilon \leq u_i \leq u_n + \epsilon$ from which $u_n - \epsilon \leq v_n \leq u_n + \epsilon$ and thus

$$|v_n - u_n| \leq \epsilon.$$

For all $n \geq \sup\{m_1, m_2\}$, it follows that $|u_n - \ell| \leq 2\epsilon$. Therefore, $u_n \to \ell$, which proves that $\mathbb{R}$ is sequentially complete. □

Let us show that the closed bounded subsets of $\mathbb{R}$ are compact which is the **Borel–Lebesgue theorem**[7].

THEOREM 4.16.– *For any subset of* $\mathbb{R}$*:*

(a) *Compact* ⇔ *closed and bounded.*

(b) *Relatively compact* ⇔ *precompact* ⇔ *bounded.*

(c) *In* (a) *and* (b)*, it is possible to replace closed, compact and relatively compact by their sequential counterparts.* ▮

Proof. Bounded ⇔ precompact. Every bounded set U of $\mathbb{R}$ is precompact because the balls $B(n\epsilon, \epsilon) = [(n-1)\epsilon, (n+1)\epsilon]$ cover it when n spans the finite set of $n \in \mathbb{Z}$ such that $|n| \leq (1/\epsilon) \sup_{x\in U} |x|$. The converse is always true (Theorem 2.25 (a)).

Other properties. They follow from the previous one and from Fréchet's Theorem 4.14 since $\mathbb{R}$, being a Banach space (Theorem 4.15), is a Fréchet space (Theorem 4.11). □

Let us show that the bounded subsets of $\mathbb{R}$ are relatively sequentially compact, which is the **Bolzano–Weierstrass theorem**[8].

---

7 **History of the Borel–Lebesgue theorem 4.16.** Émile BOREL proved that any countable open cover of a closed segment of $\mathbb{R}$ has a finite subcover, then Henri LEBESGUE proved it without assuming countability.

8 **History of the Bolzano–Weierstrass theorem 4.17.** Bernard BOLZANO pictured that any bounded sequence has a convergent subsequence but could not prove it for lack of a consistent definition of $\mathbb{R}$. Karl WEIERSTRASS established it in his teachings, which, although unpublished, had a considerable influence.

THEOREM 4.17.– *Any bounded sequence of* $\mathbb{R}$ *has a converging subsequence.* ∎

Proof. Any bounded sequence of $\mathbb{R}$ is relatively sequentially compact according to Borel–Lebesgue theorem 4.16 (b) and (c). As a result (Definition 2.24 (b)), it has a converging subsequence. □

Let us show that any real number is the limit of a sequence of rational numbers.

THEOREM 4.18.– *The set* $\mathbb{Q}$ *is dense and sequentially dense in* $\mathbb{R}$. ∎

Proof. Every $x \in \mathbb{R}$ being (Theorem 1.5) the limit of a sequence of $\mathbb{Q}$, the latter is sequentially dense (Definition 2.19 (b)), and so (Theorem 2.20) dense, in $\mathbb{R}$. □

Observe that $\mathbb{R}$ is separable.

THEOREM 4.19.– *The space* $\mathbb{R}$ *is separable and sequentially separable.* ∎

Proof. The set $\mathbb{Q}$ is countable (Theorem 1.6) and it is sequentially dense in $\mathbb{R}$ (Theorem 4.18). Therefore, $\mathbb{R}$ is sequentially separable (Definition 2.21 (b)) which leads to (Theorem 2.22) separable. □

Let us show that the compact subsets of $\mathbb{R}$ have a minimum and a maximum.

THEOREM 4.20.– *Any non-empty compact subset of* $\mathbb{R}$ *has a minimum and a maximum.* ∎

Proof. Let U be a non-empty compact subset of $\mathbb{R}$. It is bounded (Theorem 2.25 (a)), and therefore it has (Theorem 1.2) a least upper bound M and a greatest lower bound m. For all $\epsilon > 0$, there is $u \in U$ such that $M - \epsilon \leq u \leq M$ (otherwise $M - \epsilon$ would be an upper bound lower than M which would contradict the definition of M). Thus, $M \in \overline{U}$ (Theorem 2.17 (b)). In addition, $\overline{U} = U$ because U is closed (Theorem 2.26 (a)). Therefore M is an element of U, and consequently it is its maximum. Similarly, it can be shown that m is its minimum. □

4.6. Convergent sequences

Let us define the convergent series of a separated semi-normed space. We denote

$$\sum_{n=i}^{k} u_n = u_i + u_{i+1} + \cdots + u_k.$$

DEFINITION 4.21.– *Let* E *be a semi-normed space,* $\{\| \ \|_{E;\nu} : \nu \in \mathcal{N}_E\}$ *its family of semi-norms and* $(u_n)_{n\in\mathbb{N}} \subset E$.

(a) ***The series*** $\sum_{n\in\mathbb{N}} u_n$ ***converges*** *if the sequence* $(\sum_{n=1}^{k} u_n)_{k\in\mathbb{N}}$ *has a limit.*

This limit, denoted by $\sum_{n\in\mathbb{N}} u_n$*, is the* ***sum*** *of the series.*

(b) *The series* $\sum_{n\in\mathbb{N}} u_n$ *is* ***absolutely convergent*** *if, for all* $\nu \in \mathcal{N}_E$*, the sequence* $(\sum_{n=1}^{k} \|u_n\|_{E;\nu})_{k\in\mathbb{N}}$ *is majored, which is denoted by* $\sum_{n\in\mathbb{N}} \|u_n\|_{E;\nu} < \infty$.

That is if, for all $\nu \in \mathcal{N}_E$*, there exists* $c \in \mathbb{R}$ *such that, for all* $k \in \mathbb{N}$*, there is* $\sum_{n=1}^{k} \|u_n\|_{E;\nu} \leq c$. ▮

Characterization of the absolute convergence. The sequence $(\sum_{n=1}^{k} \|u_n\|_{E;\nu})_{k\in\mathbb{N}}$ is majored if and only if it has a finite limit (Theorem 1.3, because it is increasing). Therefore,

$$\sum_{n\in\mathbb{N}} u_n \textit{ is absolutely convergent in } E \quad\Leftrightarrow\quad \sum_{n\in\mathbb{N}} \|u_n\|_{E;\nu} \textit{ converges for all } \nu \in \mathcal{N}_E. \ \square$$

Comparison of the notions of convergence. *Absolutely convergent* is neither stronger nor weaker than *convergent*. As a matter of fact, it can be easily be proven that, in all separated semi-normed space not reduced to $\{0\}$:

$$\textit{Convergent} \quad \not\Rightarrow \quad \textit{absolutely convergent}. \tag{4.6}$$

Conversely, in some separated semi-normed spaces,

$$\textit{Absolutely convergent} \quad \not\Rightarrow \quad \textit{convergent}. \tag{4.7}$$

This is, for example, the case in the space $\mathcal{C}(\mathbb{R})$ of continuous functions endowed with the semi-norms of pointwise convergence $\|f\|_x = |f(x)|$ indexed by $x \in \mathbb{R}$. However, absolutely convergent series converge in every Neumann space (Theorem 4.22). □

Convexity inequality. It should be observed that, in a semi-normed space E, there is always (by iterating the inequality of Definition 2.2 (a) of a semi-norm)

$$\|u_1 + u_2 + \cdots + u_k\|_{E;\nu} \leq \|u_1\|_{E;\nu} + \|u_2\|_{E;\nu} + \cdots + \|u_k\|_{E;\nu}. \tag{4.8}$$

□

Let us show that, in a Neumann space, absolute convergence leads to convergence.

THEOREM 4.22.– *In a Neumann space, every absolutely convergent series converges.*

▮

Proof. Let $\sum_{n\in\mathbb{N}} u_n$ be an absolutely convergent series of a separated semi-normed space E.

For any semi-norm $\| \ \|_{E;\nu}$ of E, the real sequence $(\sum_{n=1}^{k} \|u_n\|_{E;\nu})_{k\in\mathbb{N}}$ is increasing and majored, therefore (Theorem 1.3) it converges and thus (Theorem 2.7 (a)) it is a Cauchy sequence. For all $\epsilon > 0$, there exists thus $k_0 \in \mathbb{N}$ such that, for every $m \geq k \geq k_0$, it follows that $\sum_{n=1}^{m} \|u_n\|_{E;\nu} - \sum_{n=1}^{k} \|u_n\|_{E;\nu} \leq \epsilon$, and then

$$\Big\|\sum_{n=1}^{m} u_n - \sum_{n=1}^{k} u_n\Big\|_{E;\nu} = \Big\|\sum_{n=k+1}^{m} u_n\Big\|_{E;\nu} \leq \sum_{n=k+1}^{m} \|u_n\|_{E;\nu} \leq \epsilon.$$

The sequence $(\sum_{n=1}^{k} u_n)_{k\in\mathbb{N}}$ is thus a Cauchy sequence. If E is a Neumann sequence, it converges (Definitions 4.10 (a) and 4.8), which means that the series $\sum_{n\in\mathbb{N}} u_n$ converges. □

Let us show that a metrizable space in which every absolutely convergent series converges is sequentially complete.

THEOREM 4.23.– *Every metrizable semi-normed space in which every absolutely convergent series converges is a Fréchet space.* ■

Proof. Let E be a metrizable semi-normed space in which every absolutely convergent series converges and let $(u_n)_{n\in\mathbb{N}}$ be a Cauchy sequence of E. The point is to prove that it converges, that is that E is sequentially complete.

Let (Theorem 4.4) $(\| \ \|_j)_{j\in\mathbb{N}}$ be an increasing sequence of semi-norms defining the topology of E. For all $j \in \mathbb{N}$, there is $m_j \in \mathbb{N}$ such that, for all $n \geq m_j$,

$$\|u_n - u_{m_j}\|_j \leq \frac{1}{2^j}.$$

Let us choose successively m_1, m_2, etc., such that $m_{i+1} \geq m_i + 1$ for all $i \in \mathbb{N}$. The semi-norm $\| \ \|_j$ increasing with j, for all $i \geq j$, there is

$$\|u_{m_{i+1}} - u_{m_i}\|_j \leq \|u_{m_{i+1}} - u_{m_i}\|_i \leq \frac{1}{2^i}.$$

Let us denote $w_1 = u_{m_1}$ and $w_i = u_{m_i} - u_{m_{i-1}}$ if $i \geq 2$. For all $j \in \mathbb{N}$ and $k \geq j$,

$$\sum_{i=1}^{k} \|w_i\|_j \leq \|w_1\|_j + \cdots + \|w_j\|_j + \frac{1}{2^j} + \frac{1}{2^{j+1}} + \cdots + \frac{1}{2^{k-1}} \leq c_j$$

where, see (1.10), p. 7, $c_j = \|w_1\|_j + \cdots + \|w_j\|_j + 2^{1-j}$. This inequality is also true for $k \leq j$, therefore the series $\sum_{i\in\mathbb{N}} w_i$ is absolutely convergent.

By assumption, such a series converges, namely $\sum_{i=1}^{n} w_i = u_{m_n}$ converges when $n \to \infty$. This subsequence $(u_{m_n})_{n\in\mathbb{N}}$ being convergent, the whole Cauchy sequence $(u_n)_{n\in\mathbb{N}}$ converges (Theorem 2.8 (b)). □

4.7. Sequential completion of a semi-normed space

The sequential completeness of E is essential for the construction of $\mathcal{D}'(\Omega; E)$ and of its subspaces $W^{m,p}(\Omega; E)$. When E does not have this property, it may be helpful to include it into a space $\mathcal{E}$ that has the property, in order to, for example, consider continuous functions with values in E as distributions with values in $\mathcal{E}$ and, thus, to be able to differentiate them.

Observe that there is still such a Neumann space $\mathcal{E}$.

THEOREM 4.24.– *Given a separated semi-normed space E, there exists $\mathcal{E}$ such that:*

$$\begin{cases} \mathcal{E} \text{ is a Neumann space,} \\ E \text{ is included and is dense in } \mathcal{E}, \\ \text{the semi-norms of } \mathcal{E} \text{ extend those of } E, \text{ and} \\ \overline{E}^{\text{seq}\,\mathcal{E}} = \mathcal{E} \text{ (that is, } \mathcal{E} \text{ is its only sequentially closed subspace containing } E). \end{cases} \quad (4.9)$$

□

Proof. There is a separated semi-normed space F containing E, whose semi-norms extend those of E, in which E is dense and which is complete (this notion is recalled on this page), see for example [KÖTHE, 64, properties (2), p. 209]. It is called a **completion** of E. It is sequentially complete like all complete spaces, see for instance [SCHWARTZ, 99, p. 251]. The required properties are then satisfied by the sequential closure $\mathcal{E} = \overline{E}^{\text{seq}\,F}$ of E in F, endowed with the restrictions of the semi-norms of F. □

Let us show that, if E is metrizable, any element of $\mathcal{E}$ is the limit of a sequence of E.

THEOREM 4.25.– *Let*

E be a metrizable semi-normed space.

Then, there exists a space $\widehat{E}^{\text{seq}}$, called a **sequential completion** *of E, such that:*

$$\begin{cases} \widehat{E}^{\text{seq}} \text{ is a Fréchet space,} \\ E \text{ is included and is dense and sequentially dense in } \widehat{E}^{\text{seq}}, \text{ and} \\ \text{the semi-norms of } \widehat{E}^{\text{seq}} \text{ extend those of } E. \end{cases} \quad (4.10)$$

□

Proof. The space $\mathcal{E}$ given by Theorem 4.24 has these properties because when E is metrizable (Definition 4.1 (a)), it is also (since their families of semi-norms are indexed over the same set) thus it is a Fréchet space (Definition 4.10 (b)) and (Theorem 4.6 (e)) dense coincide therein with sequentially dense. □

Another proof. The use of the notion of complete space (via Theorem 4.24) is avoided by introducing the vector space $\mathcal{G}$ consisting of the sets g of Cauchy sequences of E that are equivalent in the following sense: $(u_n)_{n\in\mathbb{N}}$ is equivalent to $(v_n)_{n\in\mathbb{N}}$ if the interlaced sequence $(u_1, v_1, u_2, v_2, \ldots, u_n, v_n, \ldots)$ is a Cauchy sequence. To each semi-norm $\|\ \|_{E;\nu}$ of E, a semi-norm on $\mathcal{G}$ is associated with $\|g\|_{\mathcal{G};\nu} = \lim \|u_n\|_{E;\nu}$ where $(u_n)_{n\in\mathbb{N}}$ is any sequence of g. An injection i of E in $\mathcal{G}$ is defined by $i(u) = \dot{u}$ where $\dot{u}$ is the set of Cauchy sequences equivalent to the constant sequence $(u, u, u, \ldots)$.

A space $\widehat{E}^{\text{seq}}$ verifying (4.10) is obtained by replacing each element of $\mathcal{G}$ of the form $\dot{u}$ by u and by defining its semi-norms by $\|g\|_{\widehat{E}^{\text{seq}};\nu} = \|g\|_{E;\nu}$ if $g \in E$ and $\|g\|_{\widehat{E}^{\text{seq}};\nu} = \|g\|_{\mathcal{G};\nu}$ otherwise. □

CAUTION. A separated semi-normed space E do not always have a sequential completion $\widehat{E}^{\text{seq}}$ verifying (4.10) [SIMON, to be published]. By default, it is still possible to use a space $\mathcal{E}$ verifying (4.9). □

Non-uniqueness. The space $\mathcal{E}$ is not unique for a given E but if $\mathcal{E}_1$ and $\mathcal{E}_2$ are two spaces verifying (4.9), there exists an isomorphism from one onto the other that coincides with the identity in E.

Similarly, if $\widehat{E}_1^{\text{seq}}$ and $\widehat{E}_2^{\text{seq}}$ are two spaces satisfying (4.10), there is an isomorphism from one onto the other which coincides with the identity in E according to Theorem 10.4. □

Complete spaces. A semi-normed space (or more generally a uniform space) is said to be **complete** if every Cauchy filter converges [SCHWARTZ, 99, chap. 18, § 8, definition 1, p. 251] or, which is equivalent [KÖTHE, 64, chap. 1, § 5, property (1), p. 33], if any generalized Cauchy sequence converges.

We do not make use of this notion (except in the proof of Theorem 4.24 above), because sequentially complete is enough for us, is much simpler and more general (complete results in sequentially complete) and, above all, because some useful spaces are sequentially complete but not complete. For instance:

$$\begin{cases} \text{For all Hilbert space, or reflexive Banach space, } E \text{ of infinite dimension,} \\ E\text{-weak is sequentially complete but not complete.} \end{cases} \quad (4.11)$$

Or even:

$$\begin{cases} \textit{For all Fréchet space } E \textit{ of infinite dimension,} \\ E'\text{-}*\text{weak} \textit{ is sequentially complete but not complete.} \end{cases} \tag{4.12}$$

Proof of (4.11). If E is a reflexive Banach space of infinite dimension, E-weak is sequentially complete (Theorem 17.7) but not complete [SCHAEFER, 96, exercise 21 (b), p. 195]. □

Proof of (4.12). If E is a Fréchet space, E'-$*$weak is sequentially complete (Theorem 13.8 (b)) but not complete [SCHAEFER, 96, exercise 21 (b), p. 195]. □

Completeness and metrizability. For a metrizable space, *complete* is equivalent to *sequentially complete* [SCHWARTZ, 99, Theorem XVIII, 8; 1, p. 251] or [HORVÁTH, 59, chap. 2, § 9, p. 135]. This is why some authors talk about the completeness of a Banach or Fréchet space while they use, actually, only its sequential completeness. □

Chapter 5

Hilbert Spaces

5.1. Hilbert spaces

Let us define Hilbert spaces[1].

Definition 5.1.– (a) *A **scalar product** on a vector space E is a bilinear, symmetric and definite positive mapping b from $E \times E$ into $\mathbb{R}$, which means: for all u, v and w in E and $t \in \mathbb{R}$,*

$$b(u, v + w) = b(u, v) + b(u, w), \quad b(u, tv) = tb(u, v),$$
$$b(v, u) = b(u, v),$$
$$b(v, v) > 0 \text{ if } v \neq 0_E.$$

(b) *A **pre-Hilbert space** is a vector space E endowed with a scalar product $(\ ,\ )_E$ and with the **Hilbertian norm** defined by*

$$\|u\|_E \stackrel{\text{def}}{=} (u, u)_E^{1/2}.$$

(c) *A **Hilbert space** is a sequentially complete pre-Hilbert space.* ▮

Justification.– In (b), the mapping $\|\ \|_E$ is indeed a norm according to Minkowski's inequality below (Theorem 5.3 (b)). The square root $(u, u)^{1/2}$ is defined in § 1.4.

In (a), only linearity with respect to the second variable is expressed because linearity with respect to the first one results thereof according to symmetry. □

1 **History of the notion of Hilbert space.** John von Neumann introduced separable Hilbert spaces in 1929 [76]. Previously, David Hilbert had implicitly used the ℓ^2 space of real square-integrable sequences in 1906 [57].

Let us prove a basic equality and the **parallelogram law** (b).

THEOREM 5.2.– *Let E be a pre-Hilbert space, $(\ ,\ )_E$ its scalar product and $\|\ \|_E = (\ ,\ )_E^{1/2}$. Then, for all $u \in E$ and $v \in E$:*

(a) $$\|u+v\|_E^2 = \|u\|_E^2 + \|v\|_E^2 + 2(u,v)_E.$$

(b) $$\|u+v\|_E^2 + \|u-v\|_E^2 = 2(\|u\|_E^2 + \|v\|_E^2). \blacksquare$$

Proof. (a) According to bilinearity and symmetry

$$\|u+v\|_E^2 = (u+v, u+v)_E =$$
$$= (u,u)_E + (v,v)_E + (u,v)_E + (v,u)_E = \|u\|_E^2 + \|v\|_E^2 + 2(u,v)_E.$$

(b) Following (a), $\|u-v\|_E^2 = \|u\|_E^2 + \|v\|_E^2 - 2(u,v)_E$ that is added to (a). □

Let us prove the **Cauchy–Schwarz**[2] (a) and **Minkowski**[3] (b) inequalities.

THEOREM 5.3.– *Let E be a pre-Hilbert space, $(\ ,\ )_E$ its scalar product and $\|\ \|_E = (\ ,\ )_E^{1/2}$. Then, for all $u \in E$ and $v \in E$:*

(a) $$|(u,v)_E| \le \|u\|_E \|v\|_E.$$

(b) $$\|u+v\|_E \le \|u\|_E + \|v\|_E. \blacksquare$$

Proof. (a) If there would exist u and $v \neq 0_E$ such that $|(u,v)_E| > \|u\|_E\|v\|_E$, the inequality of Theorem 5.2 (a) with tv instead of v where $t = -(u,v)_E/\|v\|_E^2$ would give

$$\|u+tv\|_E^2 = \|u\|_E^2 + t^2\|v\|_E^2 + 2t(u,v)_E = \|u\|_E^2 - \frac{(u,v)_E^2}{\|v\|_E^2} < 0$$

which would contradict the positivity of the scalar product. This proves (a) except for $v = 0_E$, in which case it results from $(u, 0_E)_E = (u, 0 0_E)_E = 0\,(u, 0_E)_E = 0$.

(b) According to the equality of Theorem 5.2 (a) and 5.3 (a),

$$\|u+v\|_E^2 = \|u\|_E^2 + \|v\|_E^2 + 2(u,v)_E \le$$
$$\le \|u\|_E^2 + \|v\|_E^2 + 2\|u\|_E\|v\|_E = (\|u\|_E + \|v\|_E)^2$$

---

2 **History of the Cauchy–Schwarz inequality.** The inequality (a) of Theorem 5.3 was proved for finite sums by Augustin CAUCHY in 1821 [25, p. 373]. It was proved for integrals in 1859 by V. BUNIAKOWSKY [21, p. 4] then in 1885 by Herman SCHWARZ [101, p. 251].

3 **History of Minkowski's inequality.** Hermann MINKOWSKI proved the inequality (b) of Theorem 5.3 for finite sums in 1896, see [74, 2nd ed., p. 115–117].

from which, by taking the square roots, the stated inequality. □

Let us compare Hilbert spaces to Banach, Fréchet and Neumann spaces.

THEOREM 5.4.– *For every vector space:*

$$\textit{Hilbert space} \;\Rightarrow\; \textit{Banach space} \;\Rightarrow\; \textit{Fréchet space} \;\Rightarrow\; \textit{Neumann space}. \;\blacksquare$$

Proof. Any Hilbert space, being normed and sequentially complete (Definition 5.1), is a Banach space (Definitions 4.10 (c)). The other implications have been addressed in Theorem 4.11. □

It should be observed that $\mathbb{R}$ is a Hilbert space.

THEOREM 5.5.– *The space $\mathbb{R}$ is a Hilbert space whose scalar product is the ordinary product.* ▮

Proof. The ordinary product is a scalar product on $\mathbb{R}$ and it generates its norm since $\|x\|_{\mathbb{R}} = |x| = (xx)^{1/2}$. Moreover, $\mathbb{R}$ is sequentially complete (Theorem 4.15). □

It should be observed that any closed subspace of a Hilbert space is a Hilbert space.

THEOREM 5.6.– *Every closed or sequentially closed topological subspace of a Hilbert space E is a Hilbert space (when it is endowed with the restriction of the scalar product of E).* ▮

Proof. A closed or sequentially closed subspace F of a Hilbert space E is sequentially complete (Theorem 4.12 (a)) since E is, and the restriction of the scalar product of E is a scalar product on F that generates its norm. □

Norm generated by a coercive form. Let us show that the topology of a Hilbert space is not changed by substituting its scalar product by any continuous symmetric and coercive bilinear form.

THEOREM 5.7.– *Let E be a Hilbert space, a be a symmetric bilinear form on E such that there are real numbers $\alpha > 0$ and $\beta > 0$ such that, for all $v \in E$,*

$$\beta \|v\|_E^2 \geq a(v,v) \geq \alpha \|v\|_E^2.$$

Then, a is a scalar product on E and

the norm defined by $|||u||| = \sqrt{a(u,u)}$ generates the topology of E. □

Proof. This follows from Definition 5.1 of a scalar product and from Definition 3.1 of equivalent families of semi-norms (applied to families reduced to a norm). □

Pre-Hilbertian norms. A **pre-Hilbertian norm** is a norm that can be defined by a scalar product (it is said that it *derives* therefrom). In the case of a sequentially complete space, it is also known as a **Hilbertian norm**. According to the **Jordan–Neumann theorem**[4], a norm is pre-Hilbertian if and only if it satisfies the parallelogram law established in Theorem 5.2 (b). □

---

4 **History of the Jordan–Neumann theorem.** Pascual JORDAN and John VON NEUMANN proved in 1935 [61] that any norm verifying the parallelogram law is pre-Hilbertian.

Variant of the definition of a pre-Hilbert space. A **pre-Hilbert space** often designates a vector space E endowed with a pre-Hilbertian norm. This is equivalent to Definition 5.1 (b) because when it exists, the scalar product is unique (for a fixed pre-Hilbert norm) and is given by:

$$(u, v)_E = \frac{1}{4}(\|u+v\|_E^2 - \|u-v\|_E^2). \ \square$$

5.2. Projection in a Hilbert space

Let us now define the projection onto a closed convex set in a Hilbert space[5].

THEOREM 5.8.– *Let*

K be a non-empty closed convex subset of a Hilbert space E,

$\| \; \|_E$ the norm and $(\; , \;)_E$ the scalar product of E, and $u \in E$. Then:

(a) *There is a unique element $p \in K$, called* **projection** *of u onto K, such that*

$$\|u-p\|_E = \min_{v \in K} \|u-v\|_E.$$

(b) *This projection is the unique element $p \in K$ such that*

$$(u-p, v-p)_E \leq 0, \quad \forall v \in K. \ \blacksquare$$

Proof. (a) Projection existence. Let

$$d = \inf_{v \in K} \|u-v\|_E$$

and let $(v_n)_{n \in \mathbb{N}}$ be a minimizing sequence, that is such that $v_n \in K$ and

$$\|u-v_n\|_E \to d. \tag{5.1}$$

The parallelogram law (Theorem 5.2 (b)) with respect to $u - v_n$ and $u - v_m$ gives

$$\|2u-(v_n+v_m)\|_E^2 + \|v_n - v_m\|_E^2 = 2(\|u-v_n\|_E^2 + \|u-v_m\|_E^2).$$

Since K is convex, $(v_n+v_m)/2 \in K$ then $\|2u-(v_n+v_m)\|_E^2 \geq 4d^2$ wherefrom, when $n \to \infty$,

$$\sup_{m \geq n} \|v_n - v_m\|_E^2 \to 0.$$

5 **History of the projection theorem 5.8.** Frigyes RIESZ established the existence of a projection onto every closed convex subset of a Hilbert space in 1934 [94, p. 36]. Ernst FISCHER had previously proved it in 1907 [42] for the closed subspaces of $L^2(0,1)$.

Thus, $(v_n)_{n\in\mathbb{N}}$ is a Cauchy sequence. Since E is sequentially complete, it converges to a limit p. Since K, being closed, is sequentially closed (Theorem 2.10),

$$p \in K.$$

Since (Theorem 2.5 (c)) $|\|u - v_n\|_E - \|u - p\|_E| \leq \|v_n - p\|_E$ which tends to 0, $\|u - v_n\|_E \to \|u - p\|_E$ and thus (5.1) leads to

$$\|u - p\|_E = d.$$

Unicity. If p' is another projection, the alternating sequence p, p', p, $p'\dots$ is minimizing and therefore, as it has just been seen, convergent from which $p' = p$.

(b) *Characterization of the projection.* Let $v \in K$, $t \in [0, 1]$ and $w = (1 - t)p + tv$. Since K is convex, $w \in K$ and consequently the projection p of u satisfies

$$\|u - p\|_E \leq \|u - w\|_E = \|(u - p) - t(v - p)\|_E$$

from which

$$\|u - p\|_E^2 \leq \|u - p\|_E^2 - 2t(u - p, v - p)_E + t^2\|v - p\|_E^2$$

and thus $2(u - p, v - p)_E \leq t\|v - p\|_E$ if $t > 0$. Hence, $(u - p, v - p)_E \leq 0$.

Conversely, let $p \in K$ be such that $(u - p, v - p)_E \leq 0$ for all $v \in K$. Since

$$\|u - v\|_E^2 = \|(u - p) - (v - p)\|_E^2 = \|u - p\|_E^2 + \|v - p\|_E^2 - 2(u - p, v - p)_E$$

it yields $\|u - v\|_E^2 \geq \|u - p\|_E^2$ for all $v \in K$. Therefore, p is the projection of u. □

Let us show that the projection is *contracting*.

THEOREM 5.9.– *Let E be a Hilbert space, K be a non-empty closed convex subset of E, u and v belong to E and p_u and p_v be their projections onto K. Then:*

$$\|p_u - p_v\|_E \leq \|u - v\|_E. \blacksquare$$

Proof. As a matter of fact,

$$\begin{aligned}
&\|u - v\|_E^2 = \\
&= \|(p_u - p_v) + (u - p_u + p_v - v)\|_E \\
&= \|p_u - p_v\|_E^2 + \|u - p_u + p_v - v\|_E^2 + 2(p_u - p_v, u - p_u + p_v - v)_E \\
&\geq \|p_u - p_v\|_E^2
\end{aligned}$$

since, due to the characterization of the projection of Theorem 5.8 (b),

$$(p_u - p_v, u - p_u + p_v - v)_E = (p_u - p_v, u - p_u)_E + (p_v - p_u, v - p_v)_E \geq 0. \ \Box$$

Let us characterize the projection onto a closed vector subspace[6].

THEOREM 5.10.– *Let*

F be a closed vector subspace of a Hilbert space E,

$(\ ,\ )_E$ the scalar product of E and $u \in E$.

Then, the projection of u onto F is the unique element $p \in F$ such that

$$(u - p, v)_E = 0, \quad \forall v \in F. \ ■$$

Proof. Let $v \in F$ and $t \in \mathbb{R}$. Then, $tv \in F$ (since F is a vector space) so the projection p of u onto F verifies (Theorem 5.8 (b))

$$0 \geq (u - p, tv - p)_E = -(u - p, p)_E + t(u - p, v)_E.$$

Hence $(u - p, v)_E = 0$ since the latter holds for all t.

Conversely, if an element p of F verifies $(u - p, v)_E = 0$ for all $v \in F$, it also verifies $(u - p, v - p)_E = 0$ because $v - p \in F$. Therefore (Theorem 5.8 (b)), this is the projection of u onto F. $\Box$

Two elements u and v of a Hilbert space E are said to be **orthogonal** if $(u, v)_E = 0$. It is said that u is **orthogonal** to a subset V of E if it is orthogonal to all its elements.

5.3. The space $\mathbb{R}^d$

Let us define the Euclidean space $\mathbb{R}^d$.

DEFINITION 5.11.– *Given an integer $d \geq 1$, the **Euclidean space** is*

$$\mathbb{R}^d \stackrel{\text{def}}{=} \{x = (x_1, \ldots, x_d) : x_i \in \mathbb{R}, \ \forall i \in [\![1, d]\!]\}.$$

*endowed with the **Euclidean norm** and the scalar product defined by*

$$|x| \stackrel{\text{def}}{=} (x_1^2 + \cdots + x_d^2)^{1/2},$$

$$x \cdot y \stackrel{\text{def}}{=} x_1 y_1 + \cdots + x_d y_d. \ ■$$

6 **History of Theorem 5.10.** The characterization of the projection of Theorem 5.10 was established in 1934 by Frigyes RIESZ [94].

Justification.– The mapping $(x, y) \mapsto x \cdot y$ is a scalar product (Definition 5.1 (a)) and subsequently $|x| = (x \cdot x)^{1/2}$ defines a norm following Minkowski's inequality (Theorem 5.3 (b)). □

Observe that $\mathbb{R}^d$ is sequentially complete and separable.

THEOREM 5.12.– *The space $\mathbb{R}^d$ is a Hilbert space which is sequentially separable and therefore separable.* ▮

Proof. The space $\mathbb{R}^d$ is a pre-Hilbert space (Definitions 5.1 (b) and 5.11) and $\mathbb{R}$ being sequentially complete (Theorem 4.15), it is sequentially complete (this can be proven as Theorem 6.7 (a) below). It is therefore a Hilbert space.

Since $\mathbb{Q}$ is sequentially dense in $\mathbb{R}$ (Theorem 4.18), $\mathbb{Q}^d$ is sequentially dense in $\mathbb{R}^d$ (this can be proven as Theorem 6.8 (b) below). In addition, $\mathbb{Q}^d$ is countable (Theorems 1.6 and 1.1 (d)). Therefore (Definition 2.21 (b)), $\mathbb{R}^d$ is sequentially separable which results (Theorem 2.22) in being separable. □

Let us show that the closed bounded subsets of $\mathbb{R}^d$ are compact which is the **Borel–Lebesgue theorem**[7].

THEOREM 5.13.– *For any subset of* $\mathbb{R}^d$*:*

(a) *Compact* $\Leftrightarrow$ *closed bounded.*

(b) *Relatively compact* $\Leftrightarrow$ *precompact* $\Leftrightarrow$ *bounded.*

(c) *In* (a) *and* (b), *it is possible to replace closed, compact and relatively compact by their sequential counterparts.* ▮

Proof. Bounded $\Leftrightarrow$ *precompact.* Every bounded set U of $\mathbb{R}^d$ is precompact because the balls $B(x, \epsilon)$ cover it when x spans the finite subset of $x \in \mathbb{R}^d$ such that $2\epsilon\sqrt{d}x \in \mathbb{Z}^d$ and $|x| \leq \sup_{u \in U} |u|$. The converse is always true (Theorem 2.25 (a)).

Other properties. They result from the equivalencies of Fréchet's theorem 4.14 since bounded is equivalent to precompact. □

Complement to Theorem 5.13 (b). In a normed space, bounded sets are relatively compact if and only if the space is of finite dimension according to **Riesz's compactness theorem**[8].

On the other hand, there are semi-normed spaces of infinite dimension in which bounded sets are relatively compact (for example $\mathcal{C}^\infty$ and $\mathcal{D}'$). They are thus not normable. □

7 **History of the Borel–Lebesgue theorem 5.13.** See note 7, p. 59.

8 **History of Riesz's compactness theorem.** Frigyes RIESZ showed in 1918 [93] that the bounded subsets of a normed space are relatively compact if and only if the space is of finite dimension.

Let us show that the bounded subsets of $\mathbb{R}^d$ are relatively sequentially compact, which is the **Bolzano–Weierstrass theorem**[9].

THEOREM 5.14.– *Any bounded sequence of* $\mathbb{R}^d$ *has a convergent subsequence.* ▮

Proof. Every bounded sequence of $\mathbb{R}^d$ is relatively sequentially compact due to Theorem 5.13 (b) and (c), thus (Definition 2.24) it has a convergent subsequence. □

Let us express the Cauchy–Schwarz and Minkowski inequalities in $\mathbb{R}^d$.

THEOREM 5.15.– *For all* $x \in \mathbb{R}^d$ *and* $y \in \mathbb{R}^d$*:*

(a) $$|x \cdot y| \leq |x||y|.$$

(b) $$|x + y| \leq |x| + |y|. \ ▮$$

Proof. These are particular cases of Theorem 5.3. □

Let us compare the Euclidean norm to the sum of the absolute values of the components and also to their upper bound.

THEOREM 5.16.– *For all* $x \in \mathbb{R}^d$ *and* $y \in \mathbb{R}^d$*:*

(a) $$(x_1^2 + \cdots + x_d^2)^{1/2} \leq |x_1| + \cdots + |x_d| \leq d^{1/2}(x_1^2 + \cdots + x_d^2)^{1/2}.$$

(b) $$d^{-1/2}(x_1^2 + \cdots + x_d^2)^{1/2} \leq \sup\{|x_1|, \cdots, |x_d|\} \leq (x_1^2 + \cdots + x_d^2)^{1/2}. \ ▮$$

Proof. (a) The first inequality is obtained by squaring the two members. The second is the result of the Cauchy–Schwartz inequality (Theorem 5.15 (a)) with respect to $(|x_1|, \ldots, |x_d|)$ and $(1, \ldots, 1)$.

(b) The first inequality is obtained by majoring all the $|x_i|$ by their least upper bound. The second is obtained by majoring this bound by the $|x_i|$ that realizes it. □

Unicity of the topology of $\mathbb{R}^d$**.** According to Theorem 5.16, each of the two norms defined by $\|x\| = |x_1| + \cdots + |x_d|$ and $|||x||| = \sup\{|x_1|, \ldots, |x_d|\}$ generates the topology of $\mathbb{R}^d$.

More generally, all the norms on $\mathbb{R}^d$, and even all the families of semi-norms that make it separated, are equivalent and thus define the same topology according to a theorem of Tychonoff[10]. □

---

9 **History of the Bolzano–Weierstrass theorem 5.14.** See note 8, p. 59.

10 **History of Tychonoff's theorem.** Andrei TYCHONOFF proved in 1935 [111, p. 769] that, on a space of finite dimension, all separated locally convex topologies are equivalent. Tychonoff's theorem usually refers to Theorem 11.10 which states that every product of compact sets is compact.

Chapter 6

Product, Intersection, Sum and Quotient of Spaces

6.1. Product of semi-normed spaces

Let us define the finite product of semi-normed spaces (not necessarily separated).

DEFINITION 6.1.– *Let E_1, …, E_d be semi-normed spaces and $\{\| \ \|_{E_i;\nu_i} : \nu_i \in \mathcal{N}_{E_i}\}$ be the family of semi-norms of E_i.*

The ***product****, or* ***Euclidean product****, of the E_i is the vector space*

$$E_1 \times E_2 \times \cdots \times E_d \stackrel{\text{def}}{=} \{u = (u_1, u_2, \ldots, u_d) : u_i \in E_i, \ \forall i \in [\![1, d]\!]\}$$

endowed with the semi-norms, said ***Euclidean****, indexed by $\nu_1 \in \mathcal{N}_{E_1}, \ldots, \nu_d \in \mathcal{N}_{E_d}$,*

$$\|u\|_{E_1\times\cdots\times E_d;\nu_1,\ldots,\nu_d} \stackrel{\text{def}}{=} (\|u_1\|^2_{E_1;\nu_1} + \cdots + \|u_d\|^2_{E_d;\nu_d})^{1/2}. \ \blacksquare$$

Justification.– The product $E_1 \times \cdots \times E_d$ satisfies Definition 2.1 of a vector space and, each mapping $(u_1, \ldots, u_d) \mapsto \|u_i\|_{E_i;\nu_i}$ being a semi-norm on the latter, their Euclidean sum $(u_1, \ldots, u_d) \mapsto (\|u_1\|^2_{E_1;\nu_1} + \cdots + \|u_d\|^2_{E_d;\nu_d})^{1/2}$ is itself also a semi-norm on the product as it will be verified in Theorem 6.2 below. □

It remains to verify that the Euclidean sum of semi-norms is a semi-norm.

THEOREM 6.2.– *Let $p_1, p_2 \ldots, p_d$ be semi-norms on the same vector space E. Then:*

(a) *A semi-norm $(\sum_{1\le i\le d} p_i^2)^{1/2}$ is defined on E by: for all $u \in E$,*

$$\Big(\sum_{1\le i\le d} p_i^2\Big)^{1/2}(u) \stackrel{\text{def}}{=} \Big(\sum_{1\le i\le d} (p_i(u))^2\Big)^{1/2}.$$

(b) *One of the* p_i *is a norm* $\Rightarrow$ $\Big(\sum_{1\le i\le d} p_i^2\Big)^{1/2}$ *is a norm.* ▮

Proof. (a) The properties of Definition 2.2 (a) of a semi-norm are satisfied because

$$\Big(\sum_{1\le i\le d}(p_i(u))^2\Big)^{1/2}\ge 0,\quad \Big(\sum_{1\le i\le d}(p_i(tu))^2\Big)^{1/2}=|t|\Big(\sum_{1\le i\le d}(p_i(u))^2\Big)^{1/2},$$

and because, according to Minkowski's inequality in $\mathbb{R}^d$ (Theorem 5.15 (b) with respect to $x=(p_1(u),\dots,p_d(u))$ and $y=(p_1(v),\dots,p_d(v))$),

$$\Big(\sum_{1\le i\le d}(p_i(u+v))^2\Big)^{1/2}\le\Big(\sum_{1\le i\le d}(p_i(u)+p_i(v))^2\Big)^{1/2}=|x+y|\le$$

$$\le|x|+|y|=\Big(\sum_{1\le i\le d}(p_i(u))^2\Big)^{1/2}+\Big(\sum_{1\le i\le d}(p_i(v))^2\Big)^{1/2}.$$

(b) If there exists $k\in[\![1,d]\!]$ such that p_k is a semi-norm then, when $u\ne 0_E$, it results that $p_k(u)>0$, hence $(\sum_{1\le i\le d}(p_i(u))^2)^{1/2}>0$ and thus $(\sum_{1\le i\le d}p_i^2)^{1/2}$ is a semi-norm. □

Note that the product of separated spaces is separated.

THEOREM 6.3.– *Every finite product* $E_1\times E_2\times\cdots\times E_d$ *of separated semi-normed spaces is a separated semi-normed space.* ▮

Proof. With the notations of Definition 6.1, if $(\|u_1\|^2_{E_1;\nu_1}+\cdots+\|u_d\|^2_{E_d;\nu_d})^{1/2}=0$ for all $\nu_1\in\mathcal{N}_{E_1}\dots,\nu_d\in\mathcal{N}_{E_d}$, then, for all $i\in[\![1,d]\!]$, it yields $\|u_i\|_{E_i;\nu_i}=0$ for all $\nu_i\in\mathcal{N}_{E_i}$ and therefore $u_i=0_{E_i}$ if E_i is separated (Definition 2.3 (c)). Then, $u=0$ which proves that $E_1\times\cdots\times E_d$ is separated. □

Let us characterize the convergent sequences of a product space.

THEOREM 6.4.– *Let* E_1, $E_2\dots$, E_d *be separated semi-normed spaces and, for all* $i\in[\![1,d]\!]$, *let* $(u_{i,n})_{n\in\mathbb{N}}\subset E_i$ *and* $u_i\in E_i$. *Then:*

$$(u_{1,n},\dots,u_{d,n})\to(u_1,\dots,u_d)\text{ in }E_1\times\cdots\times E_d\quad\Leftrightarrow$$

$$\Leftrightarrow\quad[\,u_{i,n}\to u_i\text{ in }E_i,\ \forall i\in[\![1,d]\!]\,].$$ ▮

Proof. With the notations of Definition 6.1 of a product and by Definition 2.6 (a) of the convergence of a sequence, $(u_{1,n},\dots,u_{d,n})\to(u_1,\dots,u_d)$ in $E_1\times\cdots\times E_d$ if and only if, for all $\nu_1\in\mathcal{N}_{E_1}\dots,\nu_d\in\mathcal{N}_{E_d}$,

$$(\|u_{1,n}-u_1\|^2_{E_1;\nu_1}+\cdots+\|u_{d,n}-u_d\|^2_{E_d;\nu_d})^{1/2}\to 0.$$

This implies, for all $i \in [\![1, d]\!]$, that $\|u_{i,n} - u_i\|^2_{E_i;\nu_i} \to 0$ for all $\nu_i \in \mathcal{N}_{E_i}$, which means that $u_{i,n} \to u_i$ in E_i. And vice versa. □

Observe that the topology of a product space is preserved if that of each space is.

THEOREM 6.5.– *Let $E_1, E_2 \ldots, E_d$ be separated semi-normed spaces.*

If the family of semi-norms of each E_i is replaced by an equivalent family, the family of semi-norms obtained on $E_1 \times \cdots \times E_d$ is equivalent to the initial family. ▮

Proof. If, for each $i \in [\![1, d]\!]$, the family $\{\| \; \|_{E_i;\nu_i} : \nu_i \in \mathcal{N}_i\}$ of E_i is dominated (Definition 3.1 (a)) by the family $\{||| \; |||_{E_i;\mu_i} : \mu_i \in \mathcal{M}_i\}$ then, for each $\nu_i \in \mathcal{N}_{E_i}$, there exist a finite set $M_{\nu_i} \in \mathcal{M}_i$ and $c_{\nu_i} \in \mathbb{R}$ such that, for all $u_i \in E_i$, it yields $\|u_i\|_{E_i;\nu_i} \leq c_{\nu_i} \sup_{\mu_i \in M_{\nu_i}} |||u_i|||_{E_i;\mu_i}$. Definition 6.1 of the semi-norms of the product then leads to

$$\|u\|_{E_1 \times \cdots \times E_d;\nu_1,\ldots,\nu_d} \leq c \sup_{(\mu_1,\ldots,\mu_d) \in M_{\nu_1} \times \cdots \times M_{\nu_d}} |||u|||_{E_1 \times \cdots \times E_d;\mu_1,\ldots,\mu_d}$$

where $c = \sup\{c_{\nu_1}, \ldots, c_{\nu_d}\}$. The initial family of the product is thus dominated by the new one.

If the initial family of each E_i is equivalent to the new one, that is dominates and is dominated by the new one, the same thus happens for the product. □

Let us give metrizability and normability properties of a product space.

THEOREM 6.6.– *If $E_1, E_2 \ldots, E_d$ are separated semi-normed spaces:*

(a) *The E_i are metrizable* $\Rightarrow$ *$E_1 \times \cdots \times E_d$ is metrizable.*

(b) *The E_i are normable* $\Rightarrow$ *$E_1 \times \cdots \times E_d$ is normable.*

(c) *The E_i are normed* $\Rightarrow$ $\begin{cases} E_1 \times \cdots \times E_d \textit{ is normed and} \\ \|u\|_{E_1 \times \cdots \times E_d} = (\|u_1\|^2_{E_1} + \cdots + \|u_d\|^2_{E_d})^{1/2}. \end{cases}$ ▮

Proof. (a) If each E_i is metrizable (Definition 4.1 (a)), its family of semi-norms is equivalent to a family indexed by a countable set $\mathcal{N}_i$. The family of semi-norms of $E_1 \times \cdots \times E_d$ is then equivalent (Theorem 6.5) to the family indexed by $\mathcal{N}_1 \times \cdots \times \mathcal{N}_d$ which is countable (Theorem 1.1 (d)). In addition, $E_1 \times \cdots \times E_d$ is separated (Theorem 6.3), and therefore it is metrizable.

(c) If the E_i are normed, the family of semi-norms of $E_1 \times \cdots \times E_d$ is reduced to the norm $(\|u_1\|^2_{E_1} + \cdots + \|u_d\|^2_{E_d})^{1/2}$.

(b) If each E_i is normable (Definition 4.1 (b)), its family of semi-norms is equivalent to a norm. The family of semi-norms of $E_1 \times \cdots \times E_d$ is then equivalent (Theorem 6.5) to the norm considered in the above proof of (c). □

Let us give completeness properties of a product space.

THEOREM 6.7.– *If $E_1, E_2 \ldots, E_d$ are separated semi-normed spaces:*

(a) *The E_i are Neumann spaces* $\Rightarrow$ $E_1 \times \cdots \times E_d$ *is a Neumann space.*

(b) *The E_i are Fréchet spaces* $\Rightarrow$ $E_1 \times \cdots \times E_d$ *is a Fréchet space.*

(c) *The E_i are Banach spaces* $\Rightarrow$ $E_1 \times \cdots \times E_d$ *is a Banach space.*

(d) *The E_i are Hilbert spaces* $\Rightarrow$ $\begin{cases} E_1 \times \cdots \times E_d \text{ is a Hilbert space} \\ \text{and has a scalar product} \\ (u_1, v_1)_{E_1} + \cdots + (u_d, v_d)_{E_d}. \end{cases}$ ■

Proof. (a) If $((u_{1,n}, \ldots, u_{d,n}))_{n \in \mathbb{N}}$ is a Cauchy sequence of $E_1 \times \cdots \times E_d$, then $(u_{i,n})_{n \in \mathbb{N}}$ is a Cauchy sequence of E_i so, if E_i is sequentially complete, namely if it is a Neumann space, it converges therein to a limit u_i. If this holds for all $i \in [\![1, d]\!]$, then (Theorem 6.4) $(u_{1,n}, \ldots, u_{d,n}) \to (u_1, \ldots, u_d)$ in $E_1 \times \cdots \times E_d$ that is therefore sequentially complete.

(b) If the E_i are metrizable and sequentially complete, namely if they are Fréchet spaces, the same occurs with $E_1 \times \cdots \times E_d$ according to (a) and Theorem 6.6 (a).

(c) If the E_i are normed and sequentially complete, namely if they are Banach spaces, the same occurs with $E_1 \times \cdots \times E_d$ according to (a) and Theorem 6.6 (c).

(d) This results from (c) because $(u, v) = \sum_{i=1}^{d} (u_i, v_i)_{E_i}$ defines a scalar product (Definition 5.1) on $E_1 \times \cdots \times E_d$ that verifies $\|u\|_{E_1 \times \cdots \times E_d} = (u, u)^{1/2}$. □

Let us give separation properties of a product space.

THEOREM 6.8.– *If $E_1, E_2 \ldots, E_d$ are separated semi-normed spaces:*

(a) *The E_i are separable* $\Rightarrow$ $E_1 \times \cdots \times E_d$ *is separable.*

(b) *The E_i are sequentially separable*
$\Rightarrow$ $E_1 \times \cdots \times E_d$ *is sequentially separable.* ■

Proof. (b) If E_i is sequentially separable (Definition 2.21 (b)), there is a countable subset U_i of E_i such that every $u_i \in E_i$ is the limit of a sequence $(u_{i,n})_{n \in \mathbb{N}}$ of U_i. If this holds for all $i \in [\![1,d]\!]$, then (Theorem 6.4) $(u_1,\ldots,u_d)$ is the limit of the sequence $(u_{1,n},\ldots,u_{d,n})$ of $U_1 \times \cdots \times U_d$. The latter being countable (Theorem 1.1 (d)), $E_1 \times \cdots \times E_d$ is sequentially separable.

(a) If the E_i are separable (Definition 2.21 (a)), for all $i \in [\![1,d]\!]$, there is a countable subset U_i of E_i such that $\overline{U_i} = E_i$. Then:

$$\overline{U_1 \times \cdots \times U_d} = E_1 \times \cdots \times E_d. \tag{6.1}$$

In effect, let $(u_1,\ldots,u_d) \in E_1 \times \cdots \times E_d$, N be a finite subset of $\mathcal{N}_{E_1} \times \cdots \times \mathcal{N}_{E_d}$ and $\eta > 0$. For all $i \in [\![1,d]\!]$, let $N_i = \{\nu_i : (\nu_1,\ldots,\nu_d) \in N\}$. Since $\overline{U_i} = E_i$, the characterization of the closure of Theorem 2.17 (b) shows that there exists $v_i \in U_i$ such that $\sup_{\nu_i \in N_i} \|v_i - u_i\|_{E_i;\nu_i} \leq \eta/\sqrt{d}$. According to Definition 6.1 of the semi-norms of the product,

$$\sup_{(\nu_1,\ldots,\nu_d)\in N} \|v-u\|_{E_1\times\cdots\times E_d;\nu_1,\ldots,\nu_d} \leq \Big(\sum_{1\leq i\leq d} \sup_{\nu_i\in N_i} \|v_i-u_i\|^2_{E_i;\nu_i}\Big)^{1/2} \leq \eta.$$

Therefore, $u \in \overline{U_1 \times \cdots \times U_d}$ according to the characterization of Theorem 2.17 (b), now with respect to $E_1 \times \cdots \times E_d$. This ends the proof of (6.1).

Since $U_1 \times \cdots \times U_d$ is countable (Theorem 1.1 (d)), it follows that $E_1 \times \cdots \times E_d$ is separable. □

Let us introduce two other families of semi-norms defining the product topology. The designation *Euclidean* is reserved for the semi-norms of Definition 6.1 and for the product endowed with them.

THEOREM 6.9.– *Let $E_1, E_2\ldots, E_d$ be separated semi-normed spaces and, for all $i \in [\![1,d]\!]$, let $\{\|\ \|_{E_i;\nu_i} : \nu_i \in \mathcal{N}_{E_i}\}$ be the family of semi-norms of E_i.*

The family of semi-norms of $E_1 \times E_2 \times \cdots \times E_d$ is equivalent to each of both following families of semi-norms, also indexed by $\nu_1 \in \mathcal{N}_{E_1}, \ldots, \nu_d \in \mathcal{N}_{E_d}$:

(a) $$|||(u_1,\ldots,u_d)|||_{E_1\times\cdots\times E_d;\nu_1,\ldots,\nu_d} \stackrel{\text{def}}{=} \|u_1\|_{E_1;\nu_1} + \cdots + \|u_d\|_{E_d;\nu_d}.$$

(b) $$||||(u_1,\ldots,u_d)||||_{E_1\times\cdots\times E_d;\nu_1,\ldots,\nu_d} \stackrel{\text{def}}{=} \sup\{\|u_1\|_{E_1;\nu_1},\ldots,\|u_d\|_{E_d;\nu_d}\}.$$ ∎

Proof. This indeed defines semi-norms on $E_1 \times \cdots \times E_d$ whose families are both equivalent (Definition 3.1) to the family of Definition 6.1 since, due to the inequalities of Theorem 5.16 with respect to $x = (\|u_1\|_{E_1;\nu_1},\ldots,\|u_d\|_{E_d;\nu_d})$,

$$(\|u_1\|^2_{E_1;\nu_1} + \cdots + \|u_d\|^2_{E_d;\nu_d})^{1/2} \leq \|u_1\|_{E_1;\nu_1} + \cdots + \|u_d\|_{E_d;\nu_d} \leq$$
$$\leq d^{1/2}(\|u_1\|^2_{E_1;\nu_1} + \cdots + \|u_d\|^2_{E_d;\nu_d})^{1/2}$$

and

$$d^{-1/2}(\|u_1\|^2_{E_1;\nu_1} + \cdots + \|u_d\|^2_{E_d;\nu_d})^{1/2} \leq \sup\{\|u_1\|_{E_1;\nu_1}, \cdots, \|u_d\|_{E_d;\nu_d}\} \leq$$
$$\leq (\|u_1\|^2_{E_1;\nu_1} + \cdots + \|u_d\|^2_{E_d;\nu_d})^{1/2}. \; \square$$

Usefulness of the various families of semi-norms on a product space. Euclidean semi-norms (Definition 6.1) are adapted to the Hilbertian case (Theorem 6.7 (d)) and they provide the "natural" norm of the Euclidean space $\mathbb{R}^3$ (Definition 5.11) where we live. They are essential for the study of physics problems.

The semi-norms of Theorem 6.9 are often easier to use, for example for the study of continuity issues.

Other semi-norms will be used depending on the context. So, the product E^d of a space by itself will be endowed (Definition 6.10) with semi-norms indexed only by $\mathcal{N}_E$ (instead of $\mathcal{N}_E^d$). □

Infinite product space. The product of an infinite family of separated semi-normed spaces is defined by (11.15), p. 160. Its family of semi-norms, defined by (11.16), is equivalent to that of Definition 6.1 when the family of spaces is finite. □

6.2. Product of a semi-normed space by itself

Let us define the finite product of a semi-normed space by itself.

DEFINITION 6.10.– *Let E be a separated semi-normed space, $\{\| \ \|_{E;\nu} : \nu \in \mathcal{N}_E\}$ its family of semi-norms and $d \in \mathbb{N}$.*

*The **product of** E **by itself** is the vector space*

$$E^d \stackrel{\text{def}}{=} \{u = (u_1, u_2, \ldots, u_d) : u_i \in E, \ \forall i \in [\![1, d]\!]\}$$

endowed with the semi-norms, indexed by $\nu \in \mathcal{N}_E$,

$$\|u\|_{E^d;\nu} \stackrel{\text{def}}{=} (\|u_1\|^2_{E;\nu} + \cdots + \|u_d\|^2_{E;\nu})^{1/2}. \; \blacksquare$$

Justification. It suffices to rewrite the justification of Definition 6.1 of $E_1 \times \cdots \times E_d$ with $E_1 = \cdots = E_d = E$ and $\nu_1 = \cdots = \nu_d = \nu$. □

Let us show that this product is topologically equal to the ordinary product.

THEOREM 6.11.– *Let E be a separated semi-normed space. Then:*

(a) $\qquad E^d$ *is a separated semi-normed space.*

(b) $$E^d \underset{\leftrightarrow}{=} \underbrace{E \times \cdots \times E}_{d \text{ times}}.$$

(c) *If E is normed, then E^d is normed and $\| \; \|_{E^d} = \| \; \|_{E\times\cdots\times E}$.* ▮

Proof. (a) If $(\|u_1\|^2_{E;\nu}+\cdots+\|u_d\|^2_{E;\nu})^{1/2} = 0$ for all $\nu \in \mathcal{N}_E$, then, for all $i \in [\![1, d]\!]$, it follows that $\|u_i\|_{E;\nu} = 0$ for all $\nu \in \mathcal{N}_E$ and thus $u_i = 0_E$ since E is separated. As a result, $u = 0_{E^d}$ which proves that E^d is separated.

(b) The families of the semi-norms of Definitions 6.1 and 6.10 are equivalent (Definition 3.1) because, for all $\nu \in \mathcal{N}_E$,

$$\|u\|_{E^d;\nu} = \|u\|_{E\times\cdots\times E;\nu,\dots,\nu}$$

and for all $\nu_1 \in \mathcal{N}_E \dots, \nu_d \in \mathcal{N}_E$,

$$\|u\|_{E\times\cdots\times E;\nu_1,\dots,\nu_d} = (\|u_1\|_{E;\nu_1}+\cdots+\|u_d\|_{E;\nu_d})^{1/2} \le \|u\|_{E^d;\nu_1}+\cdots+\|u\|_{E^d;\nu_d}.$$

(c) If E is normed, the family of semi-norms of E^d is reduced to the single norm $\|u\|_{E^d} \overset{\text{def}}{=} (\|u_1\|^2_E + \cdots + \|u_d\|^2_E)^{1/2}$, which is equal to $\|u\|_{E\times\cdots\times E}$. □

Let us show that this product has the properties of the ordinary product.

THEOREM 6.12.– *Let E be a separated semi-normed space.*

(a) *Each of the following properties is satisfied by E^d if it is by E:*

$\left\{\begin{array}{l}\textit{Metrizable, normed, normable, Neumann space, Fréchet space, Banach space,}\\ \textit{Hilbert space, separable, sequentially separable.}\end{array}\right.$

(b) *If E is a Hilbert space,*

$$(u, v)_{E^d} = (u, v)_{E\times\cdots\times E} = (u_1, v_1)_E + \cdots + (u_d, v_d)_E. \;▮$$

Proof. (a) The space $E \times \cdots \times E$ has each of these properties if E has it (Theorems 6.6, 6.7 and 6.8) and it is topologically equal to E^d (Theorem 6.11 (b)). The topological equalities preserving metrizability and normability (Theorem 4.3), sequential completeness (Theorem 4.9) and separability, sequential or not, (Theorem 3.4 (b)), these properties are then verified by E^d, as well as the fact of being a Neumann or Fréchet space that results thereof (Definition 4.10).

If E is normed, E^d also is (Theorem 6.11 (c)), and therefore it is a Banach space if E is.

If E is a Hilbert space, $(u, v)_{E^d} = \sum_{i=1}^d (u_i, v_i)_E$ defines a scalar product (Definition 5.1) on E^d that verifies $\|u\|_{E^d} = (u, u)^{1/2}_{E^d}$, thus E^d is a Hilbert space because it is sequentially complete, as it has just been seen, since E is.

(b) The value of the scalar product of E^d has just been established. It coincides with that of $E \times \cdots \times E$ obtained in Theorem 6.7 (d). □

6.3. Intersection of semi-normed spaces

Let us define the finite or infinite intersection of semi-normed spaces (not necessarily separated).

DEFINITION 6.13.– *Let $\{E_i : i \in \mathcal{I}\}$ be semi-normed spaces such that*

$$\begin{cases} \text{the intersection of the } E_i \text{ is not empty and their additions and} \\ \text{multiplications coincide on their intersection} \end{cases}$$

and let $\{\| \ \|_{E_i;\nu_i} : \nu_i \in \mathcal{N}_{E_i}\}$ be the family of the semi-norms of E_i.

*The **intersection** of E_i is the vector space (for the restrictions of their additions and of their multiplications)*

$$\bigcap_{i \in \mathcal{I}} E_i \stackrel{\text{def}}{=} \{u : u \in E_i, \ \forall i \in \mathcal{I}\}$$

endowed with the semi-norms, indexed by $j \in \mathcal{I}$ and $\nu_j \in \mathcal{N}_{E_j}$,

$$\|u\|_{\bigcap_{i \in \mathcal{I}} E_i;j,\nu_j} \stackrel{\text{def}}{=} \|u\|_{E_j;\nu_j}. \ \blacksquare$$

Justification.– The intersection of E_i satisfies Definition 2.1 of a vector space and each $\| \ \|_{E_i;\nu_i}$ is a semi-norm on it. □

Finite intersection. A finite intersection is also denoted by $E_1 \cap \ldots \cap E_d = \bigcap_{i \in [\![1,d]\!]}$.

It is more rewarding to endow it with the semi-norms, indexed by $\nu_1 \in \mathcal{N}_{E_1}, \ldots, \nu_d \in \mathcal{N}_{E_d}$,

$$|\!|\!|u|\!|\!|_{E_1 \cap \ldots \cap E_d;\nu_1,\ldots,\nu_d} \stackrel{\text{def}}{=} \left(\|u\|^2_{E_1;\nu_1} + \cdots + \|u\|^2_{E_d;\nu_d}\right)^{1/2}$$

because this family, equivalent to that of Definition 6.13, makes it a normed, or Banach, or Hilbert space when the E_i are such spaces, as it will be seen in Theorems 6.18 and 6.19. □

It should be observed that an intersection of spaces is separated if one of them is.

THEOREM 6.14.– *Any intersection $\bigcap_{i \in \mathcal{I}} E_i$ of semi-normed spaces is separated if one of the E_i is separated.* ▮

Proof. Assume that there exists $k \in \mathcal{I}$ such that E_k be separated (Definition 2.3 (b)). If $\|u\|_{E_j;j,\nu_j} = 0$ for all $j \in \mathcal{I}$ and $\nu_j \in \mathcal{N}_{E_j}$, then $u = 0_{E_k}$ and thus $u = 0_{E_j}$ for all $j \in \mathcal{I}$ since additions and therefore zero elements of E_j coincide on their intersection. Therefore, $\bigcap_{i \in \mathcal{I}} E_i$ is separated. □

Let us characterize the convergent sequences of an intersection of spaces.

THEOREM 6.15.– *Let $\bigcap_{i\in\mathcal{I}} E_i$ be an intersection of semi-normed spaces, $(u_n)_{n\in\mathbb{N}}$ be a sequence of $\bigcap_{i\in\mathcal{I}} E_i$ and $u \in \bigcap_{i\in\mathcal{I}} E_i$. Then:*

$$u_n \to u \text{ in } \bigcap_{i\in\mathcal{I}} E_i \quad \Leftrightarrow \quad [\, u_n \to u \text{ in } E_i,\ \forall i \in \mathcal{I}\,]. \ \blacksquare$$

Proof. With the notations of Definition 6.13 of an intersection and according to Definition 2.6 (a) of the convergence of a sequence, $u_n \to u$ in $\bigcap_{i\in\mathcal{I}} E_i$ is equivalent to $\|u_n - u\|_{E_i;\nu_i} \to 0$ for all $i \in \mathcal{I}$ and $\nu_i \in \mathcal{N}_{E_i}$, namely to $u_n \to u$ in E_i for all $i \in \mathcal{I}$. □

Observe that the topology of a product space is preserved if that of each space is.

THEOREM 6.16.– *Let $\bigcap_{i\in\mathcal{I}} E_i$ be an intersection of semi-normed spaces.*

If the family of semi-norms of each E_i is replaced by an equivalent family, the family of semi-norms obtained on their intersection is equivalent to the initial family. ■

Proof. If, for all $j \in \mathcal{I}$, the family $\{\| \ \|_{E_j;\nu_j} : \nu_j \in \mathcal{N}_j\}$ of E_j is dominated by the family $\{||| \ |||_{E_j;\mu_j} : \mu_j \in \mathcal{M}_j\}$, that is if for all $\nu_j \in \mathcal{N}_{E_j}$ there exist a finite set $M_{\nu_j} \in \mathcal{M}_j$ and $c_{\nu_j} \in \mathbb{R}$ such that $\| \ \|_{E_j;\nu_j} \le c_{\nu_j} \sup_{\mu_j \in M_{\nu_j}} ||| \ |||_{E_j;\mu_j}$, then Definition 6.13 of the semi-norms of the intersection leads to

$$\|u\|_{\bigcap_{i\in\mathcal{I}} E_i;j,\nu_j} = \|u\|_{E_j;\nu_j} \le c_{\nu_j} \sup_{\mu_j \in M_{\nu_j}} |||u|||_{E_j;\mu_j} = c_{\nu_j} \sup_{\mu_j \in M_{\nu_j}} |||u|||_{\bigcap_{i\in\mathcal{I}} E_i;j,\mu_j}.$$

The initial family of the product is thus dominated by the new one.

If the initial family of each E_i is equivalent to the new one, that is dominates and is dominated by the new one, the same thus happens for $\bigcap_{i\in\mathcal{I}} E_i$. □

Let us observe that an intersection of spaces is included in each of them.

THEOREM 6.17.– *Let $\bigcap_{i\in\mathcal{I}} E_i$ be an intersection of semi-normed spaces and $j \in \mathcal{I}$. Then:*

$$\bigcap_{i\in\mathcal{I}} E_i \subsetneqq E_j. \ \blacksquare$$

Proof. According to Definition 6.13 of the intersection, $\bigcap_{i\in\mathcal{I}} E_i \subset E_j$ and, for all $u \in \bigcap_{i\in\mathcal{I}} E_i$, we have $\|u\|_{E_j;j,\nu_j} = \|u\|_{\bigcap_{i\in\mathcal{I}} E_i;\nu_j}$. Consequently, $\bigcap_{i\in\mathcal{I}} E_i \subsetneqq E_j$ following Definition 3.6 (c) of a topological inclusion. □

Let us give metrizability and normability properties of an intersection of spaces.

THEOREM 6.18.– *Let $\bigcap_{i\in\mathcal{I}} E_i$ be an intersection of semi-normed spaces. Then:*

(a) $\begin{cases} \mathcal{I} \text{ is countable and} \\ \text{the } E_i \text{ are metrizable} \end{cases} \Rightarrow \bigcap_{i\in\mathcal{I}} E_i$ *is metrizable.*

(b) $\begin{cases} \mathcal{I} \text{ is finite and} \\ \text{the } E_i \text{ are normable} \end{cases} \Rightarrow \begin{cases} \bigcap_{i\in\mathcal{I}} E_i \text{ is normable and its topology} \\ \text{is generated by the norm} \\ \qquad |||u||| = (\sum_{i\in\mathcal{I}} \|u\|^2_{E_i})^{1/2}. \end{cases}$ ▮

Proof. (a) If each E_i is metrizable (Definition 4.1 (a)), its family of semi-norms is equivalent (Definition 3.1) to a family indexed by a countable set $\mathcal{N}_i$. The family of semi-norms of $\bigcap_{i\in\mathcal{I}} E_i$ is then equivalent (Theorem 6.16) to the family indexed by the union of the $\mathcal{N}_i$ which is countable (Theorem 1.1 (e)) if $\mathcal{I}$ is. In addition, the E_i being separated (since metrizable), the intersection $\bigcap_{i\in\mathcal{I}} E_i$ also is (Theorem 6.14) and therefore is metrizable.

(b) If the E_i are normed and if $\mathcal{I}$ is finite, then $|||u||| = (\sum_{i\in\mathcal{I}} \|u\|^2_{E_i})^{1/2}$ is a norm (Theorem 6.2) on their intersection. It is equivalent to the family of semi-norms (which are norms) of Definition 6.13 because, for all $j \in \mathcal{I}$, by denoting by n the number of elements of $\mathcal{I}$,

$$\|u\|_{\bigcap_{i\in\mathcal{I}} E_i;j} = \|u\|_{E_j} \leq |||u||| \leq \sqrt{n} \sup_{k\in\mathcal{I}} \|u\|_{E_k} = \sqrt{n} \sup_{k\in\mathcal{I}} \|u\|_{\bigcap_{i\in\mathcal{I}} E_i;k}. \quad \square$$

Let us give completeness and other properties of an intersection of spaces.

THEOREM 6.19.– *Let $\{E_i : i \in \mathcal{I}\}$ and F be separated semi-normed spaces such that, for all $i \in \mathcal{I}$,*

$$E_i \subsetneqq F.$$

Then:

(a) *The E_i are Neumann spaces* $\Rightarrow \bigcap_{i\in\mathcal{I}} E_i$ *is a Neumann space.*

(b) $\begin{cases} \mathcal{I} \text{ is countable and} \\ \text{the } E_i \text{ are Fréchet spaces} \end{cases} \Rightarrow \bigcap_{i\in\mathcal{I}} E_i$ *is a Fréchet space.*

(c) $\begin{cases} \mathcal{I} \text{ is finite and} \\ \text{the } E_i \text{ are Banach spaces} \end{cases} \Rightarrow \begin{cases} \text{The norm } |||u||| = (\sum_{i\in\mathcal{I}} \|u\|^2_{E_i})^{1/2} \\ \text{generates the topology of } \bigcap_{i\in\mathcal{I}} E_i \\ \text{and makes it a Banach space.} \end{cases}$

(d) $\begin{cases} \mathcal{I} \text{ is finite and} \\ \text{the } E_i \text{ are Hilbert spaces} \end{cases} \Rightarrow \begin{cases} \text{The norm } |||u||| = (\sum_{i\in\mathcal{I}} \|u\|^2_{E_i})^{1/2} \\ \text{generates the topology of } \bigcap_{i\in\mathcal{I}} E_i \\ \text{and makes it a Hilbert space of} \\ \text{scalar product } (u,v) = \sum_{i\in\mathcal{I}} (u,v)_{E_i}. \end{cases}$ ∎

Proof. (a) Let $(u_n)_{n\in\mathbb{N}}$ be a Cauchy sequence of $\bigcap_{i\in\mathcal{I}} E_i$. It is a Cauchy sequence in each E_j (Theorem 3.8 (a)) since $\bigcap_{i\in\mathcal{I}} E_i$ is therein topologically included (Theorem 6.17).

If each E_j is sequentially complete, $(u_n)_{n\in\mathbb{N}}$ converges therein to a limit ℓ. The latter does not depend on j because the convergence also takes place in F (Theorem 3.8 (a)) where the limit is unique (Theorem 2.7 (a)). Then (Theorem 6.15), $u_n \to \ell$ in $\bigcap_{i\in\mathcal{I}} E_i$. Therefore, $\bigcap_{i\in\mathcal{I}} E_i$ is sequentially complete.

(b) When $\mathcal{I}$ is countable, if the E_i are metrizable and sequentially complete, namely if they are Fréchet spaces, the same holds with $\bigcap_{i\in\mathcal{I}} E_i$ due to (a) and Theorem 6.18 (a).

(c) When $\mathcal{I}$ is finite, if the E_i are normed and sequentially complete, namely if they are Banach spaces, the same holds with $\bigcap_{i\in\mathcal{I}} E_i$ due to (a) and Theorem 6.18 (b).

(d) This results from (c) because $(u,v) = \sum_{i\in\mathcal{I}}(u,v)_{E_i}$ defines a scalar product (Definition 5.1) on $\bigcap_{i\in\mathcal{I}} E_i$ that verifies $|||u||| = (u,u)^{1/2}$. □

6.4. Sum of semi-normed spaces

Let us define the finite sum of separated semi-normed spaces.

DEFINITION 6.20.– *Let $E_1, E_2 \dots, E_d$ and F be separated semi-normed spaces such that, for each $i \in [\![1,d]\!]$,*

$$E_i \underset{\rightarrow}{\subseteq} F.$$

For each $i \in [\![1,d]\!]$, let

$$\begin{cases} \{\| \; \|_{E_i;\nu_i} : \nu_i \in \mathcal{N}_{E_i}\} \text{ a filtering family of} \\ \text{semi-norms generating the topology of } E_i. \end{cases}$$

*The **sum** of E_i is the vector space*

$$E_1 + E_2 + \cdots + E_d \stackrel{\text{def}}{=} \{u \in F : u = u_1 + u_2 + \cdots + u_d,\ u_i \in E_i,\ \forall i \in [\![1,d]\!]\}$$

endowed with the semi-norms, indexed by $\nu_1 \in \mathcal{N}_{E_1}$, $\nu_2 \in \mathcal{N}_{E_2} \dots$ and $\nu_d \in \mathcal{N}_{E_d}$,

$$\|u\|_{E_1+\cdots+E_d;\nu_1,\dots,\nu_d} \stackrel{\text{def}}{=} \inf_{u_i\in E_i : u_1+\cdots+u_d=u} (\|u_1\|^2_{E_1;\nu_1} + \cdots + \|u_d\|^2_{E_d;\nu_d})^{1/2}.$$ ∎

Justification.– The mapping $u \mapsto \|u\|_{E_1+\cdots+E_d;\nu_1,\ldots,\nu_d}$ is indeed a semi-norm on the space $E_1 + \cdots + E_d$. As a matter of fact, it verifies the triangle inequality

$$\|u+v\|_{E_1+\cdots+E_d;\nu_1,\ldots,\nu_d} \leq \|u\|_{E_1+\cdots+E_d;\nu_1,\ldots,\nu_d} + \|v\|_{E_1+\cdots+E_d;\nu_1,\ldots,\nu_d} \tag{6.2}$$

because, if $v_1 + \cdots + v_d = v$, then $u_1 + v_1 + \cdots + u_d + v_d = u + v$ and thus, according to Minkowski's inequality in $\mathbb{R}^d$ (Theorem 5.15 (b) with respect to $x = (\|u_1\|_{E_1;\nu_1}, \ldots, \|u_d\|_{E_d;\nu_d})$ and $y = (\|v_1\|_{E_1;\nu_1}, \ldots, \|v_d\|_{E_d;\nu_d})$,

$$\begin{aligned}
\|u+v\|_{E_1+\cdots+E_d;\nu_1,\ldots,\nu_d} &\leq (\|u_1+v_1\|^2_{E_1;\nu_1} + \cdots + \|u_d+v_d\|^2_{E_d;\nu_d})^{1/2} \leq \\
&\leq ((\|u_1\|_{E_1;\nu_1} + \|v_1\|_{E_1;\nu_1})^2 + \cdots + (\|u_d\|_{E_d;\nu_d} + \|v_d\|_{E_d;\nu_d})^2)^{1/2} = \\
&= |x+y| \leq |x| + |y| = \\
&= (\|u_1\|^2_{E_1;\nu_1} + \cdots + \|u_d\|^2_{E_d;\nu_d})^{1/2} + (\|v_1\|^2_{E_1;\nu_1} + \cdots + \|v_d\|^2_{E_d;\nu_d})^{1/2},
\end{aligned}$$

hence (6.2) by taking the infimum over all decompositions of u, then of v. The other properties of Definition 2.2 (a) of a semi-norm are verified in a similar manner.

On the other hand, for each $i \in [\![1, d]\!]$, there is indeed (Theorem 3.15) a filtering family of semi-norms that generates the topology of E_i and the topology resulting on $E_1 + \cdots + E_d$ does not depend on the choice of these families, as it will be verified in the next theorem. □

It remains to verify that, thanks to the filtration assumption, the obtained topology is independent of the choice of semi-norms.

THEOREM 6.21.– *Let $E_1 + \cdots + E_d$ be a sum of separated semi-normed spaces. Then:*

(a) *The space $E_1 + \cdots + E_d$ is separated semi-normed.*

(b) *The topology of $E_1 + \cdots + E_d$ does not depend on the choice, for every $i \in [\![1, d]\!]$, of the filtering family of semi-norms defining the topology of E_i.* ▮

Proof. For every $i \in [\![1, d]\!]$, let $\{\| \ \|_{E_i;\nu_i} : \nu_i \in \mathcal{N}_{E_i}\}$ be a filtering family of semi-norms generating the topology of E_i given by Theorem 3.15.

(a) *A majoration*. Let $\| \ \|_{F;\mu}$ be a semi-norm of F. For each $i \in [\![1, d]\!]$, the inclusion $E_i \underset{\rightarrow}{\subseteq} F$ and the assumption in Definition 6.20 that the family of semi-norms of E_i be filtering result (Theorem 3.17 (d)) in the existence of $c_i \in \mathbb{R}$ and $\nu_i \in \mathcal{N}_{E_i}$ such that, for all $u_i \in E_i$,

$$\|u_i\|_{F;\mu} \leq c_i \, \|u_i\|_{E_i;\nu_i}.$$

Let $u \in E_1 + \cdots + E_d$. Any decomposition $u = u_1 + \cdots + u_d$ thus verifies, with the second inequality of Theorem 5.16 (a), by denoting $c = \sup(c_1, \ldots, c_d)$,

$$\|u\|_{F;\mu} \le \|u_1\|_{F;\mu} + \cdots + \|u_d\|_{F;\mu} \le c\,(\|u_1\|_{E_1;\nu_1} + \cdots + \|u_d\|_{E_d;\nu_d}) \le$$
$$\le cd^{1/2}(\|u_1\|^2_{E_1;\nu_1} + \cdots + \|u_d\|^2_{E_d;\nu_d})^{1/2}.$$

By taking the infimum over all decompositions it yields, denoting $c' = cd^{1/2}$,

$$\|u\|_{F;\mu} \le c'\,\|u\|_{E_1+\cdots+E_d;\nu_1,\ldots,\nu_d}. \tag{6.3}$$

Separation. If $\|u\|_{E_1+\cdots+E_d;\nu_1,\ldots,\nu_d} = 0$ for all $\nu_1 \in \mathcal{N}_{E_1}, \ldots, \nu_d \in \mathcal{N}_{E_d}$, then $\|u\|_{F;\mu} = 0$ for all semi-norms of F following (6.3), hence $u = 0_F$ since F is separated (Definition 2.3 (c)). Therefore, $E_1 + \cdots + E_d$ is separated.

(b) For every $i \in [\![1, d]\!]$, let $\{\|\ \|_{E_i;\nu_i} : \nu_i \in \mathcal{N}_{E_i}\}$ and $\{|\!|\!|\ |\!|\!|_{E_i;\sigma_i} : \sigma_i \in \mathcal{M}_{E_i}\}$ be two filtering families of semi-norms generating the topology of E_i. Then (Theorem 3.17 (d)), there exist $c_i \in \mathbb{R}$ and $\lambda_i \in \mathcal{M}_{E_i}$ such that, for all $u_i \in E_i$,

$$\|u_i\|_{E_i;\nu_i} \le c_i\,|\!|\!|u_i|\!|\!|_{E_i;\lambda_i}. \tag{6.4}$$

Let $u \in E_1 + \cdots + E_d$. Every decomposition $u_1 + \cdots + u_d = u$ thus verifies

$$(\|u_1\|^2_{E_1;\nu_1} + \cdots + \|u_d\|^2_{E_d;\nu_d})^{1/2} \le c\,(|\!|\!|u_1|\!|\!|^2_{E_1;\lambda_1} + \cdots + |\!|\!|u_d|\!|\!|^2_{E_d;\lambda_d})^{1/2}$$

where $c = \sup(c_1, \ldots, c_d)$. By taking the infimum over all decompositions it yields,

$$\|u\|_{E_1+\cdots+E_d;\nu_1,\ldots,\nu_d} \le c\,|\!|\!|u|\!|\!|_{E_1+\cdots+E_d;\lambda_1,\ldots,\lambda_d}.$$

The second family of semi-norms of $E_1 + \cdots + E_d$ thus dominates the first. And vice versa because it is possible to exchange them. They are therefore equivalent. □

Need for the filtration assumption. Would not only filtering families of semi-norms defining the topologies of E_i be considered, Definition 6.20 would not necessarily yield a unique topology for $E_1 + \cdots + E_d$.

For example, the topology of $\mathbb{R}^2$ is generated by the semi-norms $\|x\|_1 = |x_1|$ and $\|x\|_2 = |x_2|$. It is also generated by the semi-norms $\|x\|_3 = |x_1 + x_2|$ and $\|x\|_4 = |x_1 - x_2|$. For these two families and following Definition 6.20, the space $\mathbb{R}^2 + \mathbb{R}^2$ would be endowed with the four semi-norms $\|x\|_{i,j} = \inf_{u+v=x} \|u\|_i + \|v\|_j$, where $i = 1$ or 2 and $j = 3$ or 4. Semi-norms that are all zero (for example, $\|x\|_{1,3} = 0$ is realized by $u = (0, x_1 + x_2)$ and $v = (x_1, -x_1)$). While, if $\mathbb{R}^2$ is endowed with its norm (which, on its own, is a filtering family), $\mathbb{R}^2 + \mathbb{R}^2$ is endowed with, see (6.6), p. 86, the topology of $\mathbb{R}^2$. □

CAUTION. The construction of the semi-norms of Definition 6.20 might suggest that the infimum of several semi-norms on the same vector space is a semi-norm. This is not true.

For example, semi-norms are defined on $\mathbb{R}^2$ by $\|x\|_1 = |x_1|$ and $\|x\|_2 = |x_2|$ but are not by $p(x) = \inf\{\|x\|_1, \|x\|_2\}$ because $p((0,1) + (1,0)) > p((0,1)) + p((1,0))$. □

Let us observe that a sum of convergent sequences converges.

THEOREM 6.22.– *Let $E_1, E_2 \dots, E_d$ be a sum of separated semi-normed spaces and, for all $i \in [\![1,d]\!]$, let $(u_{i,n})_{n\in\mathbb{N}} \subset E_i$ and $u_i \in E_i$. Then:*

$$u_{i,n} \to u_i \text{ in } E_i, \ \forall i \in [\![1,d]\!] \quad \Rightarrow$$
$$\Rightarrow \quad u_{1,n} + \cdots + u_{d,n} \to u_1 + \cdots + u_d \text{ in } E_1 + \cdots + E_d. \ \blacksquare$$

Proof. With the notations of Definition 6.20 of a sum,

$$\|(u_{1,n} + \cdots + u_{d,n}) - (u_1 + \cdots + u_d)\|_{E_1+\cdots+E_d;\nu_1,\dots,\nu_d}$$
$$\le (\|u_{1,n} - u_1\|^2_{E_1;\nu_1} + \cdots + \|u_{d,n} - u_d\|^2_{E_d;\nu_d})^{1/2}.$$

According to Definition 2.6 (a) of the convergence of a sequence, if $u_{i,n} \to u_i$ in E_i for all $i \in [\![1,d]\!]$, then $\|u_{i,n} - u_i\|_{E_i;\nu_i} \to 0$ for all $\nu_i \in \mathcal{N}_{E_i}$, thus the right-hand side of the above inequality, and therefore the left-hand one, tend to 0, whence the conclusion. □

CAUTION. The converse of Theorem 6.22 does not hold:

$$u_{1,n} + \cdots + u_{d,n} \to u_1 + \cdots + u_d \text{ in } E_1 + \cdots + E_d \quad \not\Rightarrow \quad u_{i,n} \to u_i \text{ in } E_i. \tag{6.5}$$

Proof of (6.5). Let $(u_n)_{n\in\mathbb{N}}$ be a sequence of a separated semi-normed space E that does not converge. Yet, $u_n + (-u_n) = 0_E$ that converges in $E + E$ since this is equal to E according to the identity (6.6) below. □

Let us observe that a sum of semi-normed spaces contains each of them.

THEOREM 6.23.– *Let $E_1, E_2 \dots, E_d$ and F be separated semi-normed spaces such that $E_i \underset{\to}{\subseteq} F$ for all $i \in [\![1,d]\!]$. Then, for any $j \in [\![1,d]\!]$:*

$$E_j \underset{\to}{\subseteq} E_1 + \cdots + E_d \underset{\to}{\subseteq} F. \ \blacksquare$$

Proof. First inclusion. For every $i \in [\![1,d]\!]$, let $\{\| \ \|_{E_i;\nu_i} : \nu_i \in \mathcal{N}_{E_i}\}$ be a filtering family of semi-norms generating the topology of E_i and let $\nu_i \in \mathcal{N}_{E_i}$. Given $u \in E_j$, by decomposing $u = u_1 + \cdots + u_d$, where $u_j = u$ and $u_i = 0_{E_i}$ otherwise, Definition 6.20 yields

$$\|u\|_{E_1+\cdots+E_d;\nu_1,\dots,\nu_d} \le \|u\|_{E_j;\nu_j}.$$

Therefore (Definition 3.6 (c)), $E_j \underset{\to}{\subseteq} E_1 + \cdots + E_d$.

Second equality. The inclusion $E_1 + \cdots + E_d \underset{\to}{\subseteq} F$ results from the inequality (6.3). □

Addition of a space to itself. For every separated semi-normed space E, Theorem 6.23 with respect to $E_i = E = F$ yields $E \underset{\to}{\subseteq} E + \cdots + E \underset{\to}{\subseteq} E$ and thus

$$E + \cdots + E \underset{\to}{=} E. \tag{6.6}$$

□

Let us observe that the families of semi-norms of a sum of spaces are filtering.

THEOREM 6.24.– *Any family of semi-norms given by Definition 6.20 of a sum of separated semi-normed spaces is filtering.* ∎

Proof. Let $E = E_1 + \cdots + E_d$ be a sum of separated semi-normed spaces. According to its Definition 6.20, each E_i is endowed with a filtering family of semi-norms, $\{\| \ \|_{E_i;\nu_i} : \nu_i \in \mathcal{N}_{E_i}\}$. Let $\|u\|_{E;\nu_1^1,\ldots,\nu_d^1}, \ldots, \|u\|_{E;\nu_1^n,\ldots,\nu_d^n}$ be n semi-norms of E. Due to the filtering property, for all $i \in [\![1, d]\!]$, there exists $\nu_i \in \mathcal{N}_{E_i}$ such that

$$\sup_{1 \le m \le n} \| \ \|_{E_i;\nu_i^m} \le \| \ \|_{E_i;\nu_i}.$$

Let $u \in E$ and $u = u_1 + \cdots + u_d$ be one of its decompositions. For each $m \in [\![1, n]\!]$,

$$(\|u_1\|^2_{E_1;\nu_1^m} + \cdots + \|u_d\|^2_{E_d;\nu_d^m})^{1/2} \le (\|u_1\|^2_{E_1;\nu_1} + \cdots + \|u_d\|^2_{E_d;\nu_d})^{1/2}$$

from which, by taking the infimum over all decompositions of u,

$$\|u\|_{E;\nu_1^m,\ldots,\nu_d^m} \le \|u\|_{E;\nu_1,\ldots,\nu_d}.$$

Thus, $\sup_{1 \le m \le n} \| \ \|_{E;\nu_1^m,\ldots,\nu_d^m} \le \| \ \|_{E;\nu_1,\ldots,\nu_d}$ which proves that the family of semi-norms of $E_1 + \cdots + E_d$ is filtering. □

Let us introduce two other families of semi-norms defining the sum topology.

THEOREM 6.25.– *Let $E_1 + E_2 + \cdots + E_d$ be a sum of separated semi-normed spaces and, for each $i \in [\![1, d]\!]$, let $\{\| \ \|_{E_i;\nu_i} : \nu_i \in \mathcal{N}_{E_i}\}$ be a filtering family of semi-norms generating the topology of E_i.*

Then, the family of semi-norms of $E_1 + \cdots + E_d$ is equivalent to each of the following two families of semi-norms, also indexed by $\nu_1 \in \mathcal{N}_{E_1}, \ldots, \nu_d \in \mathcal{N}_{E_d}$:

(a) $|||(u_1, \ldots, u_d)|||_{E_1+\cdots+E_d;\nu_1,\ldots,\nu_d} \stackrel{\text{def}}{=}$
$$\inf_{u_i \in E_i : u_1 + \cdots + u_d = u} \|u_1\|_{E_1;\nu_1} + \cdots + \|u_d\|_{E_d;\nu_d}.$$

(b) $||||(u_1, \ldots, u_d)||||_{E_1+\cdots+E_d;\nu_1,\ldots,\nu_d} \stackrel{\text{def}}{=}$
$$\inf_{u_i \in E_i : u_1 + \cdots + u_d = u} \sup\{\|u_1\|_{E_1;\nu_1}, \ldots, \|u_d\|_{E_d;\nu_d}\}.$$ ∎

Proof. We verify that they are indeed semi-norms on $E_1 + \cdots + E_d$ in a similar way to what we did in the justification of Definition 6.20 for the default semi-norms. They are equivalent (Definition 3.1) to the latter because, according to the inequalities in Theorem 5.16 with respect to $x = (\|u_1\|_{E_1;\nu_1}, \ldots, \|u_d\|_{E_d;\nu_d})$,

$$(\|u_1\|^2_{E_1;\nu_1} + \cdots + \|u_d\|^2_{E_d;\nu_d})^{1/2} \le \|u_1\|_{E_1;\nu_1} + \cdots + \|u_d\|_{E_d;\nu_d} \le$$
$$\le d^{1/2}(\|u_1\|^2_{E_1;\nu_1} + \cdots + \|u_d\|^2_{E_d;\nu_d})^{1/2}$$

and

$$d^{-1/2}(\|u_1\|^2_{E_1;\nu_1} + \cdots + \|u_d\|^2_{E_d;\nu_d})^{1/2} \leq \sup\{\|u_1\|_{E_1;\nu_1}, \cdots, \|u_d\|_{E_d;\nu_d}\} \leq$$
$$\leq (\|u_1\|^2_{E_1;\nu_1} + \cdots + \|u_d\|^2_{E_d;\nu_d})^{1/2}. \ \square$$

Let us show that a sum of metrizable or normed spaces also is.

THEOREM 6.26.– *Let $E_1 + \cdots + E_d$ be a sum of separated semi-normed spaces. Then:*

(a) *The E_i are metrizable* $\Rightarrow$ *$E_1 + \cdots + E_d$ is metrizable.*

(b) *The E_i are normed* $\Rightarrow$ *$E_1 + \cdots + E_d$ is normed.* ▮

Proof. (a) If each E_i is metrizable, it may be endowed (Theorem 4.4) with an increasing sequence of semi-norms, which is a filtering family of semi-norms. The family of semi-norms of their sum is then (Definition 6.20) indexed by $\mathbb{N}^d$ which is countable (Theorems 1.6 and 1.1 (d)). In addition, $E_1 + \cdots + E_d$ is separated (Theorem 6.21 (a)) thus (Definition 4.1 (a)) it is metrizable.

(b) If each E_i is normed, the family of semi-norms of $E_1 + \cdots + E_d$ is reduced (Definition 6.20) to a single semi-norm, which is a norm since $E_1 + \cdots + E_d$ is separated (Theorem 6.21 (a)). □

Let us give completeness properties of a sum of spaces.

THEOREM 6.27.– *Let $E_1 + \cdots + E_d$ be a sum of separated semi-normed spaces. Then:*

(a) *The E_i are Fréchet spaces* $\Rightarrow$ *$E_1 + \cdots + E_d$ is a Fréchet space.*

(b) *The E_i are Banach spaces* $\Rightarrow$ *$E_1 + \cdots + E_d$ is a Banach space.* ▮

Proof. (a) Assume that the E_i are Fréchet spaces. Then, $E_1 + \cdots + E_d$ is metrizable (Theorem 6.26 (a)) thus, in order to be a Fréchet space, it suffices (Theorem 4.23) that any absolutely convergent series converges therein.

Let $\sum_{n\in\mathbb{N}} u_n$ be an absolutely convergent sequence in $E_1 + \cdots + E_d$, which we endow with the semi-norms of Theorem 6.25 (a). This means (Definition 4.21 (b)) that, for all $\nu_1 \in \mathcal{E}_1 \ldots, \nu_d \in \mathcal{E}_d$, there exists $c_{\nu_1,\ldots,\nu_d} \in \mathbb{R}$ such that, for all $j \in \mathbb{N}$,

$$\sum_{1\leq n\leq j} |||u_n|||_{E_1+\cdots+E_d;\nu_1,\ldots,\nu_d} \leq c_{\nu_1,\ldots,\nu_d}.$$

According to the definition of these semi-norms, for each $n \in \mathbb{N}$, there is a decomposition $u_n = u_{n,1} + \cdots + u_{n,d}$, where $u_{n,i} \in E_i$ for all $i \in [\![1,d]\!]$, such that

$$\|u_{n,1}\|_{E_1;\nu_1} + \cdots + \|u_{n,d}\|_{E_d;\nu_d} \le |||u_n|||_{E_1+\cdots+E_d;\nu_1,\ldots,\nu_d} + 2^{-n}.$$

Then, see (1.10), p. 7, for all $i \in [\![1,d]\!]$ and $j \in \mathbb{N}$,

$$\sum_{1 \le n \le j} \|u_{n,i}\|_{E_i;\nu_i} \le c_{\nu_1,\ldots,\nu_d} + 1 + 2^{-1} + \cdots + 2^{-j} \le c_{\nu_1,\ldots,\nu_d} + 2.$$

The series $\sum_{n \in \mathbb{N}} u_{n,i}$ is thus absolutely convergent in E_i. The latter being sequentially complete (by assumption), this series converges (Theorem 4.22). That is to say (Definition 4.21 (a)), that there exists $\ell_i \in E_i$ such that when $j \to \infty$,

$$\sum_{1 \le n \le j} u_{n,i} \to \ell_i \quad \text{in } E_i. \tag{6.7}$$

By summing in i from 1 to d, it yields (Theorem 6.22) that $\sum_{1 \le n \le j} u_n \to \ell_1 + \cdots + \ell_d$ in $E_1 + \cdots + E_d$. Therefore, the sequence $\sum_{n \in \mathbb{N}} u_n$ indeed converges.

(b) This results from (a) and from Theorem 6.26 (b) because (Definition 4.10) a Banach space is a normed Fréchet space. □

6.5. Direct sum of semi-normed spaces

Let us define the direct sum of vector subspaces of a same separated semi-normed space.

DEFINITION 6.28.– *Let $E_1, E_2 \ldots, E_d$ and F be separated semi-normed spaces such that*

$$E_i \subsetneq F, \ \forall i \in [\![1,d]\!],$$

$$[\, u_1 + \cdots + u_d = 0_F, \ u_i \in E_i, \ \forall i \in [\![1,d]\!] \,] \ \Rightarrow \ u_i = 0_{E_i}, \ \forall i \in [\![1,d]\!] \tag{6.8}$$

and let $\{\| \ \|_{E_i;\nu_i} : \nu_i \in \mathcal{N}_{E_i}\}$ be the family of semi-norms of E_i.

*The **direct sum** of E_i is the vector space*

$$E_1 \oplus E_2 \oplus \cdots \oplus E_d \stackrel{\text{def}}{=} \{u \in F : u = u_1 + u_2 + \cdots + u_d, \ u_i \in E_i, \ \forall i \in [\![1,d]\!]\}$$

endowed with the semi-norms, indexed by $\nu_1 \in \mathcal{N}_{E_1}$, $\nu_2 \in \mathcal{N}_{E_2} \ldots$ and $\nu_d \in \mathcal{N}_{E_d}$,

$$\|u\|_{E_1 \oplus \cdots \oplus E_d;\nu_1,\ldots,\nu_d} \stackrel{\text{def}}{=} (\|u_1\|^2_{E_1;\nu_1} + \cdots + \|u_d\|^2_{E_d;\nu_d})^{1/2}. \tag{6.9}$$

■

Justification.– The equality (6.9) indeed defines a real number $\|u\|_{E_1\oplus\cdots\oplus E_d;\nu_1,\ldots,\nu_d}$ because:

$$\begin{cases}\text{every } u \in E_1 \oplus \cdots \oplus E_d \text{ has a unique decomposition}\\ u = u_1 + \cdots + u_d \text{ where } u_i \in E_i \text{ for all } i \in [\![1, d]\!].\end{cases} \quad (6.10)$$

In effect, if there were two distinct decompositions, their difference would be a non-zero decomposition of 0_F which would contradict Assumption (6.8).

The mappings $\| \ \|_{E_1\oplus\cdots\oplus E_d;\nu_1,\ldots,\nu_d}$ thus defined "coincide" with the semi-norms of the product of E_i (Definition 6.1), in the sense where

$$\|u_1 + \cdots + u_d\|_{E_1\oplus\cdots\oplus E_d;\nu_1,\ldots,\nu_d} = \|(u_1, \ldots, u_d)\|_{E_1\times\cdots\times E_d;\nu_1,\ldots,\nu_d}. \quad (6.11)$$

Therefore, they also have all the properties of semi-norms (Definition 2.2 (a)). □

Direct sum versus product. The equality (6.11) shows, see Theorem 7.29, that

$u_1 + \cdots + u_d \mapsto (u_1, \ldots, u_d)$ *is an isomorphism from* $E_1 \oplus \cdots \oplus E_d$ *onto* $E_1 \times \cdots \times E_d$. □

Let us observe that the direct sum of separated semi-normed spaces is separated.

THEOREM 6.29.– *Any direct sum* $E_1 \oplus \cdots \oplus E_d$ *of separated semi-normed spaces is a separated semi-normed space.* ▮

Proof. Let $u = u_1+\cdots+u_d \in E_1\oplus\cdots\oplus E_d$. If, with the notations of Definition 6.28, $(\|u_1\|^2_{E_1;\nu_1} + \cdots + \|u_d\|^2_{E_d;\nu_d})^{1/2} = 0$ for all $\nu_1 \in \mathcal{N}_{E_1}\ldots, \nu_d \in \mathcal{N}_{E_d}$, then, for all $i \in [\![1, d]\!]$, it yields $\|u_i\|_{E_i;\nu_i} = 0$ for all $\nu_i \in \mathcal{N}_{E_i}$ and therefore $u_i = 0_{E_i}$ since E_i is separated (Definition 2.3 (c)). Consequently, $u = 0$ which proves that $E_1 \oplus \cdots \oplus E_d$ is separated. □

Let us characterize the convergent sequences of a direct sum.

THEOREM 6.30.– *Let* $E_1\oplus\cdots\oplus E_d$ *be a direct sum of separated semi-normed spaces and for every* $i \in [\![1, d]\!]$, *let* $(u_{i,n})_{n\in\mathbb{N}} \subset E_i$ *and* $u_i \in E_i$. *Then:*

$$u_{i,n} \to u_i \text{ in } E_i, \ \forall i \in [\![1, d]\!] \quad \Leftrightarrow$$

$$\Leftrightarrow \quad u_{1,n} + \cdots + u_{d,n} \to u_1 + \cdots + u_d \text{ in } E_1 \oplus \cdots \oplus E_d. \ ▮$$

Proof. With the notations of Definition 6.28 of a direct sum and following Definition 2.6 (a) of the convergence of a sequence, $u_{1,n} + \cdots + u_{d,n} \to u_1 + \cdots + u_d$ in $E_1 \oplus \cdots \oplus E_d$ when $n \to \infty$ if and only if, for all $\nu_1 \in \mathcal{N}_{E_1}\ldots, \nu_d \in \mathcal{N}_{E_d}$,

$$(\|u_{1,n} - u_1\|^2_{E_1;\nu_1} + \cdots + \|u_{d,n} - u_d\|^2_{E_d;\nu_d})^{1/2} \to 0.$$

This implies, for all $i \in [\![1, d]\!]$, that $\|u_{i,n} - u_i\|_{E_i;\nu_i} \to 0$ for all $\nu_i \in \mathcal{N}_{E_i}$, which means that $u_{i,n} \to u_i$ in E_i. And vice versa. □

Let us observe that the direct sum topology is maintained when each topology is.

THEOREM 6.31.– *Let $E_1 \oplus \cdots \oplus E_d$ be a direct sum of separated semi-normed spaces.*

If the family of semi-norms of each E_i is replaced by an equivalent family, the family of semi-norms obtained on their direct sum is equivalent to the initial family. ▮

Proof. If, for all $i \in [\![1, d]\!]$, the family $\{\| \ \|_{E_i;\nu_i} : \nu_i \in \mathcal{N}_i\}$ of E_i is dominated by the family $\{|\!|\!| \ |\!|\!|_{E_i;\mu_i} : \mu_i \in \mathcal{M}_i\}$, that is if for all $\nu_i \in \mathcal{N}_i$ there exist a finite set $M_{\nu_i} \in \mathcal{M}_i$ and $c_{\nu_i} \in \mathbb{R}$ such that $\|u_i\|_{E_i;\nu_i} \leq c_{\nu_i} \sup_{\mu_i \in M_{\nu_i}} |\!|\!|u_i|\!|\!|_{E_i;\mu_i}$ for all $u_i \in E_i$, then from Definition 6.28 of semi-norms of the direct sum it follows that

$$\|u\|_{E_1 \oplus \cdots \oplus E_d;\nu_1,\ldots,\nu_d} \leq c \sup_{(\mu_1,\ldots,\mu_d) \in M_{\nu_1} \times \cdots \times M_{\nu_d}} |\!|\!|u|\!|\!|_{E_1 \oplus \cdots \oplus E_d;\mu_1,\ldots,\mu_d}$$

where $c = \sup\{c_{\nu_1}, \ldots, c_{\nu_d}\}$. The initial family of the direct sum is thus dominated by the new one.

If the initial family of each E_i is equivalent to the new one, that is dominates and is dominated by the new one, the same thus happens for the direct sum. □

Let us verify that the direct sum coincides with the ordinary sum, which leads to the fact that it has all the properties established in section 6.4.

THEOREM 6.32.– *Let $E_1 \oplus \cdots \oplus E_d$ be a direct sum of separated semi-normed spaces and $i \in [\![1, d]\!]$. Then:*

(a) $$E_1 \oplus \cdots \oplus E_d \underset{\leftrightarrows}{=} E_1 + \cdots + E_d.$$

(b) *E_i is a topological subspace of $E_1 \oplus \cdots \oplus E_d$.* ▮

Proof. (a) The vector spaces $E_1 \oplus \cdots \oplus E_d$ and $E_1 + \cdots + E_d$ are equal according to their Definitions 6.28 and 6.20.

Every family of semi-norms being equivalent (Theorem 3.15) to a filtering family, for each $i \in [\![1, d]\!]$, the family of semi-norms of E_i may be replaced by a filtering family $\{\| \ \|_{E_i;\nu_i} : \nu_i \in \mathcal{N}_{E_i}\}$ which (Theorems 6.31 and 6.21) preserves the topologies of $E_1 \oplus \cdots \oplus E_d$ and $E_1 + \cdots + E_d$. Given $u \in E_1 \oplus \cdots \oplus E_d$, its decomposition $u = u_1 + \cdots + u_d$ is unique according to (6.10), thus Definitions 6.28 and 6.20 yield

$$\|u\|_{E_1 \oplus \cdots \oplus E_d;\nu_1,\ldots,\nu_d} = \left(\|u_1\|^2_{E_1;\nu_1} + \cdots + \|u_d\|^2_{E_d;\nu_d}\right)^{1/2} = \|u\|_{E_1 + \cdots + E_d;\nu_1,\ldots,\nu_d}.$$

These families, being both indexed by $\nu_1 \in \mathcal{N}_{E_1}, \ldots, \nu_d \in \mathcal{N}_{E_d}$, are identical thus $E_1 \oplus \cdots \oplus E_d$ is topologically equal (Definition 3.6 (b)) to $E_1 + \cdots + E_d$.

(b) For all $u_i \in E_i$, the equality (6.9) with $u_j = 0_{E_j}$ for $j \neq i$ yields

$$\|u_i\|_{E_1 \oplus \cdots \oplus E_d;\nu_1,\ldots,\nu_d} = \|u_i\|_{E_i;\nu_i}. \tag{6.12}$$

The family of the semi-norms of E_i is therefore equivalent to the family of restrictions to E_i of the semi-norms of $E_1 \oplus \cdots \oplus E_d$ (these families coincide if we count only once the semi-norms of $E_1 \oplus \cdots \oplus E_d$ whose restrictions are equal to E_i according to (6.12), namely those obtained by varying the ν_j for $j \neq i$). Thus, E_i is a topological subspace (Definition 3.9) of $E_1 \oplus \cdots \oplus E_d$. □

Let us give basic properties of a direct sum.

THEOREM 6.33.– *Let $E_1 \oplus \cdots \oplus E_d$ be a direct sum of separated semi-normed spaces. Then:*

(a) *The E_i are metrizable* $\Rightarrow$ *$E_1 \oplus \cdots \oplus E_d$ is metrizable.*

(b) *The E_i are normed* $\Rightarrow$ *$E_1 \oplus \cdots \oplus E_d$ is normed.* ▮

Proof. Being verified by $E_1 + \cdots + E_d$ (Theorems 6.26 (a) and (b)), these properties also are by $E_1 \oplus \cdots \oplus E_d$ since the latter is topologically equal (Theorem 6.32 (a)). □

Let us give completeness and other properties of a direct sum.

THEOREM 6.34.– *Let $E_1 \oplus \cdots \oplus E_d$ be a direct sum of separated semi-normed spaces. Then:*

(a) *The E_i are Neumann spaces* $\Rightarrow$ *$E_1 \oplus \cdots \oplus E_d$ is a Neumann space.*

(b) *The E_i are Fréchet spaces* $\Rightarrow$ *$E_1 \oplus \cdots \oplus E_d$ is a Fréchet space.*

(c) *The E_i are Banach spaces* $\Rightarrow$ *$E_1 \oplus \cdots \oplus E_d$ is a Banach space.*

(d) *The E_i are Hilbert spaces* $\Rightarrow$ $\begin{cases} E_1 \oplus \cdots \oplus E_d \text{ is a Hilbert space} \\ \text{and its scalar product is} \\ (u_1, v_1)_{E_1} + \cdots + (u_d, v_d)_{E_d}. \end{cases}$ ▮

Proof. (a) If $(u_{1,n} + \cdots + u_{d,n})_{n\in\mathbb{N}}$ is a Cauchy sequence of the space $E_1 \oplus \cdots \oplus E_d$, then $(u_{i,n})_{n\in\mathbb{N}}$ is a Cauchy sequence of E_i, therefore if E_i is sequentially complete, it converges therein to a limit u_i. If this holds for all $i \in [\![1, d]\!]$, then (Theorem 6.30) $u_{1,n} + \cdots + u_{d,n} \to u_1 + \cdots + u_d$ in $E_1 \oplus \cdots \oplus E_d$ that is therefore sequentially complete.

(b) *and* (c). These properties result from (a) and Theorem 6.33 since (Definition 4.10) a Fréchet space is a metrizable Neumann space, and a Banach space is a normed Neumann space.

(d) This results from (c) because $(u, v) = \sum_{i=1}^{d} (u_i, v_i)_{E_i}$ defines a scalar product (Definition 5.1) on $E_1 \times \cdots \times E_d$ that verifies $\|u\|_{E_1 \oplus \cdots \oplus E_d} = (u, u)^{1/2}$. □

6.6. Quotient space

Let us define the quotient of a vector space by a vector subspace.

DEFINITION 6.35.– *Let E be a vector space and F a vector subspace of E. For all $u \in E$, let us denote*

$$\widetilde{u} \stackrel{\text{def}}{=} \{x \in E : x - u \in F\}.$$

*The **quotient space** of E by F is the vector space*

$$E/F \stackrel{\text{def}}{=} \{\widetilde{u} : u \in E\}$$

endowed with the addition and the multiplication defined, for all $u \in E$, $v \in E$ and $t \in \mathbb{R}$, by

$$\widetilde{u} + \widetilde{v} \stackrel{\text{def}}{=} \widetilde{u + v}, \quad t\widetilde{u} \stackrel{\text{def}}{=} \widetilde{tu}. \tag{6.13}$$

■

Sum of sets. This definition of $\widetilde{u} + \widetilde{v}$ is a particular case of the sum of subsets of a vector space defined by (3.4), p. 46. □

Justification of Definition 6.35. Preliminary. First, let us show that:

$$x \in \widetilde{u} \quad \Leftrightarrow \quad \widetilde{x} = \widetilde{u}. \tag{6.14}$$

Let $x \in \widetilde{u}$. If $y \in \widetilde{x}$, then $y - x \in F$ from which $y - u = (y - x) + (x - u) \in F$ because F is a vector subspace of E and therefore $y \in \widetilde{u}$. Conversely, if $y \in \widetilde{u}$, then $y - x = (y - u) + (u - x) \in F$ and therefore $y \in \widetilde{x}$. As a result, $\widetilde{x} = \widetilde{u}$.

Conversely, if $\widetilde{x} = \widetilde{u}$, then x is an element of $\widetilde{x}$ (by definition) and therefore of $\widetilde{u}$.

Definitions of the sum and of the product. Let us show that $\widetilde{u} + \widetilde{v}$ and $t\widetilde{u}$ depend only on the sets $\widetilde{u}$ and $\widetilde{v}$, not on the elements u and v with which they are associated.

If $\widetilde{x} = \widetilde{u}$ and $\widetilde{y} = \widetilde{v}$, then $x \in \widetilde{u}$ and $y \in \widetilde{v}$ according to (6.14), thus $x - u \in F$ and $y - v \in F$ hence $(x + y) - (u + v) = (x - u) + (y - v) \in F$, and so $x + y \in \widetilde{u + v}$ hence, again according to (6.14), $\widetilde{x + y} = \widetilde{u + v}$, namely $\widetilde{x} + \widetilde{y} = \widetilde{u} + \widetilde{v}$.

If $\widetilde{x} = \widetilde{u}$, then $x \in \widetilde{u}$ according to (6.14), thus $x - u \in F$ hence $tx - tu = t(x - u) \in F$, and so $tx \in \widetilde{tu}$ from which again according to (6.14), $\widetilde{tx} = \widetilde{tu}$, that is $t\widetilde{x} = t\widetilde{u}$.

Vector space. The addition and multiplication properties required in Definition 2.1 for E/F to be a vector space result from the similar properties of the operations of E. □

Let us norm the quotient of a vector space by the kernel of a semi-norm.

THEOREM 6.36.– *Let E be a semi-normed space, p a semi-norm on E,*

$$\mathrm{N}_p \overset{\text{def}}{=} \{z \in E : p(z) = 0\}$$

*the **kernel** of p. Then, N_p is a vector subspace of E and a norm is defined on the quotient space*

$$E/\mathrm{N}_p = \{\widetilde{u} : u \in E\} \quad \textit{where} \quad \widetilde{u} = \{x \in E : p(x-u) = 0\}$$

by: for all $u \in E$,

$$\|\widetilde{u}\|_{E/\mathrm{N}_p} \overset{\text{def}}{=} p(u). \blacksquare$$

Proof. The set N_p is a vector subspace of E because, if $y \in \mathrm{N}_p$, $z \in \mathrm{N}_p$ and $t \in \mathbb{R}$, then $y + z \in \mathrm{N}_p$ since $p(y+z) \le p(y) + p(z) = 0$ and $ty \in \mathrm{N}_p$ since $p(ty) = |t|\, p(y) = 0$.

The value of $\|\widetilde{u}\|_{E/\mathrm{N}_p}$ depends only on the set $\widetilde{u}$, not on the element u with which it is associated. It turns out that if $\widetilde{v} = \widetilde{u}$, then $v \in \widetilde{u}$ according to (6.14) thus $p(v-u) = 0$ from which $p(v) = p(u)$ since $|p(v) - p(u)| \le p(v-u)$ (Theorem 2.5 (c)).

The mapping $\| \ \|_{E/\mathrm{N}_p}$ is a semi-norm (Definition 2.2 (a)) on the vector space E/N_p (Definition 6.35) because

$$\|\widetilde{u}\|_{E/\mathrm{N}_p} = p(u) \ge 0,$$

$$\|t\widetilde{u}\|_{E/\mathrm{N}_p} = \|\widetilde{tu}\|_{E/\mathrm{N}_p} = p(tu) = |t|\, p(u) = |t|\, \|\widetilde{u}\|_{E/\mathrm{N}_p},$$

$$\|\widetilde{u} + \widetilde{v}\|_{E/\mathrm{N}_p} = \|\widetilde{u+v}\|_{E/\mathrm{N}_p} = p(u+v) \le p(u) + p(v) = \|\widetilde{u}\|_{E/\mathrm{N}_p} + \|\widetilde{v}\|_{E/\mathrm{N}_p}.$$

This is a norm because $\|\widetilde{u}\|_{E/\mathrm{N}_p} = 0$ leads to $p(u) = 0$ from which $0_E \in \widetilde{u}$ and thus $\widetilde{u} = \widetilde{0_E}$ according to (6.14). That is, $\widetilde{u} = 0_{E/\mathrm{N}_p}$. $\square$

PART 2

CONTINUOUS MAPPINGS

Chapter 7

Continuous Mappings

7.1. Continuous mappings

Let us define continuity and variants[1].

DEFINITION 7.1.– *Let E and F be two separated semi-normed spaces, $\{\| \|_{E;\nu} : \nu \in \mathcal{N}_E\}$ and $\{\| \|_{F;\mu} : \mu \in \mathcal{N}_F\}$ their families of semi-norms, $X \subset E$ and T a mapping from X into F.*

(a) *T is **continuous at a point** u of X if, for all $\mu \in \mathcal{N}_F$ and $\epsilon > 0$, there exist a finite set $N \subset \mathcal{N}_E$ and $\eta > 0$ such that:*

$$v \in X, \ \sup_{\nu \in N} \|v - u\|_{E;\nu} \leq \eta \quad \Rightarrow \quad \|T(v) - T(u)\|_{F;\mu} \leq \epsilon.$$

*T is **continuous** if it is at any point of X.*

(b) *T is **uniformly continuous** if, for all $\mu \in \mathcal{N}_F$ and $\epsilon > 0$, there exist a finite set $N \subset \mathcal{N}_E$ and $\eta > 0$ such that:*

$$u \in X, \ v \in X, \ \sup_{\nu \in N} \|v - u\|_{E;\nu} \leq \eta \quad \Rightarrow \quad \|T(v) - T(u)\|_{F;\mu} \leq \epsilon.$$

---

1 **History of the notion of continuous mapping.** Augustin CAUCHY defined sequential continuity for a real function on a line segment, in 1821, in [25]. Bernard BOLZANO also contributed to the emerging of this notion.

History of the notion of uniformly continuous mapping. Eduard HEINE defined the notion of uniform continuity for functions defined on (a subset of) $\mathbb{R}^d$, in 1870, in [55]. This notion had already been implicitly used by CAUCHY in 1823 to define the integral of a real function [26, p. 122-126], then explicitly by Peter DIRICHLET.

(c) *T is* ***sequentially continuous at a point*** *u of X if, for every sequence $(u_n)_{n\in\mathbb{N}}$ of X:*

$$u_n \to u \text{ in } E \quad \Rightarrow \quad T(u_n) \to T(u) \text{ in } F.$$

T is ***sequentially continuous*** *if it is at any point of X.* ▮

Let us compare these notions of continuity.

THEOREM 7.2.– *For any mapping from a subset of a separated semi-normed space into a separated semi-normed space:*

(a) *Uniformly continuous* $\Rightarrow$ *continuous* $\Rightarrow$ *sequentially continuous.*

(b) *Continuous at a point* $\Rightarrow$ *sequentially continuous at this point.* ▮

Proof. (b) Let us use the notations of Definition 7.1. Assume that T is continuous and $u_n \to u$. Then, $\|T(u_n) - T(u)\|_{F;\mu} \leq \epsilon$ as soon as n is large enough so that $\sup_{\nu\in N} \|u_n - u\|_{E;\nu} \leq \eta$. Thus $T(u_n) \to T(u)$, hence T is sequentially continuous.

(a) The first implication is trivial. The second is the result of (b). □

Converses. *Uniformly continuous* coincides with *continuous* for any linear mapping (Theorem 7.14 (a)) and also for any mapping defined on a compact set (Theorem 8.10), but this is not general. For example, the mapping $x \mapsto 1/x$ of $(0, 1)$ in itself is continuous but not uniformly continuous.

Continuous coincides with *sequentially continuous* for the mappings that are defined on a subset of a metrizable space (Theorem 9.1), but this is not the case for every separated semi-normed space. For example, since the set U defined by (2.3), p. 18, is not closed, there exists $u \in \overline{U} \setminus U$ and therefore the mapping equal to zero in U and to 1 at u is not continuous; however, it is sequentially continuous because, U being sequentially closed, there is no sequence of U converging to u. □

Let us show that the characterizations of continuity and uniform continuity can be simplified with a filtering family: the upper envelope of a finite number of semi-norms is replaced by a single semi-norm.

THEOREM 7.3.– *Let E and F be two separated semi-normed spaces, $\{\| \; \|_{E;\nu} : \nu \in \mathcal{N}_E\}$ and $\{\| \; \|_{F;\mu} : \mu \in \mathcal{N}_F\}$ their families of semi-norms, $X \subset E$, T a mapping from X into F and $u \in X$. If*

the family of semi-norms $\{\| \; \|_{E;\nu} : \nu \in \mathcal{N}_E\}$ is filtering,

then:

(a) *T is continuous at the point u if and only if, for all $\mu \in \mathcal{N}_F$ and $\epsilon > 0$, there exist $\nu \in \mathcal{N}_E$ and $\eta > 0$ such that:*

$$v \in X, \; \|v - u\|_{E;\nu} \leq \eta \quad \Rightarrow \quad \|T(v) - T(u)\|_{F;\mu} \leq \epsilon.$$

(b) *T is uniformly continuous if and only if, for all $\mu \in \mathcal{N}_F$ and $\epsilon > 0$, there exist $\nu \in \mathcal{N}_E$ and $\eta > 0$ such that:*

$$u \in X,\ v \in X,\ \|v - u\|_{E;\nu} \leq \eta \quad \Rightarrow \quad \|T(v) - T(u)\|_{F;\mu} \leq \epsilon. \ \blacksquare$$

Proof. (a) If T is continuous at the point u, there is (Definition 7.1 (a)) a finite set $N \subset \mathcal{N}_E$ such that $\sup_{\lambda \in N} \|v - u\|_{E;\lambda} \leq \eta$ yields $\|T(v) - T(u)\|_{F;\mu} \leq \epsilon$. On the other hand, according to Definition 3.14 of a filtering family, there is $\nu \in \mathcal{N}_E$ such that $\sup_{\lambda \in N} \|\ \|_{E;\lambda} \leq \|\ \|_{E;\nu}$. Then, $\|v-u\|_{E;\nu} \leq \eta$ indeed yields $\|T(v)-T(u)\|_{F;\mu} \leq \epsilon$.

Conversely, this property satisfies Definition 7.1 (a) of continuity with $N = \{\nu\}$

(b) A similar procedure is used for uniform continuity. □

Let us observe that uniformly continuous mappings preserve Cauchy sequences.

THEOREM 7.4.– *Let E and F be two separated semi-normed spaces, $X \subset E$, T a mapping from X into F, $(u_n)_{n \in \mathbb{N}} \subset X$ and $(v_n)_{n \in \mathbb{N}} \subset X$. If*

T is uniformly continuous

then:

(a) $\quad (u_n)_{n \in \mathbb{N}}$ *is Cauchy in E* $\Rightarrow$ $(T(u_n))_{n \in \mathbb{N}}$ *is Cauchy in F.*

(b) $\quad u_n - v_n \to 0_E$ *in E* $\Rightarrow$ $T(u_n) - T(v_n) \to 0_F$ *in F.* ∎

Proof. Let us use the notations of Definition 7.1 of the notions of continuity and let $\mu \in \mathcal{N}_F$ and $\epsilon > 0$. Assume that T is uniformly continuous and let N be the finite subset of $\mathcal{N}_E$ and $\eta > 0$ given by Definition 7.1.

(a) If $(u_n)_{n \in \mathbb{N}}$ is a Cauchy sequence (Definition 2.6 (b)), for each $\nu \in N$ there exists $m_\nu \in \mathbb{N}$ such that $n' \geq n \geq m_\nu$ yields $\|u_{n'} - u_n\|_{E;\nu} \leq \eta$. Let $m = \sup_{\nu \in N} m_\nu$ (it exists and it is finite following Theorem 1.2 (c)). If $n' \geq n \geq m$, Definition 7.1 (b) of uniform continuity leads to $\|T(u_{n'}) - T(u_n)\|_{F;\mu} \leq \epsilon$. Therefore, $(u_n)_{n \in \mathbb{N}}$ is a Cauchy sequence.

(b) If $u_n - v_n \to 0_E$ (Definition 2.6 (a)), for each $\nu \in N$ there exists $m_\nu \in \mathbb{N}$ such that $n \geq m_\nu$ yields $\|u_n - v_n\|_{E;\nu} \leq \eta$. By denoting $m = \sup_{\nu \in N} m_\nu$, as soon as $n \geq m$, it follows that $\|T(u_n) - T(v_n)\|_{F;\mu} \leq \epsilon$. Thus, $T(u_n) - T(v_n) \to 0_F$. □

7.2. Continuity and change of topology or restriction

Let us show that the various notions of continuity are preserved by topological equalities.

THEOREM 7.5.– *For any mapping from a subset of a separated semi-normed space E into a separated semi-normed space F:*

$$\left\{\begin{array}{l}\textit{Continuity, uniform continuity and sequential continuity are preserved when the}\\ \textit{family of semi-norms of } E \textit{ and/or that of } F \textit{ is replaced by an equivalent family.}\end{array}\right.$$ ∎

Proof. This results from Theorems 7.6 and 7.7 below according to Definition 3.6 of topological equalities and inclusions. □

Let us give the effect of changing the space of values on a continuous mapping.

THEOREM 7.6.– *Let E, F_1 and F_2 be separated semi-normed spaces such that:*

$$F_1 \subsetneqq F_2$$

and let $X \subset E$. Then, for any mapping from X into F_1:

(a) *Continuous from X into F_1* $\Rightarrow$ *continuous from X into F_2.*

(b) *Uniformly continuous from X into F_1*
$\Rightarrow$ *uniformly continuous from X into F_2.*

(c) *Sequentially continuous from X into F_1*
$\Rightarrow$ *sequentially continuous from X into F_2.* ∎

Proof. Properties (a) and (b) follow from Definitions 3.6 (c) of topological inclusion and 7.1 of continuity notions.

Property (c) results from the fact that every sequence converging in F_1 converges in F_2 (Theorem 3.8 (a)). □

Let us now show the effect of changing the space of definition.

THEOREM 7.7.– *Let E_1, E_2 and F be separated semi-normed spaces such that:*

$$E_1 \subsetneqq E_2.$$

Let $X \subset E_1$ and denote by X_{E_1} the subset X of the semi-normed space E_1 and X_{E_2} that of E_2. Then, every mapping from X into F verifies:

(a) *Continuous from X_{E_2} into F $\Rightarrow$ continuous from X_{E_1} into F.*

(b) *Uniformly continuous from X_{E_2} into F*
$\Rightarrow$ *uniformly continuous from X_{E_1} into F.*

(c) *Sequentially continuous from X_{E_2} into F*
$\Rightarrow$ *sequentially continuous from X_{E_1} into F.* ▌

Different formulation. In terms of topology, X_{E_1} is the set X endowed with the topology of E_1. □

Proof of Theorem 7.7. Properties (a) and (b) follow from Definitions 3.6 (c) of topological inclusion and 7.1 of continuity notions.

Property (c) results from the fact that every sequence converging in F_1 converges in F_2 (Theorem 3.8 (a)). □

Other proofs of Theorems 7.6 and 7.7. These properties result from those of composite mappings (Theorem 7.9) since the identity from E_1 into E_2 is uniformly continuous (Theorems 7.16 (a) and 7.14 (a)). □

Let us observe that the restrictions of continuous mappings are continuous.

THEOREM 7.8.– *Let E and F be separated semi-normed spaces, $X \subset E$, T a mapping from X into F and $U \subset X$. Then, denoting by $T|_U$ the restriction of T to U:*

(a) *T is continuous $\Rightarrow$ $T|_U$ is continuous.*

(b) *T is uniformly continuous $\Rightarrow$ $T|_U$ is uniformly continuous.*

(c) *T is sequentially continuous $\Rightarrow$ $T|_U$ is sequentially continuous.* ▌

Proof. This follows from Definition 7.1 of these notions of continuity. □

A mapping is said **continuous on** a subset U of its domain of definition if its restriction to U is continuous. Uniform continuity and sequential continuity on U are defined in a similar way.

CAUTION. The terminology "*continuous on*" can cause confusion because:

$$\textit{Continuous on a subset } U \quad \not\Rightarrow \quad \textit{continuous at every point of } U. \tag{7.1}$$

However, "*continuous at every point of U*" always yields to "*continuous on U*". If U is open, these two notions coincide. When they do not coincide and when there is a risk of confusing them, it is better to use "*continuity of the restriction to U*" instead of "*continuity on U*".

Proof of (7.1). The mapping equal to 1 in $[0, 1]$ and to 0 in the rest of $\mathbb{R}$ is continuous on $[0, 1]$ (it is constant therein) but it is not continuous at points 0 and 1 (it "jumps" there). □

7.3. Continuity of composite mappings

Given mappings T from X into Y and S from Y into G, the **composite mapping** $S \circ T$ is defined from X into G by

$$(S \circ T)(u) = S(T(u)).$$

Indifferently, it can be denoted by $S \circ T$ or $S(T)$. Let us show that any composite of continuous mappings is continuous.

THEOREM 7.9.– *Let E, F and G be separated semi-normed spaces, $X \subset E, Y \subset F$ and let T be a mapping from X into Y and S a mapping from Y into G. Then:*

(a) $\quad$ *S and T are continuous $\Rightarrow$ $S \circ T$ is continuous.*

(b) $\quad$ *S and T are uniformly continuous $\Rightarrow$ $S \circ T$ is uniformly continuous.*

(c) $\quad$ *S and T are sequentially continuous $\Rightarrow$ $S \circ T$ is sequentially continuous.* ∎

Proof. (a) *Continuity.* Let us use the notations of Definition 7.1 (a) of continuity, and let $\| \ \|_{G;\lambda}$ be a semi-norm of G and $\sigma > 0$. If S is continuous at a point $T(u)$, there exist a finite set $M \subset \mathcal{N}_F$ and $\epsilon > 0$ such that:

$$w \in Y, \ \sup_{\mu \in M} \|w - T(u)\|_{F;\mu} \leq \epsilon \quad \Rightarrow \quad \|S(w) - S(T(u))\|_{G;\lambda} \leq \sigma.$$

If in addition, T is continuous at u, let us denote by N_μ and η_μ the quantities provided by Definition 7.1 (a) and let $N = \bigcup_{\mu \in M} N_\mu$ and $\eta = \min_{\mu \in M} \eta_\mu$. Then:

$$v \in X, \ \sup_{\nu \in N} \|v - u\|_{E;\nu} \leq \eta \quad \Rightarrow \quad \sup_{\mu \in M} \|T(v) - T(u)\|_{F;\mu} \leq \epsilon \quad \Rightarrow$$

$$\Rightarrow \quad \|S(T(v)) - S(T(u))\|_{G;\lambda} \leq \sigma.$$

Since N is finite and $\eta > 0$, this proves that $S \circ T$ is continuous.

(b) *Uniform continuity.* A similar procedure is utilized.

(c) *Sequential continuity.* Consider a sequence $(u_n)_{n \in \mathbb{N}}$ of X and $u \in X$ such that $u_n \to u$ in E. If S and T are sequentially continuous, $T(u_n) \to T(u)$ and thus $S(T(u_n)) \to S(T(u))$, which proves that $S \circ T$ is sequentially continuous. □

7.4. Continuous semi-norms

Let us observe that the semi-norms of a separated semi-normed space are continuous.

THEOREM 7.10.– *Every semi-norm $\| \ \|_{E;\nu}$ of a separated semi-normed space E is uniformly continuous, and thus continuous and sequentially continuous.* ▮

Proof. The inequality $|\|v\|_{E;\nu} - \|u\|_{E;\nu}| \leq \|v - u\|_{E;\nu}$ (Theorem 2.5 (c)) yields uniform continuity (Definition 7.1 (b)). The latter always leads to continuity and sequential continuity (Theorem 7.2 (a)). □

Let us characterize continuous semi-norms.

THEOREM 7.11.– *Let E be a separated semi-normed space, $\{\| \ \|_{E;\nu} : \nu \in \mathcal{N}_E\}$ its family of semi-norms and p a semi-norm over E. Then:*

$$\text{(a) } p \text{ is continuous} \quad \Leftrightarrow \quad \begin{cases} \text{there exist a finite set } N \subset \mathcal{N}_E \text{ and } c \geq 0 \text{ such that,} \\ \text{for all } u \in E, \\ \qquad\qquad p(u) \leq c \sup_{\nu \in N} \|u\|_{E;\nu} \end{cases}$$

$$\Leftrightarrow \quad p \text{ is uniformly continuous.}$$

(b) *If the family of semi-norms $\{\| \ \|_{E;\nu} : \nu \in \mathcal{N}_E\}$ is filtering:*

$$p \text{ is continuous} \quad \Leftrightarrow \quad \begin{cases} \text{there exist } \nu \in \mathcal{N}_E \text{ and } c \geq 0 \text{ such that, for all } u \in E, \\ \qquad\qquad p(u) \leq c\,\|u\|_{E;\nu}. \end{cases}$$ ▮

Proof. (a) *Continuity $\Rightarrow$ majoration.* If p is continuous at 0_E, Definition 7.1 (a) of continuity gives, since $p(0_E) = 0$ (Theorem 2.5 (b)), the existence of a finite set $N \subset \mathcal{N}_E$ and of $\eta > 0$ such that:

$$\sup_{\nu \in N} \|u\|_{E;\nu} \leq \eta \quad \Rightarrow \quad |p(u)| \leq 1.$$

It follows that $p(u) \leq (1/\eta) \sup_{\nu \in N} \|u\|_{E;\nu}$ for all $u \in E$, as we will verify in Theorem 7.12 below.

Majoration $\Rightarrow$ uniform continuity. The stated majoration yields (Theorem 2.5 (c))

$$|p(v) - p(u)| \leq p(v - u) \leq c \sup_{\nu \in N} \|v - u\|_{E;\nu}$$

hence uniform continuity.

Uniform continuity $\Rightarrow$ continuity. This is always true (Theorem 7.2 (a)).

(b) If the family of semi-norms E is filtering (Definition 3.15), there exists $\lambda \in \mathcal{N}_E$ such that $\sup_{\nu \in N} \| \ \|_{E;\nu} \leq \| \ \|_{E;\lambda}$ therefore the majoration in (a) leads to that in (b).

Conversely, the majoration in (b) is that in (a) for $N = \{\nu\}$. □

The following property remains to be proved.

THEOREM 7.12.– *Let p and q be two semi-norms on the same vector space E or, more generally, two real* **positively homogeneous** *mappings, namely such that, for every $u \in E$ and $t \geq 0$,*

$$p(u) \geq 0,\ p(tu) = tp(u), \quad q(u) \geq 0,\ q(tu) = tq(u).$$

If there are two real numbers $c > 0$ and $a \geq 0$ such that, for all $u \in E$,

$$p(u) \leq c \quad \Rightarrow \quad q(u) \leq a$$

then, for all $u \in E$,

$$q(u) \leq \frac{a}{c} p(u). \blacksquare$$

Proof. Assume that p and q are real positively homogeneous mappings, which is the case of semi-norms according to their Definition 2.2 (a), such that $p(u) \leq c$ leads to $q(u) \leq a$.

Let $v \in E$ be such that $p(v) > 0$ and let $t = c/p(v)$. Then, $p(tv) = tp(v) = c$ thus $q(tv) = tq(v) = cq(v)/p(v) \leq a$ from which $q(v) \leq (a/c)\, p(v)$.

This is also true when $p(v) = 0$ because otherwise it would yield $q(v) > 0$ and, for $t = 2a/q(v)$, it would give $p(tv) = tp(v) = 0$ and $q(tv) = tq(v) = 2a$. □

7.5. Continuous linear mappings

Let us define linear mappings[2].

DEFINITION 7.13.– *A mapping L from a vector space E into a vector space is* **linear** *if: for all $u \in E$, $v \in E$ and $t \in \mathbb{R}$,*

$$L(u + v) = L(u) + L(v), \quad L(tu) = t\, L(u). \blacksquare$$

When there is no ambiguity, we sometimes denote Lu instead of $L(u)$ when L is linear.

Let us characterize continuous linear mappings[3].

THEOREM 7.14.– *Let E and F be separated semi-normed spaces, $\{\| \ \|_{E;\nu} : \nu \in \mathcal{N}_E\}$ and $\{\| \ \|_{F;\mu} : \mu \in \mathcal{N}_F\}$ their families of semi-norms, and*

L a linear mapping from E into F.

2 **History of the notion of linear mapping.** Giuseppe PEANO gave the definition of a linear mapping in 1888 [83, chap. IX].

3 **History of the characterization of continuous linear mappings of Theorem 7.14 (a).** In 1911, Frigyes RIESZ [92] observed that continuous linear mappings over $\mathcal{C}(I)$ are characterized by $|Lu| \leq c\, \|u\|$, I being a compact interval.

Then:

$$\text{(a) } L \textit{ is continuous} \quad \Leftrightarrow \quad \begin{cases} \textit{for all } \mu \in \mathcal{N}_F, \textit{ there exist a finite set } N \subset \mathcal{N}_E \\ \textit{and } c \geq 0 \textit{ such that: for all } u \in E, \\ \qquad \|Lu\|_{F;\mu} \leq c \sup_{\nu \in N} \|u\|_{E;\nu}. \end{cases}$$

$$\Leftrightarrow \quad L \textit{ is uniformly continuous.}$$

(b) *If the family of semi-norms* $\{\| \ \|_{E;\nu} : \nu \in \mathcal{N}_E\}$ *is filtering:*

$$L \textit{ is continuous} \quad \Leftrightarrow \quad \begin{cases} \textit{for all } \mu \in \mathcal{N}_F, \textit{ there exist } \nu \in \mathcal{N}_E \textit{ and } c \geq 0 \\ \textit{such that: for all } u \in E, \\ \qquad \|Lu\|_{F;\mu} \leq c \|u\|_{E;\nu}. \end{cases}$$ ∎

Proof. (a) *Continuity* $\Rightarrow$ *majoration.* Linearity yields $L(0_E) = 0_F$ and thus the continuity at 0_E results in (Definition 7.1), for all $\mu \in \mathcal{N}_F$, the existence of a finite set $N \subset \mathcal{N}_E$ and of $\eta > 0$ such that:

$$\sup_{\nu \in N} \|u\|_{E;\nu} \leq \eta \quad \Rightarrow \quad \|Lu\|_{F;\mu} \leq 1.$$

From which (Theorem 7.12) the stated upper bound with $c = 1/\eta$.

Majoration $\Rightarrow$ *uniform continuity.* The stated majoration gives, with linearity,

$$\|Lv - Lu\|_{F;\mu} \leq c \sup_{\nu \in N} \|v - u\|_{E;\nu}.$$

Uniform continuity $\Rightarrow$ *continuity.* This is always true (Theorem 7.2 (a)).

(b) If the family of semi-norms E is filtering (Definition 3.14), the majoration in (a) leads to that in (b) because, for every finite subset N of $\mathcal{N}_E$, there exists $\lambda \in \mathcal{N}_E$ such that $\sup_{\nu \in N} \| \ \|_{E;\nu} \leq \| \ \|_{E;\lambda}$.

Conversely, the majoration in (b) is that in (a) for $N = \{\nu\}$. □

Discontinuous linear mappings. There are linear mappings that are neither continuous nor even sequentially continuous, even in a normed space. For example, if E is the space of real sequences $\mathcal{A} = (a_i)_{i \in \mathbb{N}}$ having a finite number of non-zero terms endowed with the norm $\|\mathcal{A}\|_E = \sup_{i \in \mathbb{N}} |a_i|$, the linear mapping L defined from E into $\mathbb{R}$ by $L(\mathcal{A}) = \sum_{i \in \mathbb{N}} a_i$ is not sequentially continuous. As a matter of fact, the sequence $\mathcal{A}_n \in E$ whose n first elements are equal to 1/n and whose following ones are zero verify $\|\mathcal{A}_n\|_E = 1/n$ and $L(\mathcal{A}_n) = 1$, therefore $\mathcal{A}_n \to 0_E$ but $L(\mathcal{A}_n) \not\to L(0_E)$ (since $L(0_E)$ is zero). □

Let us show that continuous linear mappings transform bounded sets into bounded sets.

THEOREM 7.15.– *Any continuous or sequentially continuous linear mapping from a separated semi-normed space into another transforms bounded sets into bounded sets.* ▮

Proof. Let L be a sequentially continuous linear mapping from a separated semi-normed space E into another F, and B a bounded subset of E.

If $L(B)$ were not bounded, there would exist a semi-norm F and, for each $n \in \mathbb{N}$, an element $u_n \in B$ such that:

$$\|L(u_n)\|_{F;\mu} \geq n.$$

For each semi-norm of E, we would have $\sup_{n\in\mathbb{N}} \|u_n\|_{E;\nu} \leq c_\nu < \infty$ since B is bounded. Then $v_n = u_n/n$ would verify $\|v_n\|_{E;\nu} \leq c_\nu/n$, and thus $v_n \to 0_E$. Since $\|L(v_n)\|_{F;\mu} \geq 1$, this would contradict sequential continuity. Therefore, $L(B)$ is bounded.

This still holds if L is continuous because it is then sequentially continuous (Theorem 7.2 (a)). □

Converse of Theorem 7.15. The converse holds for any linear mapping defined on a metrizable space: if it transforms bounded sets into bounded sets, it is continuous (Theorem 9.2). This does not hold in every separated semi-normed space, see the observation "*The case of a non-metrizable space*", p. 131. □

A bijection is said to be **bicontinuous** if it is continuous together with its inverse mapping.

Let us characterize equalities and topological inclusions.

THEOREM 7.16.– *Let E and F be two separated semi-normed spaces. Then:*

(a) $E \underset{\to}{\subseteq} F \quad \Leftrightarrow \quad E \subset F$ *and the identity of E in F is continuous.*

(b) $E \underset{\leftrightarrow}{=} F \quad \Leftrightarrow \quad E = F$ *and the identity of E in F is bicontinuous.* ▮

Proof. According to Definition 3.6 (c) and (b) of topological inclusion and equality, this results from the characterization of the continuous linear mappings of Theorem 7.14 (a) applied to the identity from E into F. □

An **isomorphism** (from a separated semi-normed space into another) refers to a bicontinuous linear bijection. Let us observe that its inverse mapping is also linear according to the following property.

THEOREM 7.17.– (a) *The image of any linear mapping is a vector space.*

(b) *The inverse mapping of any linear bijection is linear.* ▮

Proof. (a) Let E and F be vector spaces, L be a linear mapping from E into F, $x \in L(E)$, $y \in L(E)$ and $t \in \mathbb{R}$. There exist $u \in E$ and $v \in E$ such that $x = Lu$ and $y = Lv$. Then, $x + y = L(u + v)$ and $tx = L(tu)$ are elements of $L(E)$. The latter is therefore a vector subspace (Definition 3.5) of F and, *a fortiori*, a vector space.

(b) If L is bijective, the above equalities yield $L^{-1}(x+y) = u+v = L^{-1}(x)+L^{-1}(y)$ and $L^{-1}(tx) = tu = tL^{-1}(x)$, and thus L^{-1} is linear. □

Let us show that isomorphisms preserve sequential completeness.

THEOREM 7.18.– *Any separated semi-normed space isomorphic to a Neumann space is a Neumann space.* ▮

Proof. Let E and F be separated semi-normed spaces, L be an isomorphism from E onto F and $(v_n)_{n\in\mathbb{N}}$ be a Cauchy sequence F. The mapping L^{-1} is linear (Theorem 7.17 (b)) and continuous (by assumption), and therefore (Theorem 7.14 (a)) it is uniformly continuous. Then (Theorem 7.4 (a)) $(L^{-1}(v_n))_{n\in\mathbb{N}}$ is a Cauchy sequence of E.

If E is a Neumann space (Definition 4.10 (a)), $L^{-1}(v_n)$ has a limit u therein. Since L is sequentially continuous (Theorem 7.2 (a)), $v_n = L(L^{-1}(v_n)) \to L(u)$ in F. This proves that F is sequentially complete, that is a Neumann space. □

7.6. Continuous multilinear mappings

Let us define multilinear mappings.

DEFINITION 7.19.– *A mapping T over a product of vector spaces is **multilinear** if its **partial mappings** $u_i \mapsto T(u_1, \ldots, u_d)$ are linear.*

*In the case there are two variables, it is said to be **bilinear**.* ▮

Let us characterize continuous multilinear mappings.

THEOREM 7.20.– *Let $E_1, E_2 \ldots, E_d$ and F be separated semi-normed spaces whose families of semi-norms are indexed by $\mathcal{N}_{E_1}, \mathcal{N}_{E_2} \ldots, \mathcal{N}_{E_d}$ and $\mathcal{N}_F$, and let*

T be a multilinear mapping from $E_1 \times E_2 \times \cdots \times E_d$ into F.

Then:

(a) *T is continuous*

$$\Leftrightarrow \begin{cases} \textit{for all } \mu \in \mathcal{N}_F, \textit{ there exist finite sets } N_1 \subset \mathcal{N}_{E_1}, \dots, N_d \subset \mathcal{N}_{E_d} \\ \textit{and } c \geq 0 \textit{ such that: for all } (u_1, \dots, u_d) \in E_1 \times \dots \times E_d, \\ \qquad \|T(u_1, \dots, u_d)\|_{F;\mu} \leq c \sup_{\nu_1 \in N_1} \|u_1\|_{E_1;\nu_1} \cdots \sup_{\nu_d \in N_d} \|u_d\|_{E_d;\nu_d}. \end{cases}$$

(b) *If the family of semi-norms of each E_i is filtering:*

T is continuous

$$\Leftrightarrow \begin{cases} \textit{for all } \mu \in \mathcal{N}_F, \textit{ there exist } \nu_1 \in \mathcal{N}_{E_1}, \dots, \nu_d \in \mathcal{N}_{E_d} \textit{ and } c \geq 0 \textit{ such that:} \\ \textit{for all } (u_1, \dots, u_d) \in E_1 \times \dots \times E_d, \\ \qquad \|T(u_1, \dots, u_d)\|_{F;\mu} \leq c\|u_1\|_{E_1;\nu_1} \cdots \|u_d\|_{E_d;\nu_d}. \end{cases}$$ ∎

Proof. (a) *Continuity ⇒ majoration.* Let us endow $E_1 \times \dots \times E_d$ with the semi-norms of Theorem 6.9 (a), namely $|||u|||_\nu = \|u_1\|_{E_1;\nu_1} + \dots + \|u_d\|_{E_d;\nu_d}$ indexed by $\nu \in \mathcal{N}_{E_1} \times \dots \times \mathcal{N}_{E_d}$.

Linearity with respect to u_1 yields $T(0_{E_1}, \dots, 0_{E_d}) = 0_F$. Thus, the continuity of T at this point results (Definition 7.1 (a)) in that, for every $\mu \in \mathcal{N}_F$, there exist a finite set $N \subset \mathcal{N}_{E_1} \times \dots \times \mathcal{N}_{E_d}$ and $\eta > 0$ such that:

$$\sup_{(\nu_1,\dots,\nu_d)\in N} \|u_1\|_{E_1;\nu_1} + \dots + \|u_d\|_{E_d;\nu_d} \leq \eta \quad \Rightarrow \quad \|T(u_1, \dots, u_d)\|_{F;\mu} \leq 1.$$

By denoting $N_i = \{\nu_i : (\nu_1, \dots, \nu_d) \in N\}$, this is especially satisfied if, for all $i \in [\![1, d]\!]$,

$$\sup_{\nu_i \in N_i} \|u_i\|_{E_i,\nu_i} \leq \frac{\eta}{d}.$$

This implies the stated majoration with $c = (d/\eta)^d$ as we will verify in Theorem 7.21 below.

Majoration ⇒ continuity. The multilinearity of T gives

$$T(u_1, \dots, u_d) - T(v_1, \dots, v_d) = T(u_1 - v_1, u_2, \dots, u_d) + \\ + T(v_1, u_2 - v_2, \dots, u_d) + \dots + T(v_1, \dots, v_{d-1}, u_d - v_d).$$

From the stated majoration, it results that

$$\|T(u_1, \dots, u_d) - T(v_1, \dots, v_d)\|_{F;\mu} \leq \\ \leq cb^{d-1} \sup_{\nu_1 \in N_1, \dots, \nu_d \in N_d} \|u_1 - v_1\|_{E_1;\nu_1} + \dots + \|u_d - v_d\|_{E_d;\nu_d} \quad (7.2)$$

where $b = \sup_{1 \leq i \leq d,\, \nu_i \in N_i} \|u_i\|_{E_i;\nu_i} + \|v_i\|_{E_i;\nu_i}$. Therefore, T is continuous.

(b) If, for all $i \in [\![1, d]\!]$, the family of semi-norms of E_i is filtering (Definition 3.14), there exists $\lambda_i \in \mathcal{N}_{E_i}$ such that $\sup_{\nu_i \in N_i} \| \ \|_{E_i;\nu_i} \leq \| \ \|_{E_i;\lambda_i}$ therefore the majoration in (a) leads to that in (b).

Conversely, the majoration of (b) is that of (a) for $N_i = \{\nu_i\}$. □

The following property remains to be proved.

THEOREM 7.21.– *Let q be a real mapping over a product $E_1 \times \cdots \times E_d$ of vector spaces and, for each $i \in [\![1, d]\!]$, let p_i be a real mapping over E_i such that, for all $u_i \in E_i$ and $t_i \geq 0$:*

$$q(u_1, \ldots, u_d) \geq 0, \quad q(t_1 u_1, \ldots, t_d u_d) = t_1 \cdots t_d \, q(u_1, \ldots, u_d),$$
$$p_i(u_i) \geq 0, \quad p_i(t_i u_i) = t_i p(u_i).$$

If there exist real numbers $c_1 > 0, \ldots, c_d > 0$ and $a \geq 0$ such that, for all $(u_1, \ldots, u_d)$ in $E_1 \times \cdots \times E_d$,

$$[\,\forall i \in [\![1, d]\!], \ p_i(u_i) \leq c_i\,] \quad \Rightarrow \quad q(u_1, \ldots, u_d) \leq a \tag{7.3}$$

then, for all $(u_1, \ldots, u_d) \in E_1 \times \cdots \times E_d$,

$$q(u_1, \ldots, u_d) \leq \frac{a}{c_1 \cdots c_d} p_1(u_1) \cdots p_d(u_d). \ \blacksquare$$

Proof. For all $i \in [\![1, d]\!]$, let $v_i \in E_i$ be such that $p_i(v_i) > 0$ and let $t_i = c_i / p_i(v_i)$. Then, $p_i(t_i v_i) = t_i p_i(v_i) = c_i$ therefore, if (7.3) is verified,

$$q(t_1 v_1, \ldots, t_d v_d) = \frac{c_1}{p_1(v_1)} \cdots \frac{c_d}{p_d(v_d)} q(v_1, \ldots, v_d) \leq a$$

hence the stated majoration.

This is still true if some $p_i(v_i)$ are zero because then $q(v_1, \ldots, v_d) = 0$. In effect, for the i such that $p_i(v_i) = 0$, it follows that $p_i(t_i v_i) = t_i p_i(v_i) = 0$ for any given t_i such that, by choosing $t_i = c_i / p_i(v_i)$ for the other values of i, (7.3) leads to $q(t_1 v_1, \ldots, t_d v_d) = t_1 \cdots t_d \, q(v_1, \ldots, v_d) \leq a$. This would be impossible if $q(v_1, \ldots, v_d)$ were not zero because some t_i are as large as desired. □

Let us show that continuous multilinear mappings are uniformly continuous on bounded sets.

THEOREM 7.22.– *Any continuous multilinear mapping from a product of separated semi-normed spaces into a separated semi-normed space is uniformly continuous on any bounded set.* ▮

Proof. Let us use the notations of Theorem 7.20 and assume that T is continuous multilinear. As it has been seen in the proof of the characterization (a) of the aforementioned theorem, T then satisfies the majoration (7.2). The space $E_1 \times \cdots \times E_d$ being endowed with the semi-norms $|||u|||_\nu = \|u_1\|_{E_1;\nu_1} + \cdots + \|u_d\|_{E_d;\nu_d}$ of Theorem 6.9 (a), this majoration writes, by denoting $N = N_1 \times \cdots \times N_d$,

$$\|T(u) - T(v)\|_{F;\mu} \leq cb^{d-1} \sup_{\nu \in N} |||u - v|||_\nu$$

where $b \leq b' = \sup_{\nu \in N} |||u|||_\nu + |||v|||_\nu$.

Let U be a bounded set (Definition 2.9 (a)) of $E_1 \times \cdots \times E_d$. Then, $\sup_{u \in U} |||u|||_\nu$ is finite for each ν. Since N is finite because the N_i are, then $b' < \infty$ whence uniform continuity (Definition 7.1 (b)) of T. □

Complement to Theorem 7.22. In contrast to a linear mapping (Theorem 7.14 (a)):

$$\begin{cases} \textit{A multilinear mapping (non-zero, having several variables)} \\ \textit{is never uniformly continuous over the entire space.} \end{cases} \tag{7.4}$$

Proof. If T is multilinear, then $T(tu + u) - T(tu) = ((t+1)^d - t^d)T(u) \geq tdT(u)$ if $t \geq 0$ and $d \geq 2$ and therefore is not bounded independently of t if $T(u) \neq 0$. □

Non-continuity of common bilinear mappings. Except in normed spaces, a large number of bilinear mappings that could be thought as continuous are not. For instance (Theorem 13.22), the bilinear form of duality $\langle\ ,\ \rangle_{E' \times E}$ is not continuous over $E' \times E$ if E is not normable; on the other hand, it is sequentially continuous (it is even **quasi-continuous**, that is continuous on bounded subsets).

The mapping $(L, M) \mapsto M \circ L$ is also quasi-continuous (Theorem 12.12 (e)), but the author of these lines does not know any counterexample showing that it is not continuous.

We will see in Volume II that the product is sequentially continuous from $\mathcal{C}^\infty(\Omega) \times \mathcal{D}'(\Omega; E)$ into $\mathcal{D}'(\Omega; E)$ and that the mapping $\langle\ ,\ \rangle_{\mathcal{D}'(\Omega;E) \times \mathcal{D}(\Omega)}$ is sequentially continuous from $\mathcal{D}'(\Omega; E) \times \mathcal{D}(\Omega)$ into E but that these mappings are not continuous. □

Let us show that continuous multilinear mappings transform bounded sets into bounded sets.

THEOREM 7.23.– *Any continuous or sequentially continuous multilinear mapping from a product of separated semi-normed spaces into a separated semi-normed space transforms bounded sets into bounded sets.* ▮

Proof. Let T be a sequentially continuous linear mapping from a product $E_1 \times \cdots \times E_d$ of separated semi-normed spaces into a separated semi-normed space F, and let B be a bounded subset of $E_1 \times \cdots \times E_d$.

If $T(B)$ were not bounded (Definition 2.9 (a)), there would exist a semi-norm $\|\ \|_{F;\mu}$ of F and, for each $n \in \mathbb{N}$, an element $(u_{n,1}, \ldots, u_{n,d}) \in B$ such that:

$$\|T(u_{n,1}, \ldots, u_{n,d})\|_{F;\mu} \geq n^d.$$

For each semi-norm $\| \ \|_\nu$ of $E_1 \times \cdots \times E_d$, since B is bounded, there would be $\sup_{n \in \mathbb{N}} \|(u_{n,1}, \ldots, u_{n,d})\|_\nu \leq c_\nu < \infty$. By denoting $v_{n,i} = u_{n,i}/n$, it would follow that

$$\|(v_{n,1}, \ldots, v_{n,d})\|_\nu \leq \frac{c_\nu}{n}, \qquad \|T(v_{n,1}, \ldots, v_{n,d})\|_{F;\mu} \geq 1$$

and thus $(v_{n,1}, \ldots, v_{n,d}) \to 0$ and $T(v_{n,1}, \ldots, v_{n,d}) \not\to 0$ which would contradict sequential continuity. Therefore, $T(B)$ is bounded.

This still holds if T is continuous because it is then sequentially continuous (Theorem 7.2 (a)). □

Converse of Theorem 7.23. The converse is true for any multilinear mapping defined on a product of metrizable spaces: if it transforms bounded sets into bounded sets, it is continuous (Theorem 9.3). □

7.7. Some continuous mappings

Let us show that the addition and the multiplication of a separated semi-normed space are continuous[4].

THEOREM 7.24.– *Let E be a separated semi-normed space. Then:*

(a) $(u, v) \mapsto (u + v)$ *is continuous linear from $E \times E$ into E.*

(b) $(t, u) \mapsto tu$ *is continuous bilinear from $\mathbb{R} \times E$ into E.* ▮

Proof. (a) Linearity results from Definition 2.1 of a vector space. According to the characterization of the continuous linear mappings of Theorem 7.14 (a), continuity is then the result of the following inequality:

$$\|u + v\|_{E;\nu} \leq \|u\|_{E;\nu} + \|v\|_{E;\nu} \leq 2(\|u\|^2_{E;\nu} + \|v\|^2_{E;\nu})^{1/2} = 2\|(u, v)\|_{E^2;\nu}$$

(we did use Definition 2.2 (a) of a semi-norm, the inequality $a \leq (a^2 + b^2)^{1/2}$ and Definition 6.10 of the product of a semi-normed space by itself).

(b) Bilinearity results from Definition 2.1. According to the characterization of the continuous multilinear mappings of Theorem 7.20, continuity is then the result of the inequality $\|tu\|_{E;\nu} \leq |t| \, \|u\|_{E;\nu}$ (Definition 2.2 (a) of a semi-norm). □

Let us show that the scalar product of a Hilbert space is continuous.

THEOREM 7.25.– *Let E be a Hilbert space. Then:*

$(u, v) \mapsto (u, v)_E$ *is continuous bilinear from $E \times E$ into $\mathbb{R}$.* ▮

4 **History of Theorem 7.24 (b).** Stefan BANACH proved the continuity of the scalar multiplication for a Fréchet space.

Proof. The Cauchy–Schwarz inequality $|(u,v)_E| \le \|u\|_E\|v\|_E$ (Theorem 5.3 (a)) leads to continuity according to the characterization of the continuous multilinear mappings of Theorem 7.20 (b). □

Let us show that the square, the inverse and the product of real numbers continuously depend on these.

THEOREM 7.26.– (a) *$x \mapsto x^2$ is continuous from $\mathbb{R}$ into itself.*

(b) *$x \mapsto 1/x$ is continuous from $(0,\infty)$ into itself.*

(c) *$(x,y) \mapsto xy$ is continuous from $\mathbb{R} \times \mathbb{R}$ into $\mathbb{R}$.* ▮

Proof. (a) If $|x-y| \le 1$, then $|x^2-y^2| = |x-y||x+y| \le |x-y|(2|y|+1)$, hence (Definition 7.1 (a)) the continuity of $x \mapsto x^2$ at the point y.

(b) If $y > 0$ and $x \ge y/2$, then $|1/x - 1/y| = |x-y|/xy \le 2|x-y|/y^2$. This particularly holds if $|x-y| \le y/2$, hence the continuity of $x \mapsto 1/x$ at the point y.

(c) If $|y-t| \le 1$, then $|xy-st| = |y(x-s)+s(y-t)| \le (|t|+1+|s|)(|x-s|+|y-t|)$ hence the continuity of $(x,y) \mapsto xy$ at the point (s,t). □

Let us observe that a mapping that has several components is continuous if its components are.

THEOREM 7.27.– *Let E and $F_1, \ldots, F_d$ be separated semi-normed spaces, X be an open subset of E, T be a mapping from X into $F_1 \times \cdots \times F_d$ and $u \in X$. Then, by denoting by T_i the mapping from X into E_i defined by $T_i(v) = (T(v))_i$:*

$$T \text{ is continuous at } u \quad \Leftrightarrow \quad \text{all } T_i \text{ are continuous at } u. \;▮$$

Proof. Let us endow (Theorem 3.15) the space E with a filtering family of semi-norms $\{|||\ |||_{E;\nu} : \nu \in \mathcal{N}_E\}$ defining its topology, which preserves continuity (Theorem 7.5). Let $\|\ \|_{F_1\times\cdots\times F_d;\mu_1,\ldots,\mu_d}$ be a semi-norm of $F_1 \times \cdots \times F_d$ (Definition 6.1) and $\epsilon > 0$.

If T is continuous, there exist (Theorem 7.3 (a)) $\nu \in \mathcal{N}_E$ and $\eta > 0$ such that $v \in X$ and $|||v-u|||_{E;\nu} \le \eta$ lead to

$$\|T(v)-T(u)\|_{F_1\times\cdots\times F_d;\mu_1,\ldots,\mu_d} =$$
$$= \left(\|T_1(v)-T_1(u)\|^2_{F_1;\mu_1} + \cdots + \|T_d(v)-T_d(u)\|^2_{F_d;\mu_d}\right)^{1/2} \le \epsilon.$$

For each $i \in [\![1,d]\!]$, it follows that $\|T_i(v)-T_i(u)\|_{F_i;\mu_i} \le \epsilon$ thus T_i is continuous.

Conversely, if each T_i is continuous, there exist $\nu_i \in \mathcal{N}_E$ and $\eta_i > 0$ such that $|||v-u|||_{E;\nu_i} \le \eta_i$ leads to $\|T_i(v)-T_i(u)\|_{F_i;\mu_i} \le \epsilon/\sqrt{d}$. The family of semi-norms of E being filtering, there is $\nu \in \mathcal{N}_E$ such that $\sup_{1\leq i\leq d} ||| \ |||_{E;\nu_i} \leq ||| \ |||_{E;\nu}$. Then $\min_{1\leq i\leq d} \eta_i$ exists (Theorem 1.2 (c)) and is > 0, and $|||v - u|||_{E;\nu} \leq \min_{1\leq i\leq d} \eta_i$ results in $\|T(v) - T(u)\|_{F_1\times\cdots\times F_d;\mu_1,\ldots,\mu_d} \leq \epsilon$. Therefore, T is continuous. □

Let us show that the projection from a product of spaces onto one of them is continuous.

THEOREM 7.28.– *Let $E_1, E_2, \ldots, E_d$ be separated semi-normed spaces, $i \in [\![1, d]\!]$, and P_i be the projection defined on $E_1 \times \cdots \times E_d$ by $P_i(u_1, \ldots, u_d) = u_i$. Then:*

$$P_i \text{ is continuous linear from } E_1 \times \cdots \times E_d \text{ onto } E_i. \ \blacksquare$$

Proof. Let $\{\| \ \|_{E_j;\nu_j} : \nu_j \in \mathcal{N}_{E_j}\}$ be the family of semi-norms of E_i. Given $\nu_i \in \mathcal{N}_{E_i}$, following Definition 6.1 of the semi-norms of a product,

$$\|P_i(u)\|_{E_i;\nu_i} = \|u_i\|_{E_i;\nu_i} \leq$$
$$\leq (\|u_1\|^2_{E_1;\nu_1} + \cdots + \|u_d\|^2_{E_d;\nu_d})^{1/2} = \|u\|_{E_1\times\cdots\times E_d;\nu_1,\ldots,\nu_d}$$

where, for $j \neq i$, ν_j is whatever element of $\mathcal{N}_{E_j}$. Linearity being trivial, continuity results therefrom according to the characterization of continuous linear applications of Theorem 7.14 (a). □

Let us show that the direct sum of separated semi-normed spaces is isomorphic to their product.

THEOREM 7.29.– *Let $E_1\oplus\cdots\oplus E_d$ be a direct sum of separated semi-normed spaces. Then:*

$$\begin{cases} \textit{The mapping } u_1 + \cdots + u_d \mapsto (u_1, \ldots, u_d) \textit{ is an} \\ \textit{isomorphism from } E_1 \oplus \cdots \oplus E_d \textit{ onto } E_1 \times \cdots \times E_d. \ \blacksquare \end{cases}$$

Proof. The mapping $u_1 + \cdots + u_d \mapsto (u_1, \ldots, u_d)$ is a bijection since the decomposition $u = u_1 \cdots + u_d$ is unique, see (6.10), p. 90, and it is bicontinuous according to the equality (6.11) of semi-norms. □

Chapter 8

Images of Sets Under Continuous Mappings

8.1. Images of open and closed sets

Let us give properties of preimages of open or closed sets that characterize continuity.

THEOREM 8.1.– *Let E and F be two separated semi-normed spaces, $X \subset E$, and T a mapping from X into F. Then:*

T is continuous

$$\Leftrightarrow \begin{cases} \text{for any open subset } W \text{ of } F, \text{ there exists an open subset } \mathcal{O} \text{ of } E \text{ such that} \\ \qquad T^{-1}(W) = \mathcal{O} \cap X, \end{cases}$$

$$\Leftrightarrow \begin{cases} \text{for any closed subset } W \text{ of } F, \text{ there exists a closed subset } \mathcal{F} \text{ of } E \text{ such that} \\ \qquad T^{-1}(W) = \mathcal{F} \cap X. \ \blacksquare \end{cases}$$

Proof. First equivalence, direct part. Let $\{\| \ \|_{E;\nu} : \nu \in \mathcal{N}_E\}$ and $\{\| \ \|_{F;\mu} : \mu \in \mathcal{N}_F\}$ be the families of semi-norms of E and F, $W \subset F$ and $u \in T^{-1}(W)$. Remember that $T^{-1}(W) = \{v \in X : T(v) \in W\}$.

If W is open (Definition 2.9 (b)) in F, there exists a finite set $M \subset \mathcal{N}_F$ and $\epsilon > 0$ such that:

$$w \in F, \ \sup_{\mu \in M} \|w - T(u)\|_{F;\mu} \leq \epsilon \quad \Rightarrow \quad w \in W. \tag{8.1}$$

If T is continuous (Definition 7.1 (a)), for each $\mu \in M$, there exists a finite set $N_\mu \subset \mathcal{N}_E$ and $\eta_\mu > 0$ such that:

$$v \in X, \sup_{\nu \in N_\mu} \|v - u\|_{E;\nu} \leq \eta_\mu \quad \Rightarrow \quad \|T(v) - T(u)\|_{F;\mu} \leq \epsilon. \tag{8.2}$$

Then, $N = \bigcup_{\mu \in M} N_\mu$ is finite and $\eta = \min_{\mu \in M} \eta_\mu$ exists (Theorem 1.2 (c)) and is strictly positive. Let

$$\mathcal{O}_u = \{v \in E : \sup_{\nu \in N} \|v - u\|_{E;\nu} < \eta\}, \quad \mathcal{O} = \bigcup_{u \in T^{-1}(W)} \mathcal{O}_u.$$

According to (8.2) and (8.1), $T(\mathcal{O}_u \cap X) \subset W$ thus $\mathcal{O}_u \cap X \subset T^{-1}(W)$ for each u of $T^{-1}(W)$, from which $\mathcal{O} \cap X \subset T^{-1}(W)$. Conversely, $u \in \mathcal{O}_u \subset \mathcal{O}$ for each u of $T^{-1}(W)$, from which $T^{-1}(W) \subset \mathcal{O}$, and by definition $T^{-1}(W) \subset X$. Therefore, it indeed yields $T^{-1}(W) = \mathcal{O} \cap X$. In addition, each semi-ball $\mathcal{O}_u$ is open (Definition 2.18 (b)) and thus their union $\mathcal{O}$ is indeed open (Theorem 2.11 (a)).

Converse of the first equivalence. Let $u \in X$, $\mu \in \mathcal{N}_F$ and $\epsilon > 0$. The semi-ball $W = \{w \in F : \|w - T(u)\|_{F;\mu} < \epsilon\}$ is open (Definition 2.18 (b)). If there exists an open set $\mathcal{O}$ of E such that $T^{-1}(W) = \mathcal{O} \cap X$, then $u \in \mathcal{O}$ and thus, according to Definition 2.9 (b) of an open set, there exists a finite set $N \subset \mathcal{N}_E$ and $\eta > 0$ such that $v \in E$ and $\sup_{\nu \in N} \|v - u\|_{E;\nu} \leq \eta$ lead to $v \in \mathcal{O}$; if in addition $v \in X$, then $v \in T^{-1}(W)$ and thus $T(v) \in W$, that is $\|T(v) - T(u)\|_{F;\mu} < \epsilon$. Consequently, T is indeed continuous.

Equivalence of the properties of the preimages of an open set and of a closed set. Let W be a closed subset of F. Its complement $V = F \setminus W$ is open (Definition 2.9 (c)) therefore, if $T^{-1}(V) = \mathcal{O} \cap X$ with $\mathcal{O}$ open, then $T^{-1}(W) = X \setminus T^{-1}(V) = \mathcal{F} \cap X$ where $\mathcal{F} = E \setminus \mathcal{O}$ is closed in E. And conversely. □

Let us give properties of the preimage of a set under continuous mapping defined on an open or closed set.

THEOREM 8.2.– *Let E and F be separated semi-normed spaces, $X \subset E$ and T a mapping from X into F.*

(a) *If X is open, then:*

T is continuous $\Leftrightarrow$ for every open subset W of F, $T^{-1}(W)$ is open in E.

(b) *If X is closed, then:*

T is continuous $\Leftrightarrow$ for every closed subset W of F, $T^{-1}(W)$ is closed in E. ∎

Proof. (a) This results from the first characterization of Theorem 8.1 because, if X is open, $\mathcal{O} \cap X$ also is, as any finite intersection of open sets is (Theorem 2.11 (b)).

(b) This results from the first characterization of Theorem 8.1 because, if X is closed, $\mathcal{F} \cap X$ also is, as any finite intersection of open sets is (Theorem 2.12 (a)). □

Characterization of the continuity by the open sets for the induced topology. Induced topology on a subset X of a topological space E (Definition 2.14) refers to the topology defined by:

$$\textit{The open sets of } X \textit{ are the intersections with } X \textit{ of the open sets of } E. \tag{8.3}$$

The characterizations of Theorem 8.1 then write as follows, for a mapping T from X into F:

$$T \textit{ is continuous} \quad \Leftrightarrow \quad \textit{for every open set } W \textit{ of } F, T^{-1}(W) \textit{ is an open subset of } X. \tag{8.4}$$

□

Continuous mapping in topological spaces. Property (8.4) is the **definition of the continuity** of a mapping from a topological space X into a topological space F. Definition 7.1 (a) of continuity in a subset X of a separated semi-normed space E therefore coincides with that of continuity in the topological space X endowed with the topology induced by E. □

8.2. Images of dense, separable and connected sets

Let us give properties of the image of the closure of a set under a continuous mapping.

THEOREM 8.3.– *Let E and F be separated semi-normed spaces, $X \subset E$, T a mapping from X into F and $U \subset X$. Then:*

(a) $$T \textit{ is continuous} \quad \Rightarrow \quad T(\overline{U}^E \cap X) \subset \overline{T(U)}^F.$$

(b) $$T \textit{ is sequentially continuous} \quad \Rightarrow \quad T(U^{\lim \operatorname{seq} E} \cap X) \subset T(U)^{\lim \operatorname{seq} F}.$$ ∎

Proof. (a) The set $W = \overline{T(U)}^F$ being closed in F, if T is continuous, there is (Theorem 8.1) a closed set $\mathcal{F}$ of E such that $T^{-1}(W) = \mathcal{F} \cap X$. Then, $U \subset T^{-1}(W) \subset \mathcal{F}$ thus $\overline{U}^E \subset \overline{\mathcal{F}}^E = \mathcal{F}$ which yields $T(\overline{U}^E \cap X) \subset T(\mathcal{F} \cap X) = T(T^{-1}(W)) = W$.

(b) Let $u \in U^{\lim \operatorname{seq} E} \cap X$, namely (Definition 2.15 (d)) u is an element of X and is the limit of a sequence $(u_n)_{n \in \mathbb{N}}$ of U; if T is sequentially continuous, then $T(u)$ is the limit of the sequence $(T(u_n))_{n \in \mathbb{N}}$ of $T(U)$, that is $T(u) \in T(U)^{\lim \operatorname{seq} F}$. □

Image of the interior of a set. If T is a continuous mapping from E into F and $U \subset E$:

$$T(\mathring{U}^E) \textit{ is neither always smaller nor always greater than } (T(U))^{\circ F}.$$

PROOF.– If $U = E = \mathbb{R}$ and $T(x) = 0$ for all $x \in \mathbb{R}$, then $T(\mathring{U}) = T(\mathbb{R}) = \{0\}$ which is greater than $(T(U))^\circ = \{0\}^\circ = \emptyset$.

On the contrary, if $E = \mathbb{R}$, $U = \mathbb{R} \setminus \mathbb{Q}$ and $T(x) = -x$ if $x \leq 0$ and $T(x) = \pi x$ if $x \geq 0$, then $T(\mathring{U}) = T(\emptyset) = \emptyset$ which is smaller than $(T(U))^\circ = \{x \geq 0\}^\circ = \{x > 0\}$. □

Let us give density properties of images of sets under continuous mapping.

THEOREM 8.4.– *Let E and F be separated semi-normed spaces, $X \subset E$, T a mapping from X into F and $V \subset U \subset X$. Then:*

(a) $$\begin{cases} T \text{ is continuous and} \\ V \text{ is dense in } U \end{cases} \Rightarrow \quad T(V) \text{ is dense in } T(U).$$

(b) $$\begin{cases} T \text{ is sequentially continuous and} \\ V \text{ sequentially dense in } U \end{cases} \Rightarrow \quad T(v) \text{ is sequentially dense in } T(U).$$ ▮

Proof. (a) If V is dense in U, that is (Definition 2.19 (a)) if $V \subset U \subset \overline{V}^E$, then $T(V) \subset T(U) \subset T(\overline{V}^E \cap X)$ which, if T is continuous, is included in $\overline{T(V)}^F$ (Theorem 8.3 (a)), which means that $T(V)$ is dense in $T(U)$.

(b) If V is sequentially dense in U, that is (Definition 2.19 (b)) if any $u \in U$ is the limit of a sequence $(v_n)_{n \in \mathbb{N}}$ of V, and if T is sequentially continuous (Definition 7.1 (c)), then $T(u)$ is the limit of the sequence $(T(v_n))_{n \in \mathbb{N}}$ of $T(V)$, which is thus sequentially dense in $T(U)$. □

Let us show that the image of a separable set under a continuous mapping is separable.

THEOREM 8.5.– *Let E and F be separated semi-normed spaces, $X \subset E$, T a mapping from X into F and $U \subset X$. Then:*

(a) $$\begin{cases} T \text{ is continuous and} \\ U \text{ is separable} \end{cases} \Rightarrow \quad T(U) \text{ is separable.}$$

(b) $$\begin{cases} T \text{ is sequentially continuous and} \\ U \text{ is sequentially separable} \end{cases} \Rightarrow \quad T(U) \text{ is sequentially separable.}$$ ▮

Proof. (a) If U is separable (Definition 2.21 (a)), there exists a dense countable set V in U. Then $T(V)$ is countable (Theorem 1.1 (c)), and if T is continuous, it is dense in $T(U)$ (Theorem 8.4 (a)), which is thus separable.

(b) If U is sequentially separable (Definition 2.21 (b)), there exists a countable set V that is sequentially dense in U. Then $T(V)$ is countable (Theorem 1.1 (c)), and if T is sequentially continuous, it is sequentially dense in $T(U)$ (Theorem 8.4 (b)), which is thus sequentially separable. □

Let us show that the image of a connected set under a continuous mapping is connected.

THEOREM 8.6.– *Let E and F be separated semi-normed spaces, $X \subset E$, T a mapping from X into F and $U \subset X$. Then:*

$$\begin{cases} T \text{ is continuous and} \\ U \text{ is connected} \end{cases} \Rightarrow \quad T(U) \text{ is connected.} ∎$$

Proof. Let us start with the case $U = X$. Let $\mathcal{O}_1$ and $\mathcal{O}_2$ be two open subsets of F such that:

$$T(X) \subset \mathcal{O}_1 \cup \mathcal{O}_2, \quad T(X) \cap \mathcal{O}_1 \cap \mathcal{O}_2 = \emptyset.$$

The mapping T being continuous, there exist (Theorem 8.1) two open sets $\mathcal{U}_i$ of E such that $T^{-1}(\mathcal{O}_i) = X \cap \mathcal{U}_i$. Then:

$$X \subset T^{-1}(\mathcal{O}_1) \cup T^{-1}(\mathcal{O}_2) \subset \mathcal{U}_1 \cup \mathcal{U}_2,$$
$$X \cap T^{-1}(\mathcal{O}_1) \cap T^{-1}(\mathcal{O}_2) = X \cap \mathcal{U}_1 \cap \mathcal{U}_2 = \emptyset.$$

If in addition X is connected in E (Definition 2.30 (a)), then $X \cap \mathcal{U}_i = \emptyset$ for $i = 1$ or 2. For this value of i, $T(X) \cap \mathcal{O}_i \subset T(X \cap \mathcal{U}_i) = \emptyset$. Therefore, $T(X)$ is connected in F.

Let us now examine the general case. If T is continuous, so is its restriction $T|_U$ (Theorem 7.8 (a)) so, if in addition U is connected, $T|_U(U)$, that is $T(U)$, is also as it has just been seen. □

8.3. Images of compact sets

Let us show that the image of a compact set under a continuous mapping is compact[1].

THEOREM 8.7.– *Let E and F be separated semi-normed spaces, $X \subset E$, T a mapping from X into F and $U \subset X$. Then:*

(a) $$\begin{cases} T \text{ is continuous and} \\ U \text{ is compact} \end{cases} \Rightarrow \quad T(U) \text{ is compact.}$$

(b) $$\begin{cases} T \text{ is sequentially continuous and} \\ U \text{ is sequentially compact} \end{cases} \Rightarrow \quad T(U) \text{ is sequentially compact.}$$

1 **History of Theorem 8.7 (c).** John VON NEUMANN proved in 1935 [78, Theorem 9, p. 8] that uniformly continuous mappings preserve precompactness.

(c) $\begin{cases} T \textit{ is uniformly continuous and} \\ U \textit{ is precompact} \end{cases} \Rightarrow T(U) \textit{ is precompact.}$

(d) $\begin{cases} T \textit{ is continuous,} \\ X \textit{ is closed and} \\ U \textit{ is relatively compact} \end{cases} \Rightarrow T(U) \textit{ is relatively compact.}$

(e) $\begin{cases} T \textit{ is sequentially continuous,} \\ X \textit{ is sequentially closed and} \\ U \textit{ is relatively sequentially compact} \end{cases} \Rightarrow \begin{cases} T(U) \textit{ is relatively sequentially} \\ \textit{compact.} \end{cases}$ ∎

Proof. (a) Let $\{\mathcal{W}_i\}_{i\in\mathcal{I}}$ be a family of open subsets of F covering $T(U)$.

If T is continuous, for every i, there exists (Theorem 8.1) an open set $\mathcal{O}_i$ of E such that $T^{-1}(\mathcal{W}_i) = X \cap \mathcal{O}_i$. Then:

$$U \subset \bigcup_{i\in\mathcal{I}} T^{-1}(\mathcal{W}_i) \subset \bigcup_{i\in\mathcal{I}} \mathcal{O}_i.$$

If in addition U is compact (Definition 2.23 (a)) in E, there is thus a finite set $J \subset \mathcal{I}$ such that $U \subset \bigcup_{i\in J} \mathcal{O}_i$. Then, since $U \subset X$,

$$T(U) \subset \bigcup_{i\in J} T(X \cap \mathcal{O}_i) \subset \bigcup_{i\in J} \mathcal{W}_i.$$

Therefore, $T(U)$ is indeed compact in F.

(b) Let $(w_n)_{n\in\mathbb{N}}$ be a sequence of $T(U)$ and, for every n, let $v_n \in U$ be such that $w_n = T(v_n)$.

If U is sequentially compact (Definition 2.24 (a)) in E, there exists a subsequence $(v_{m_n})_{n\in\mathbb{N}}$ such that $v_{m_n} \to v$ in E and $v \in U$. If in addition T is sequentially continuous, then $w_{m_n} \to T(v)$ in F and $T(v) \in T(U)$. Thus, $T(U)$ is indeed sequentially compact in F.

(c) Let $\| \ \|_{F;\mu}$ be a semi-norm of F and $\epsilon > 0$. Let us endow (Theorem 3.15) E with a filtering family of semi-norms $\{\| \ \|_{E;\nu} : \nu \in \mathcal{N}_E\}$ generating its topology, which preserves (Theorems 3.4 (b) and 7.5) precompactness and uniform continuity.

If T is uniformly continuous, there exist (Theorem 7.3 (b)) $\nu \in \mathcal{N}_E$ and $\eta > 0$ such that:

$$u \in X,\ v \in X,\ \|v - u\|_{E;\nu} \leq \eta \quad \Rightarrow \quad \|T(v) - T(u)\|_{F;\mu} \leq \epsilon.$$

If in addition U is precompact in E, there exists (Theorem 2.28) a finite set $A \subset U$ such that the semi-balls $B_{E;\nu}(a,\eta) = \{v \in E : \|v-a\|_{E;\nu} \le \eta\}$ centered at points $a \in A$ cover it. Then,

$$T(U) \subset \bigcup_{a \in A} T(B_{E;\nu}(a,\eta)) \subset \bigcup_{a \in A} B_{F;\mu}(T(a),\epsilon).$$

Therefore, $T(U)$ is indeed precompact (Definition 2.23 (c)) in F.

(d) If U is relatively compact (Definition 2.23 (b)), $\overline{U}$ is compact. If in addition X is closed, then $\overline{U} \subset X$. Furthermore, if T is continuous, then $T(\overline{U})$ is compact according to (a) thus (Theorem 2.26 (c)) its subset $T(U)$ is indeed relatively compact.

(e) Let $(w_n)_{n\in\mathbb{N}}$ be a sequence of $T(U)$ and, for every n, let $v_n \in U$ be such that $w_n = T(v_n)$.

If U is relatively sequentially compact (Definition 2.24 (b)), there exists a subsequence $(v_{m_n})_{n\in\mathbb{N}}$ such that $v_{m_n} \to v$ in E. If in addition X is sequentially closed, then $v \in X$. If furthermore T is sequentially continuous, then the subsequence $(w_{m_n})_{n\in\mathbb{N}}$ converges (to $T(v)$), thus $T(U)$ is indeed relatively sequentially compact. □

8.4. Images under continuous linear mappings

Let us give properties of images under continuous linear mappings.

THEOREM 8.8.– *Let E and F be separated semi-normed spaces and*

L a continuous linear mapping from E into F.

Let $U \subset E$, $V \subset E$ and $W \subset F$. Then:

(a) *If U has one of the following properties in E, $L(U)$ has the same property in F:*

bounded, compact, relatively compact, precompact, sequentially compact, relatively sequentially compact, separable, sequentially separable or connected.

(b) *If U has one of the following properties with respect to V in E, $L(U)$ has the same property with respect to $L(V)$ in F:*

dense or sequentially dense.

(c) $L(\overline{U}^E) \subset \overline{L(U)}^F$, $L(U^{\mathrm{lim\ seq}\,E}) \subset L(U)^{\mathrm{lim\ seq}\,F}$.

(d) *If W has one of the following properties in F, $L^{-1}(W)$ has the same property in E:*

open, closed or sequentially closed. ∎

Let us give the properties being satisfied as soon as the mapping is sequentially continuous.

THEOREM 8.9.– *Let E and F be separated semi-normed spaces and*

L a sequentially continuous linear mapping from E into F.

Let $U \subset E$, $V \subset E$ and $W \subset F$. Then:

(a) *If U has one of the following properties in E, $L(U)$ has the same property in F:*

bounded, sequentially compact, relatively sequentially compact or sequentially separable.

(b) *If U is sequentially dense in V, $L(U)$ is sequentially dense in $L(V)$.*

(c) $L(U^{\lim \operatorname{seq} E}) \subset L(U)^{\lim \operatorname{seq} F}$.

(d) *If W is sequentially closed in F, $L^{-1}(W)$ is sequentially closed in E.* ∎

Proof of Theorem 8.9. (a) These properties are respectively proven in Theorems 7.15 (bounded), 8.7 (b) (sequentially compact), 8.7 (e) (relatively sequentially compact), which is applicable here since E is sequentially closed (Theorem 2.13), and 8.5 (b) (sequentially compact).

(b) This property (sequential density) is proved in Theorem 8.4 (b).

(c) Theorem 8.3 (b) gives $L(U^{\lim \operatorname{seq} E} \cap E) = L(U^{\lim \operatorname{seq} E}) \subset L(U)^{\lim \operatorname{seq} F}$.

(d) Let $(u_n)_{n\in\mathbb{N}} \subset L^{-1}(W)$ and $u \in E$ such that $u_n \to u$ in E. Then, $L(u_n) \in W$ and $L(u_n) \to L(u)$ in F. If W is sequentially closed, then $L(u) \in W$, hence u is an element of $L^{-1}(W)$ which is thus sequentially closed. □

Proof of Theorem 8.8. We first remark that the properties stated in Theorem 8.9 are satisfied here since L, being continuous, is sequentially continuous (Theorem 7.2 (a)).

(a) These properties are respectively proven in Theorems 8.7 (a) (compact), 8.7 (d) (relatively compact) which is applicable here since E is closed (Theorem 2.13), 8.7 (c) (precompact), 8.5 (a) (separable), 8.6 (connected) and 8.9 (a) (other properties).

(b) The density property is proved in Theorem 8.4 (a).

(c) Theorem 8.3 (a) gives $L(\overline{U}^E \cap E) = L(\overline{U}^E) \subset \overline{L(U)}^F$.

(d) The properties of open and closed sets are respectively proved in Theorem 8.2 (a) and (b) which are applicable because E is open and closed (Theorem 2.13). □

8.5. Continuous mappings in compact sets

A mapping T from X into F is said to be **bounded** if the image of X, that is $T(X) = \{T(u) : u \in X\}$, is bounded in F.

Let us show that any continuous mapping over a compact set is uniformly continuous which is **Heine's theorem**[2].

THEOREM 8.10.– *Every continuous mapping from a compact subset of a separated semi-normed space into a separated semi-normed space is uniformly continuous and bounded.* ∎

Proof. Let E and F be separated semi-normed spaces, $\{\| \ \|_{E;\nu} : \nu \in \mathcal{N}_E\}$ and $\{\| \ \|_{F;\mu} : \mu \in \mathcal{N}_F\}$ their families of semi-norms, X a compact subset of E and T a continuous mapping from X into F.

Uniform continuity. Let $\epsilon > 0$ and $\mu \in \mathcal{N}_F$. By Definition 7.1 (a) of continuity, for all $u \in X$, there exists a finite set $N_u \subset \mathcal{N}_E$ and $\eta_u > 0$ such that $v \in X$ and $\sup_{\nu \in N_u} \|v - u\|_{E;\nu} \leq \eta_u$ lead to $\|T(v) - T(u)\|_{F;\mu} \leq \epsilon/2$. Let

$$\mathcal{O}_u = \{w \in E : \sup_{\nu \in N_u} \|w - u\|_{E;\nu} < \eta_u/2\}.$$

The open sets $\mathcal{O}_v$ cover the compact set X thus there exists a finite subset $\mathcal{U}$ of X such that $\{\mathcal{O}_u\}_{u \in \mathcal{U}}$ covers X. Then, $N = \bigcup_{u \in \mathcal{U}} N_u$ is finite and $\eta = \min_{u \in \mathcal{U}} \eta_u$ does exist (Theorem 1.2 (c)) and is strictly positive.

Let $v \in X$ and $w \in X$ be such that $\sup_{\nu \in N} \|v - w\|_{E;\nu} \leq \eta/2$. There exists $u \in \mathcal{U}$ such that $v \in \mathcal{O}_u$, from which

$$\|T(v) - T(u)\|_{F;\mu} \leq \epsilon/2.$$

In addition, it follows that $\sup_{\nu \in N_u} \|w - v\|_{E;\nu} \leq \sup_{\nu \in N} \|w - v\|_{E;\nu} \leq \eta/2 \leq \eta_u/2$ and $\sup_{\nu \in N_u} \|v - u\|_{E;\nu} \leq \eta_u/2$ from which $\sup_{\nu \in N_u} \|w - u\|_{E;\nu} \leq \eta_u$ and thus

$$\|T(w) - T(u)\|_{F;\mu} \leq \epsilon/2.$$

2 **History of Heine's theorem 8.10.** Eduard HEINE proved in 1872, in [56], that any real function on a closed and bounded interval (that is a compact interval) of $\mathbb{R}$ is uniformly continuous.

Consequently, $\|T(v) - T(w)\|_{F;\mu} \leq \epsilon$, which proves that T is uniformly continuous.

Bounded character. For all $v \in \mathcal{O}_u \cap X$, there is $\|T(v)\|_{F;\mu} \leq \|T(u)\|_{F;\mu} + \epsilon$, from which

$$\sup_{v \in X} \|T(v)\|_{F;\mu} \leq \sup_{u \in \mathcal{U}} \|T(u)\|_{F;\mu} + \epsilon < \infty.$$

This proves that T is bounded. □

Other proof of the bounded character in Theorem 8.10. If T is continuous on a compact set X, then $T(X)$ is compact (Theorem 8.7 (a)) and is thus bounded (Theorem 2.25 (a)). □

Let us show that any continuous bijection over a compact set is bicontinuous.

THEOREM 8.11.– *Any continuous bijection T from a compact subset of a separated semi-normed space into a subset of a separated semi-normed space has a continuous inverse mapping T^{-1}.* ▮

Proof. Let E and F be separated semi-normed spaces, X a compact subset of E, $W \subset F$ and T a continuous bijection from X into W.

Let us denote $S = T^{-1}$. Its definition domain is $W = T(X)$ which is compact (Theorem 8.7 (a)) and thus closed (Theorem 2.26 (a)). The continuity of S is then equivalent, according to the characterization of Theorem 8.2 (b), to the fact that the preimage $S^{-1}(V)$ of every closed subset V of E be closed in F.

This is the case because $S^{-1}(V) = T(V \cap X)$ and, X being compact and therefore closed, $V \cap X$ is closed (Theorem 2.12 (a)) and included in a compact set and thus is compact (Theorem 2.26 (a) and (c)) so that its image $T(V \cap X)$ is compact (Theorem 8.7 (a)) and thus closed. □

8.6. Continuous real mappings

Let us show that any real continuous mapping over a compact set reaches its extrema.

THEOREM 8.12.– *Let X be a nonempty subset of a separated semi-normed space E and T a mapping from X into $\mathbb{R}$. If*

T is continuous and X is compact

or if

T is sequentially continuous and X is sequentially compact

then, there exist $v \in X$ and $w \in X$ such that:

$$T(v) = \inf_{u \in X} T(u), \quad T(w) = \sup_{u \in X} T(u).$$ ▮

Proof. *Assumption* (a). If T is continuous and X is compact, then $T(X)$ is compact (Theorem 8.7 (a)) in $\mathbb{R}$ and non-empty (by assumption). Subsequently, there is (Theorem 4.20) a minimum $m = T(v)$ and a maximum $M = T(w)$. They are then the greatest lower and the least upper bounds.

Assumption (b). If T is sequentially continuous and X is sequentially compact, then $T(X)$ is sequentially compact (Theorem 8.7 (b)) which, in $\mathbb{R}$, coincides with compact (Theorem 4.6 (b)) since $\mathbb{R}$ is normed and thus metrizable (Theorem 4.2). Consequently, the conclusion is as above. □

Let us show that any continuous real function (we reserve the term **function** to a mapping defined on a subset of $\mathbb{R}$ or of $\mathbb{R}^d$) on a real interval takes all intermediate values between its values at the extremities, which is the **intermediate value theorem**[3].

THEOREM 8.13.– *Let a function f from $[a, b]$ into $\mathbb{R}$, where $-\infty < a < b < \infty$, and t be such that:*

$$f \textit{ is continuous or sequentially continuous,}$$

$$f(a) \le t \le f(b) \quad \textit{or} \quad f(a) \ge t \ge f(b).$$

Then, there exists $x \in [a, b]$ such that:

$$f(x) = t. \blacksquare$$

Proof. It suffices to assume that f is sequentially continuous since it is the case (Theorem 7.2 (a)) if it is continuous. Then, if $t = f(a)$ or $t = f(b)$, $x = a$ or b are suitable.

If $f(a) < t < f(b)$, let $X = \{u \in [a, b] : f(u) \le t\}$ and $x = \sup X$. For all $n \in \mathbb{N}$, let $x_n \in X$ such that $x_n \to x$. Then, $f(x_n) \le t$ and $f(x_n) \to f(x)$ when $n \to \infty$, from which $f(x) \le t$ (and thus $x < b$). Conversely, let $y_n = x + 1/n$. When n is large enough it follows that $y_n < b$. Then, $f(y_n)$ is defined, $f(y_n) \ge t$ and $f(y_n) \to f(x)$ when $n \to \infty$, from which $f(x) \ge t$. Therefore, $f(x) = t$.

If $f(a) > t > f(b)$, we may conclude with the previous case for the function $x \mapsto -f(x)$. □

8.7. Compacting mappings

Let us define compacting mappings.

DEFINITION 8.14.– *Let E and F be separated semi-normed spaces and T a mapping from E into F.*

---

3 **History of the intermediate value theorem 8.13.** Bernard BOLZANO proved this result in 1817 [13].

(a) *T is **compacting** if, for all bounded set B of E, its image $T(B) = \{T(u) : u \in B\}$ is relatively compact in F.*

(b) *T is **sequentially compacting** if, for any bounded set B of E, $T(B)$ is relatively sequentially compact in F.*

In other words, if every sequence $(u_n)_{n\in\mathbb{N}}$ of E that is bounded contains a subsequence $(u_{\sigma(n)})_{n\in\mathbb{N}}$ for which $(T(u_{\sigma(n)}))_{n\in\mathbb{N}}$ converges in F. ▌

Terminology. Compacting is a **new definition** (it coincides with *compact* in a normed space, see (8.6), p. 127, but not in every semi-normed space), consequently the reader should thereof recall the meaning if he makes use of it. □

Let us show that compacting mappings are preserved by topological equalities.

THEOREM 8.15.– *For any mapping from a separated semi-normed space E into another F:*

$$\left\{\begin{array}{l}\textit{The compacting or sequentially compacting character is preserved when the}\\ \textit{family of semi-norms of } E \textit{ and/or that of } F \textit{ is replaced by an equivalent family.}\end{array}\right.$$ ▌

Proof. This is a consequence from Theorems 8.16 and 8.17 below, according to Definition 3.6 of topological equality and inclusion. □

Let us give the effect of changing the space of values on a compacting mapping.

THEOREM 8.16.– *Let E, F_1 and F_2 be separated semi-normed spaces such that:*

$$F_1 \subsetneq F_2.$$

Then, for any mapping from E into F_1:

(a) *Compacting from E into F_1 $\Rightarrow$ compacting from E into F_2.*

(b) *Sequentially compacting from E into F_1*
$\Rightarrow$ *sequentially compacting from E into F_2.* ▌

Proof. According to Definitions 8.14 (a) and (b) of compacting mappings, these properties result from the fact that any compact or sequentially compact subset of F_1 has the same property in F_2 (Theorem 3.8 (b)). □

Let us now show the effect of changing the space of definition.

THEOREM 8.17.– *Let E_1, E_2 and F be separated semi-normed spaces such that:*

$$E_1 \subsetneq E_2.$$

Then, for any mapping from E_2 into F:

(a) *Compacting from E_2 into F* $\Rightarrow$ *compacting from E_1 into F*

(b) *Sequentially compacting from E_2 into F*

$\Rightarrow$ *sequentially compacting from E_1 into F.* ▮

Proof. According to Definitions 8.14 (a) and (b) of compacting mappings, these properties result from the fact that any bounded subset of E_1 is bounded in E_2 (Theorem 3.8 (b)). □

Motivation for introducing the notion of (sequentially) compacting mapping. The conventional notion of *compact* mapping (see here below) is too strong for our needs. For example, the identity of $\mathcal{D}'(\mathbb{R})$ is compacting (and sequentially compacting, namely every bounded sequence of $\mathcal{D}'(\mathbb{R})$ has a converging subsequence) while it is not compacting (because, in an infinite dimensional space, there is no relatively compact open set). □

Linear compact mappings. A linear mapping L from a separated semi-normed space E into a separated semi-normed space F is called **compact**[4] if [SCHWARTZ, 99, Definition, p. 419]:

There exists a non-empty open set U of E such that $L(U)$ be relatively compact in F.

Let us observe that, for every linear mapping:

$$\textit{Compact} \quad \Rightarrow \quad \textit{compacting} \tag{8.5}$$

(because for any bounded set B and any open set U of E, there exists a real t such that $B \subset tU$).

And that:

$$\textit{If } E \textit{ is normed:} \quad [\,\textit{Compact} \Leftrightarrow \textit{compacting}\,] \tag{8.6}$$

(because there exists then a bounded open set in E and therefore the inverse of (8.5) holds). □

4 **History of the notion of compact mapping.** This notion was introduced, on $\mathcal{C}(I)$, I being for example a compact interval, by Frigyes RIESZ [93] in 1918, under the name of **completely continuous** mapping.

1. **History of the notion of compact mapping.** This notion was introduced, on $C(I)$, I being for example a compact interval, by Frigyes RIESZ [93] in 1918 under the name of completely continuous mapping.

Chapter 9

Properties of Mappings in Metrizable Spaces

9.1. Continuous mappings in metrizable spaces

Let us show that in a metrizable space, continuity and the compacting character coincide with their sequential equivalents.

THEOREM 9.1.– *Let E and F be separated semi-normed spaces, $X \subset E$ and $u \in X$. Then:*

(a) *If E is metrizable: for any mapping from X into F,*

$$\begin{array}{rcl} \textit{Continuous} & \Leftrightarrow & \textit{sequentially continuous;} \\ \textit{Continuous at } u & \Leftrightarrow & \textit{sequentially continuous at } u. \end{array}$$

(b) *If F is metrizable: for any mapping from X into F,*

$$\textit{Compacting} \quad \Leftrightarrow \quad \textit{sequentially compacting.} \blacksquare$$

Proof. (a) Continuous at $u \Rightarrow$ sequentially continuous at u. This is always true (Theorem 7.2 (b)).

Sequentially continuous at $u \Rightarrow$ continuous at u. Here, E is metrizable and therefore it can be endowed with (Theorem 4.4) an increasing sequence of semi-norms $(\| \ \|_{E;n})_{n\in\mathbb{N}}$ defining its topology, which neither modifies (Theorem 7.5) the continuity nor the sequential continuity of T.

Let T be a mapping from X into F sequentially continuous at u. If it were not continuous at this point, there would exist (Definition 7.1 (a)) a semi-norm $\| \ \|_{F;\mu}$ of F, $\epsilon > 0$ and, for all $n \in \mathbb{N}$, $u_n \in X$ such that:

$$\|u_n - u\|_{E;n} \leq \frac{1}{n}, \quad \|T(u_n) - T(u)\|_{F;\mu} \geq \epsilon.$$

For all $k \in \mathbb{N}$ and $n \geq k$, it would follow that $\|u_n - u\|_{E;k} \leq \|u_n - u\|_{E;n} \leq 1/n$, from which $u_n \to u$. Since $T(u_n) \not\to T(u)$ this would contradict the sequential continuity of T, which proves that T is continuous at u.

Continuous $\Leftrightarrow$ *sequentially continuous.* According to Definition 7.1 (a) and (c) of these notions of continuity, this follows from the fact that it holds at any point u.

(b) *Compacting* $\Leftrightarrow$ *sequentially compacting.* Following Definition 8.14 of these notions, this results from the fact that F being here metrizable, relatively compact therein is equivalent to relatively sequentially compact (Theorem 4.6 (b)). □

Let us show that any linear mapping over a metrizable space that transforms bounded sets into bounded sets is continuous, which is the **Banach–Mazur–Orlicz–Mackey theorem**[1].

THEOREM 9.2.– *For every linear mapping from a metrizable semi-normed space into a separated semi-normed space:*

$$\textit{Continuous} \quad \Leftrightarrow \quad \textit{transforms bounded sets into bounded sets}. \ \blacksquare$$

Proof. Continuous linear $\Rightarrow$ *transforms bounded sets into bounded sets.* This is always true (Theorem 7.15).

Transforms bounded sets into bounded sets $\Rightarrow$ *continuous.* Let L be a linear mapping from a metrizable semi-normed space E into a separated semi-normed space F that transforms bounded sets into bounded sets. Let (Theorem 4.4) $(\| \ \|_{E;n})_{n\in\mathbb{N}}$ be an increasing sequence of semi-norms defining the topology of E.

If L were not continuous, the characterization of continuous linear mappings of Theorem 7.14 (b) would result in the existence of a semi-norm $\| \ \|_{F;\mu}$ of F and, for all $n \in \mathbb{N}$, of $u_n \in E$ such that:

$$\|L(u_n)\|_{F;\mu} \geq n \,\|u_n\|_{E;n}, \quad \|L(u_n)\|_{F;\mu} \neq 0.$$

1 **History of Banach–Mazur–Orlicz–Mackey theorem 9.2.** The result has been proved by Stefan BANACH in Banach spaces in 1922 [7, p. 151] and then in normed spaces in 1931 [9, p. 151]. It was extended to Fréchet spaces by Stanislaw MAZUR and Wladyslaw ORLICZ [71, p. 153]. John WEHAUSEN observed in 1938 that the result is still valid for mappings with values in any separated semi-normed space [115, p. 161]. George MACKEY proved the result in metrizable spaces in 1946 [69, Theorems 8 and 10 p. 527 and Corollary p. 528].

Then, $v_n = nu_n/\|L(u_n)\|_{F;\mu}$ would verify

$$\|v_n\|_{E;n} \leq 1, \quad \|L(v_n)\|_{F;\mu} = n. \tag{9.1}$$

Let $k \in \mathbb{N}$. For all $n \geq k$, it would follow that $\|v_n\|_{E;k} \leq \|v_n\|_{E;n} \leq 1$ since $\| \ \|_{E;n}$ increases with n, therefore

$$\sup_{n\in\mathbb{N}} \|v_n\|_{E;k} \leq \sup\{\|v_0\|_{E;k}, \ldots, \|v_{k-1}\|_{E;k}, 1\} < \infty.$$

This is so for all $k \in \mathbb{N}$, thus $\{v_n\}_{n\in\mathbb{N}}$ would be bounded but its image $\{L(v_n)\}_{n\in\mathbb{N}}$ would not according to (9.1). This would contradict our assumption, therefore L is continuous. □

Other proof when E is normed. Let E be a normed space, F a separated semi-normed space and L a linear mapping from E into F transforming bounded sets into bounded sets. Then, for any semi-norm of F,

$$\sup_{v\in E:\|v\|_E\leq 1} \|Lv\|_{F;\mu} = c < \infty.$$

Thus, for all $u \in E$, $u \neq 0$,

$$\frac{\|Lu\|_{F;\mu}}{\|u\|_E} = \left\| L\left(\frac{u}{\|u\|_E}\right)\right\|_{F;\mu} \leq c$$

which results in the continuity of L. □

The case of a non-metrizable space. The property stated in Theorem 9.2 holds for some non-metrizable spaces. In fact, it characterizes *bornological* spaces according to the Mackey–Bourbaki theorem [15, Theorem 3, p. 11].

It does not hold for all separated semi-normed space. For example, if F is a normed space of infinite dimension and $E = F$-weak, the identity of E in F transforms bounded sets into bounded sets (Banach–Mackey theorem 16.1) but it is not continuous (otherwise, according to Theorem 7.14 (a), there would exist n elements $f'_i \in F'$ and $c \geq 0$ such that $\|f\|_F \leq c\sup_{1\leq i\leq n} |\langle f'_i, f\rangle|$ for all $f \in F$ which, according to Theorem 17.17, would cause that E' be of dimension $\leq n$, and thus E and F also). □

Let us extend the Banach–Mazur–Orlicz–Mackey theorem to multilinear mappings.

THEOREM 9.3.– *For every multilinear mapping from a product of metrizable semi-normed spaces into a separated semi-normed space:*

Continuous $\Leftrightarrow$ *transforms bounded sets into bounded sets.*

$\Leftrightarrow$ *transforms bounded sets products into bounded sets.* ▌

Proof. *Continuous multilinear* $\Rightarrow$ *transforms bounded sets into bounded sets.* This is always true (Theorem 7.23).

Transforms bounded sets into bounded sets $\Rightarrow$ *continuous.* Let T be a multilinear mapping from a product $E_1 \times \cdots \times E_d$ of metrizable semi-normed spaces into a separated semi-normed space F that transforms bounded sets into bounded sets. For each $i \in [\![1, d]\!]$, let (Theorem 4.4) $(\| \ \|_{E_i;n})_{n\in\mathbb{N}}$ be an increasing sequence of semi-norms defining the topology of E_i.

If T were not continuous, the characterization of the continuous multilinear mappings of Theorem 7.20 (b) would result in the existence of a semi-norm $\| \ \|_{F;\mu}$ of F and, for all $n \in \mathbb{N}$, of $(u_{n,1}, \dots, u_{n,d}) \in E_1 \times \cdots E_d$ such that:

$$\|T(u_{n,1}, \dots, u_{n,d})\|_{F;\mu} \geq n \, \|u_{n,1}\|_{E_1;n} \cdots \|u_{n,d}\|_{E_d;n}$$

and $\|u_{n,i}\|_{E_i;n} \neq 0$ for all $i \in [\![1, d]\!]$. The $v_{n,i} = u_{n,i}/\|u_{n,i}\|_{E_i;n}$ would then verify

$$\|T(v_{n,1}, \dots, v_{n,d})\|_{F;\mu} = \frac{\|T(u_{n,1}, \dots, u_{n,d})\|_{F;\mu}}{\|u_{n,1}\|_{E_1;n} \cdots \|u_{n,d}\|_{E_d;n}} \geq n. \qquad (9.2)$$

Let $k \in \mathbb{N}$. For all $n \geq k$, it would follow that $\|v_{n,i}\|_{E_i;k} \leq \|v_{n,i}\|_{E_i;n} \leq 1$ since $\| \ \|_{E_i;n}$ increases with n, and thus

$$\sup_{n\in\mathbb{N}} \|v_{n,i}\|_{E_i;k} \leq \sup\{\|v_{1,i}\|_{E_i;k}, \dots, \|v_{k-1,i}\|_{E_i;k}, 1\} \stackrel{\text{def}}{=} c_{i,k} < \infty.$$

For all $n \in \mathbb{N}$ and for every semi-norm of $E_1 \times \cdots E_d$ (Definition 6.1), that is for all $(k_1, \dots, k_d) \in \mathbb{N}^d$, it would follow that

$$\|(v_{n,1}, \dots, v_{n,d})\|_{E_1\times\cdots E_d;k_1,\dots,k_d} = (\|v_{n,1}\|^2_{E_1;k_1} + \cdots + \|v_{n,d}\|^2_{E_d;k_d})^{1/2} \leq$$

$$\leq (c^2_{1,k_1} + \cdots + c^2_{d,k_d})^{1/2} < \infty.$$

The set $\{(v_{n,1}, \dots, v_{n,d})\}_{n\in\mathbb{N}}$ would thus be bounded in $E_1 \times \cdots \times E_d$, but its image $\{T(v_{n,1}, \dots, v_{n,d})\}_{n\in\mathbb{N}}$ would not according to (9.2). This would contradict our hypothesis, thus T is continuous.

Continuous multilinear ⇒ transforms bounded set products into bounded sets. Again, this is always true according to Theorem 7.23, because any product $B_1 \times \cdots \times B_d$ of bounded sets is bounded (according to Definitions 6.1 of the semi-norms of a product and 2.9 (a) of bounded sets).

Transforms bounded sets products into bounded sets ⇒ continuous. Any bounded set B of the space $E_1 \times \cdots \times E_d$ is included in the product of $B_i = \{u_i : (u_1, \dots, u_d) \in B\}$, and each B_i is bounded in E_i, because

$$\sup_{u_i\in B_i} \|u_i\|_{E_i;\nu_i} \leq \sup_{(u_1,\dots,u_d)\in B} (\|u_1\|^2_{E_1;\nu_1} + \cdots + \|u_d\|^2_{E_d;\nu_d})^{1/2} < \infty.$$

If T transforms every bounded set product into a bounded set, it then transforms bounded sets into bounded sets and therefore it is continuous as it has been proved above. □

Let us show that in a metrizable space, the composite under sequentially continuous linear mapping preserves uniform continuity.

THEOREM 9.4.– *Let F and G be separated semi-normed spaces,*

X a subset of a metrizable separated semi-normed E,

T a uniformly continuous mapping from X into F, and

L a sequentially continuous linear mapping from F into G.

Then:

$L \circ T$ is uniformly continuous. ▮

Proof. Let us endow (Theorem 4.4) E with an increasing sequence of semi-norms $(\| \, \|_{E;k})_{k\in\mathbb{N}}$. If $L \circ T$ were not uniformly continuous, there would exist (Definition 7.1 (b)) a semi-norm $\| \, \|_{G;\lambda}$ of G and $\epsilon > 0$ such that, for any $n \in \mathbb{N}$, there would exist u_n and v_n in X verifying

$$\|u_n - v_n\|_{E;n} \leq \frac{1}{n}, \quad \|(L \circ T)(u_n) - (L \circ T)(v_n)\|_{G;\lambda} \geq \epsilon. \tag{9.3}$$

For all $k \in \mathbb{N}$ and $n \geq k$, it would follow that $\|u_n - v_n\|_{E;k} \leq \|u_n - v_n\|_{E;n} \leq 1/n$. Since T is uniformly continuous, it would thereof result, for each semi-norm of F, that $\|T(u_n) - T(v_n)\|_{F;\mu} \to 0$ when $n \to \infty$. And thus that $T(u_n) - T(v_n) \to 0$. Since $L(T(u_n) - T(v_n)) \not\to 0$ (according to (9.3)), this would contradict the sequential continuity of L. Which proves that $L \circ T$ is uniformly continuous. □

9.2. Banach's fixed point theorem

A **fixed point** of a mapping T is a point u such that:

$$T(u) = u.$$

Let us show that any strict contraction of a Banach space has a fixed point, which is **Banach's fixed point theorem**[2].

THEOREM 9.5.– *Let*

X be an open subset of a Banach space E,

T a mapping from X into itself and $k \geq 0$ such that, for any u and v in X:

$$\|T(u) - T(v)\|_E \leq k\,\|u - v\|_E,$$

$$k < 1.$$

Then:

T has a fixed point. ▮

---

2 **History of Banach's theorem 9.5.** Stefan BANACH stated this result in 1922 [7]. The proof method, known as **successive approximations**, is often attributed to Charles Émile PICARD who used it in 1890 [87], although it was previously introduced by Giuseppe PEANO in 1888 [82].

Proof. Starting from a point $u_0 \in X$, let us define by induction, for all $n \in \mathbb{N}^*$,

$$u_{n+1} = T(u_n) \tag{9.4}$$

For $n \geq 1$,

$$\|u_{n+1} - u_n\|_E = \|T(u_n) - T(u_{n-1})\|_E \leq k\,\|u_n - u_{n-1}\|_E.$$

Hence

$$\|u_{n+1} - u_n\|_E \leq k^n \|u_1 - u_0\|_E.$$

By decomposing $u_{n+j} - u_j = \sum_{i=0}^{n-1} u_{j+i+1} - u_{j+i}$, we get, see (1.10), p. 7,

$$\|u_{n+j} - u_j\|_E \leq \sum_{i=0}^{n-1} k^{j+i} \|u_1 - u_0\|_E \leq \frac{k^j}{1-k}\|u_1 - u_0\|_E. \tag{9.5}$$

Therefore, $(u_n)_{n\in\mathbb{N}}$ is a Cauchy sequence (because $k^j \to 0$ when $j \to \infty$, see Theorem 1.4). It has a limit u since E is sequentially complete (according to Definition 4.10 (c) of a Banach space). At the limit, (9.4) gives $u = T(u)$. Therefore, u is a fixed point. □

Error estimation. By iterating (9.4), it yields $u_j = T^j(u_0)$. By passing to the limit when $n \to \infty$ in (9.5), we obtain the following bound of the error achieved by considering this point as an approximation of u:

$$\|u - u_j\|_E \leq \frac{k^j}{1-k}\|u_1 - u_0\|_E. \;\square$$

Generalization of Theorem 9.5. In Theorem 9.5, the hypothesis "T is a mapping from X into itself" can be replaced by:

$$\begin{cases} T \text{ is a mapping from } X \text{ into } E \text{ and there exist } a \in X \text{ and } m \geq 0 \text{ and such that:} \\ \quad \|T(a) - a\|_E < m(1-k), \quad \|a - w\|_E \geq m, \;\; \forall w \in E \setminus X. \;\square \end{cases}$$

Proof. The proof of Theorem 9.5 is reused starting with $u_0 = a$. We verify by induction on n that $u_n \in X$ observing that (9.5) gives $\|u_n - u_0\| \leq \|T(a) - a\|_E/(1-k) < m$. The limit u verifies this majoration and is thus also an element of X. □

9.3. Baire's theorem

Let us express **Baire's theorem**[3].

THEOREM 9.6.– *In a Fréchet space, any countable union of closed sets with empty interiors has an empty interior.* ∎

3 **History of Baire's theorem 9.6.** William OSGOOD proved that any countable union of closed sets with empty interiors of $\mathbb{R}$ has an empty interior in 1897 [80]. René BAIRE independently proved the same result for closed subsets of $\mathbb{R}^d$ in 1899 [6]. Felix HAUSDORFF extended it to the closed subsets of a Fréchet space in 1914 [53].

Proof. This follows from Theorem 9.7 hereafter by endowing E with an increasing sequence of semi-norms (Theorem 4.4), in which case the fact of having an empty interior results in containing no semi-ball (Theorem 3.17 (b)). □

Let us express Baire's theorem in terms of semi-balls (Definition 2.18).

THEOREM 9.7.– *Let E be a Fréchet space, $(\| \ \|_k)_{k\in\mathbb{N}}$ an increasing sequence of semi-norms generating its topology and, for all $n \in \mathbb{N}$, let V_n be a closed subset of E containing no semi-ball, that is no set $B_k(z,r) = \{u \in E : \|u - z\|_k \leq r\}$ where $z \in E$, $r > 0$ and $k \in \mathbb{N}$. Then:*

$$\bigcup_{n\in\mathbb{N}} V_n \text{ contains no semi-ball.}$$ ∎

Proof. By denoting $U_n = E \setminus V_n$, the point is to show that

$$\bigcap_{n\in\mathbb{N}} U_n \text{ does meet every semi-ball.} \tag{9.6}$$

Let B be an arbitrary semi-ball. Starting from $\mathcal{B}_0 = B$ and using induction, a sequence of nested semi-balls $\mathcal{B}_n = B_{m_n}(z_n, r_n)$ will be built. By assumption, V_n does not contain $\mathring{\mathcal{B}}_n$ (otherwise it would contain $B_{m_n}(z_n, r_n/2)$), thus there is $z_{n+1} \in U_n \cap \mathring{\mathcal{B}}_n$. It verifies

$$\|z_{n+1} - z_n\|_{m_n} \leq r_n. \tag{9.7}$$

Since U_n and $\mathring{\mathcal{B}}_n$ are open, $U_n \cap \mathring{\mathcal{B}}_n$ also is (Theorem 2.11 (b)), and therefore it contains (Theorem 3.17 (a)) a semi-ball $\mathcal{B}_{n+1}$ centered in z_{n+1}. Then:

$$\mathcal{B}_{n+1} \subset U_n \cap \mathcal{B}_n. \tag{9.8}$$

It is always possible to choose $m_{n+1} \geq n+1$ and $r_{n+1} \leq r_n/2$ which results in $r_n \leq 2^{-n} r$. Let $k \in \mathbb{N}$ and $n' \geq n \geq k$. Since $\| \ \|_k$ increases with k, the inequality (9.7) leads to $\|z_{n+1} - z_n\|_k \leq 2^{-n} r$, hence

$$\|z_{n'} - z_n\|_k \leq (2^{-n} + \cdots + 2^{-n'+1}) r \leq 2^{1-n} r$$

(according to the majoration of the sum of a geometric series, see (1.10), p. 7).

The sequence $(z_n)_{n\in\mathbb{N}}$ is thus a Cauchy sequence. Since E is sequentially complete (following Definition 4.10 (b) of a Fréchet space), it has a limit z. By iterating (9.8), $z_{n'} \in \mathcal{B}_{n+1}$ is obtained for all $n' \geq n+1$ from which, at the limit when $n' \to \infty$, $z \in \mathcal{B}_{n+1}$ and thus $z \in U_n \cap B$. Since this holds for every n,

$$z \in \Big(\bigcap_{n\in\mathbb{N}} U_n\Big) \cap B \tag{9.9}$$

which proves (9.6). □

9.4. Open mapping theorem

Let us state the **open mapping theorem**.

THEOREM 9.8.– *Any continuous and surjective linear mapping from a Fréchet space into a Fréchet space transforms open sets into open sets.* ∎

Terminology. Theorem 9.8 is known as *open mapping theorem* because a mapping that transforms open sets into open sets is said to be **open**. □

The proof will make use of the following result.

LEMMA 9.9.– *Let E be a semi-normed space whose semi-norms are indexed by $\mathcal{N}_E$, F a Fréchet space endowed with an increasing sequence of semi-norms and T a linear and surjective mapping from E into F.*

Then, for every $\nu \in \mathcal{N}_E$, there exist a semi-norm $\| \ \|_{F;m}$ of F and $\eta > 0$ such that, with Definition 2.18 (a) of semi-balls:

$$B_{F;m}(0_F, \eta) \subset \overline{T(B_{E;\nu}(0_E, 1))}.$$ ∎

Observation. In [BOURBAKI, 17, Chap. I, § 3, No. 3, Lemma 1, p. 23], this result is stated with the additional assumption that "T is continuous". The latter is not necessary, and moreover it is not used in the proof of [17]. □

Proof of Lemma 9.9. The index ν being fixed, for all $n \in \mathbb{N}$, let us denote

$$B_n = B_{E;\nu}(0_E, n) = \{v \in E : \|v\|_{E;\nu} \leq n\}.$$

Then, $\bigcup_{n\in\mathbb{N}} B_n = E$. Since T is surjective, it thereof results that $\bigcup_{n\in\mathbb{N}} T(B_n) = F$ and, *a fortiori*,

$$\bigcup_{n\in\mathbb{N}} \overline{T(B_n)} = F.$$

Baire's Theorem 9.7 shows that there exists $i \in \mathbb{N}$ such that $\overline{T(B_i)}$ contains a semi-ball. That is to say, there exist $w \in F$, $m \in \mathbb{N}$ and $r > 0$ such that:

$$B_{F;m}(w, r) \subset \overline{T(B_i)}.$$

In particular, $w \in \overline{T(B_i)}$. Since T is linear, it yields $T(B_i) = T(-B_i) = -T(B_i)$, and thus $-w \in \overline{T(B_i)}$. Therefore, it results that

$$B_{F;m}(0_F, r) = B_{F;m}(w, r) - w \subset \overline{T(B_i)} + \overline{T(B_i)}.$$

According to Theorem 3.18, it follows that $\overline{T(B_i)} + \overline{T(B_i)} \subset \overline{T(B_i) + T(B_i)}$. On the other hand, $T(B_i) + T(B_i) = T(B_i + B_i) \subset T(B_{2i})$. Consequently,

$$B_{F;m}(0_F, r) \subset \overline{T(B_{E;\nu}(0_E, 2i))}.$$

This leads to the stated result, with $\eta = r/2i$. □

Proof of Theorem 9.8. Let T be a continuous and surjective linear mapping from a Fréchet space E into a Fréchet space F, and U an open subset of E. By means of Theorem 4.4, E and F are endowed with increasing sequences of semi-norms denoted respectively by $(\|\ \|_{E;m})_{m\in\mathbb{N}}$ and $(\|\ \|_{F;n})_{n\in\mathbb{N}}$.

First step. Let $n \in \mathbb{N}$. According to Lemma 9.9, there exist $m_n \in \mathbb{N}$ and $\eta_n > 0$ such that $B_{F;m_n}(0_F, \eta_n) \subset \overline{T(B_{E;n}(0_E, 1))}$. Observe that we can choose $\eta_n \leq 1$. We can also choose $m_n \geq n$ since $\|\ \|_{F;m}$ increases with m.

Since T is linear, for all $t > 0$, it follows that

$$B_{F;m_n}(0_F, t\eta_n) \subset \overline{T(B_{E;n}(0_E, t))}.$$

The characterization of the closure of Theorem 3.17 (c) shows that, for all $w \in F$ such that $\|w\|_{F;m_n} \leq t\eta_n$ and for all $\ell \in \mathbb{N}$ and $\epsilon > 0$, there exists $z \in E$ such that:

$$\|z\|_{E;n} \leq t \quad \text{and} \quad \|w - T(z)\|_{F;\ell} \leq \epsilon. \tag{9.10}$$

Second step. We intend now to show that

$$B_{F;m_n}(0_F, t\eta_n) \subset T(B_{E;n}(0_E, 2t)). \tag{9.11}$$

Let thus $y \in F$ be such that $\|y\|_{F;m_n} \leq t\eta_n$. According to (9.10), there exists $z_1 \in E$ such that:

$$\|z_1\|_{E;n} \leq t \quad \text{and} \quad \|y - T(z_1)\|_{F;m_{n+1}} \leq \frac{t}{2}\eta_{n+1}.$$

Now, by using (9.10) with $w = y - T(z_1)$, we obtain $z_2 \in E$ such that:

$$\|z_2\|_{E;n+1} \leq \frac{t}{2} \quad \text{and} \quad \|y - T(z_1) - T(z_2)\|_{F;m_{n+2}} \leq \frac{t}{4}\eta_{n+2}.$$

A sequence of $z_i \in E$ is thus built by induction, such that:

$$\|z_i\|_{E;n+i-1} \leq \frac{t}{2^{i-1}} \quad \text{and} \quad \|y - T(z_1 + \cdots + z_i)\|_{F;m_{n+i}} \leq \frac{t}{2^i}\eta_{n+i}.$$

Let us denote $x_i = z_1 + \cdots + z_i$. Since the sequences of semi-norms of E and F are increasing and since $m_n \geq n$ and $\eta_n \leq 1$, it results therefrom that, if $k \leq n + i - 1$, then:

$$\|z_i\|_{E;k} \leq \frac{t}{2^{i-1}} \quad \text{and} \quad \|y - T(x_i)\|_{F;k} \leq \frac{t}{2^i}. \tag{9.12}$$

For all $k \in \mathbb{N}$, as soon as $j \geq i \geq k + 1$, it follows that

$$\|x_i - x_j\|_{E;k} = \|z_{i+1} + \cdots + z_j\|_{E;k} \leq t\Big(\frac{1}{2^i} + \cdots + \frac{1}{2^{j-1}}\Big) \leq \frac{t}{2^{i-1}}$$

(according to the majoration of the sum of a geometric series, see (1.10), p. 7).

The sequence $(x_i)_{i\in\mathbb{N}}$ is thus a Cauchy sequence. It has a limit x because E is sequentially complete (according to Definition 4.10 (b) of a Fréchet space). Then, $T(x_i) \to T(x)$ since T, being continuous, is sequentially continuous (Theorem 7.2 (a)). On the other hand, $T(x_i) \to y$ according to (9.12), therefore

$$y = T(x).$$

Finally, since (9.13) holds for $k = n$ and $i \geq 1$, it follows that (again by (1.10))

$$\|x_i\|_{E;n} = \|z_1 + \cdots + z_i\|_{E;n} \leq t\Big(1 + \frac{1}{2} + \cdots + \frac{1}{2^{i-1}}\Big) \leq 2t.$$

At the limit, $\|x\|_{E;n} \leq 2t$ is obtained (since $\| \ \|_{E;n}$ is sequentially continuous, see Theorem 7.10). Subsequently, $y \in T(B_{E;n}(0_E, 2t))$ which establishes (9.11).

Third step. Let U be an open subset of E and $u \in U$. According to the characterization of an open set of Theorem 3.17 (a), there exist $\epsilon > 0$ and $n \in \mathbb{N}$ such that:

$$B_{E;n}(u, \epsilon) \subset U.$$

In other words, $B_{E;n}(0_E, \epsilon) + u \subset U$. Since T is linear, it results thereof that

$$T(B_{E;n}(0_E, \epsilon)) + T(u) \subset T(U).$$

With (9.11) for $t = \epsilon/2$, this leads to $B_{F;m_n}(0_F, \epsilon\eta_n/2) + T(u) \subset T(U)$ that is

$$B_{F;m_n}\Big(T(u), \frac{\epsilon\eta_n}{2}\Big) \subset T(U).$$

Since $T(u)$ is an arbitrary point of $T(U)$, this implies (Theorem 3.17 (a)) that $T(U)$ is open in F. □

9.5. Banach–Schauder's continuity theorem

Let us show that continuous linear bijections on Fréchet spaces are isomorphism, which is the **Banach–Schauder theorem**[4].

THEOREM 9.10.– *If a linear bijection from a Fréchet space into a Fréchet space is continuous, its inverse mapping also is continuous.* ▮

Proof. Let T be a continuous linear bijection from a Fréchet space E onto a Fréchet space F, $\| \ \|_{E;\nu}$ a semi-norm of E, $y \in F$ and $\epsilon > 0$. The set

$$U = \{v \in E : \|v - T^{-1}(y)\|_{E;\nu} < \epsilon\}$$

4 **History of Banach–Schauder theorem 9.10.** Julius SCHAUDER proved this result for Banach spaces in 1930 [97]. Stefan BANACH generalized it to Fréchet spaces in 1932 [10, Theorem 5, p. 41].

is open and thus the open mapping theorem 9.8 shows that $T(U)$ is open in F. As a result, there exist (Definition 2.9 (b)) a finite subset M of the set indexing the semi-norms of F and $\eta > 0$ such that:

$$\sup_{\mu \in M} \|w - y\|_{F;\mu} \leq \eta$$

yields $w \in T(U)$ and thus $T^{-1}(w) \in U$ that is $\|T^{-1}(w) - T^{-1}(y)\|_{E;\nu} < \epsilon$. Consequently, T^{-1} is continuous from F into E. □

Other proof of Theorem 9.10. When T is one-to-one, the property (9.11) of the proof of Theorem 9.8 yields $T^{-1}(B_{F;m_n}(0_F, t\eta_n)) \subset B_{E;n}(0_E, 2t)$. Therefore, the inequality $\|w\|_{F;m_n} \leq t\eta_n$ yields $\|T^{-1}(w)\|_{E;n} \leq 2t$. Therefrom, it results that

$$\|T^{-1}(w)\|_{E;n} \leq \frac{2}{\eta_n} \|w\|_{F;m_n}.$$

Hence (Theorem 7.14 (a)) T^{-1} is continuous. □

Let us show that if two families of semi-norms on the same space make them Fréchet spaces and if one dominates the other, they are equivalent.

THEOREM 9.11.– *Given two Fréchet spaces E_1 and E_2:*

$$[\, E_1 = E_2 \text{ and } E_1 \underset{\rightarrow}{\subseteq} E_2 \,] \quad \Rightarrow \quad E_1 \underset{\leftrightarrow}{=} E_2. \ \blacksquare$$

Proof. If $E_1 = E_2$, the identity from E_1 into E_2 is bijective and if $E_1 \underset{\rightarrow}{\subseteq} E_2$, it is continuous; in addition, if E_1 and E_2 are Fréchet spaces, its inverse mapping is also continuous according to the Banach–Schauder theorem 9.10, and thus $E_1 \underset{\leftrightarrow}{=} E_2$. □

The case of normed spaces. In particular, Theorem 9.11 gives the following property:

$$\left\{ \begin{array}{l} \textit{If each of two norms make the same vector space sequentially complete} \\ \textit{and if one dominates the other, these two norms are equivalent.} \ \square \end{array} \right.$$

9.6. Closed graph theorem

The **graph** of a mapping T from a set E into a set F is, by definition, the subset $\{(u, T(u)) : u \in E\}$ of $E \times F$.

Let us state the **closed graph theorem**.

THEOREM 9.12.– *Any linear mapping from a Fréchet space into a Fréchet space whose graph is closed is continuous.* ▮

Proof. Let T be a linear mapping from a Fréchet space E into another F. The space $E \times F$ is also a Fréchet space (Theorem 6.7 (b)) and the graph G of T is thereof a vector subspace because

$$(u, T(u)) + (v, T(v)) = (u + v, T(u) + T(v)) = (u + v, T(u + v)),$$
$$t(u, T(u)) = (tu, tT(u)) = (u, T(tu)).$$

Let us endow G with the restrictions of the semi-norms of $E \times F$ that makes it a topological subspace (Definition 3.9). If G is closed in $E \times F$, it is likewise a Fréchet space (Theorem 4.12 (b)).

The projections $(u, v) \mapsto u$ and $(u, v) \mapsto v$ being continuous linear from $E \times F$ respectively into E and F (Theorem 7.28), their restrictions P_E and P_F to G are continuous linear from G into E and F respectively. Moreover, P_E is one-to-one, therefore its inverse mapping P_E^{-1} is continuous from E into G according to the Banach–Schauder theorem 9.10. The composite mapping $P_F \circ P_E^{-1}$ is subsequently (Theorem 7.9 (a)) continuous from E into F.

This proves the theorem since $P_F \circ P_E^{-1} = T$ (because $P_E^{-1}(u) = (u, T(u))$). □

Chapter 10

Extension of Mappings, Equicontinuity

10.1. Extension of equalities by continuity

Let us give properties of the extension of an equality by continuity.

THEOREM 10.1.– *Let E and F be separated semi-normed spaces, $U \subset X \subset E$ and T_1 and T_2 mappings from X into F. Then:*

(a) $\left.\begin{cases} T_1 \textit{ and } T_2 \textit{ are continuous from } X \textit{ into } F, \textit{ and} \\ T_1 = T_2 \textit{ in } U \end{cases}\right\} \quad \Rightarrow \quad T_1 = T_2 \textit{ in } \overline{U} \cap X.$

(b) $\left.\begin{cases} T_1 \textit{ and } T_2 \textit{ are sequentially continuous from } X \textit{ into } F, \\ T_1 = T_2 \textit{ in } U, \textit{ and} \\ X \textit{ is sequentially closed} \end{cases}\right\} \quad \Rightarrow$

$$\Rightarrow \quad T_1 = T_2 \textit{ in } \overline{U}^{\mathrm{seq}}.$$

(c) *If $F = \mathbb{R}$, it is possible to replace $T_1 = T_2$ by $T_1 \leq T_2$ in* (a) *and* (b). ❚

Proof. Let $\{\| \; \|_{E;\nu} : \nu \in \mathcal{N}_E\}$ and $\{\| \; \|_{F;\mu} : \mu \in \mathcal{N}_F\}$ be the families of semi-norms of E and F, and $T = T_2 - T_1$.

(a) Suppose that the assumptions be verified and let $u \in X$ be such that $T(u) \neq 0_F$. The point is to verify that $u \notin \overline{U}$. Let $\mu \in \mathcal{N}_F$ be such that $\|T(u)\|_{F;\mu} = a > 0$. Since T is continuous (Definition 7.1 (a)), there exist a finite set $N \subset \mathcal{N}_E$ and $\eta > 0$ such that:

$$\sup_{\nu \in N} \|v - u\|_{E;\nu} \leq \eta \quad \Rightarrow \quad \|T(v) - T(u)\|_{F;\mu} \leq \frac{a}{2} \quad \Rightarrow \quad v \notin U$$

(the second implication results from the fact that, then, $\|T(v)\|_{F;\mu} \geq a/2$ from which $T(v) \neq 0_F$). This yields $u \notin \overline{U}$ according to the characterization of the closure of Theorem 2.17 (b).

(b) Let $V = \{v \in X : T(v) = 0_F\}$. Under the assumptions made, V is sequentially closed because, if $(u_n)_{n\in\mathbb{N}} \subset V$ and $u_n \to u$ in E, then $u \in X$ (since X is sequentially closed) and $T(u_n) \to T(u)$, hence $T(u) = 0_F$ and therefore $u \in V$.

Since (Definition 2.15 (c)) $\overline{U}^{\mathrm{seq}}$ is the smallest sequentially closed set containing U, it is included in V. Thus, $T = 0$ in $\overline{U}^{\mathrm{seq}}$.

(c) We conclude by replacing $T_1 = T_2$ by $T_1 \leq T_2$ in the proofs of (a) and (b). □

Let us complete these extension properties in terms of density.

THEOREM 10.2.– *Let E and F be separated semi-normed spaces, $U \subset X \subset E$ and T_1 and T_2 mappings from X into F. Then:*

$$\text{(a)}\quad \left.\begin{cases} T_1 \text{ and } T_2 \text{ are continuous from } X \text{ into } F, \\ T_1 = T_2 \text{ in } U, \text{ and} \\ U \text{ is dense in } X \end{cases}\right\} \quad\Rightarrow\quad T_1 = T_2 \text{ in } X.$$

$$\text{(b)}\quad \left.\begin{cases} T_1 \text{ are } T_2 \text{ are sequentially continuous from } X \text{ into } F, \\ T_1 = T_2 \text{ in } U, \text{ and} \\ U \text{ is sequentially dense in X} \end{cases}\right\} \quad\Rightarrow\quad T_1 = T_2 \text{ in } X.$$

■

Proof. (a) This results from Theorem 10.1 (a) because $X \subset \overline{U}$ if U is dense in X (Definition 2.19 (a)).

(b) If U is sequentially dense in X (Definition 2.19 (b)), any point $u \in X$ is the limit of a sequence of points $u_n \in U$; if in addition the T_i are sequentially continuous, $T_i(u_n) \to T_i(u)$; if finally $T_1(u_n) = T_2(u_n)$, then $T_1(u) = T_2(u)$. □

Non-equivalence of Theorems 10.1 (b) and 10.2 (b). The property (b) of Theorem 10.2 is weaker than that of Theorem 10.1 because U is not always sequentially dense in $\overline{U}^{\mathrm{seq}}$, see (2.5), p. 24.

On the other hand, the properties (a) of Theorems 10.1 and 10.2 are equivalent because U is always dense in $\overline{U}$. □

10.2. Continuous extension of mappings

Let us first explicitly build the **continuous extension** of a continuous Neumann space-valued mapping to the set $U^{\mathrm{lim\ seq}}$ of the limits of its domain U.

THEOREM 10.3.– *Let E be a separated semi-normed space, $X \subset E$ and*

F a Neumann space,

T a uniformly continuous mapping from X into F.

Then:

(a) *For every* $u \in X^{\text{lim seq}}$, *there exists a unique point* $\overline{T}(u)$ *of* F *such that:*

$$\overline{T}(u) = \lim_{n \to \infty} T(u_n)$$

where $(u_n)_{n \in \mathbb{N}}$ *is any sequence of* X *such that* $u_n \to u$ *(in* E*).*

(b) $\overline{T}$ *is uniformly continuous from* $X^{\text{lim seq}}$ *into* F.

(c) $\left\{\begin{array}{l} \overline{T} \textit{ has no other uniformly continuous, continuous, or} \\ \textit{sequentially continuous extension from } X^{\text{lim seq}} \textit{ into } F. \end{array}\right.$ ∎

Proof. (a) *Limit existence.* Let $u \in X^{\text{lim seq}}$. According to its Definition 2.15 (d), there exists a sequence $(u_n)_{n \in \mathbb{N}}$ of X such that $u_n \to u$. It is (Theorem 2.7 (a)) a Cauchy sequence in E. Since T is uniformly continuous, $(T(u_n))_{n \in \mathbb{N}}$ is then a Cauchy sequence (Theorem 7.4 (a)) in F, which is sequentially complete. It has therefore a limit denoted by $\overline{T}(u)$.

Limit unicity. Let $(v_n)_{n \in \mathbb{N}}$ be another sequence of X converging to u and let $\widetilde{T}(u)$ be the limit of $T(v_n)$. Since $v_n - u_n \to 0$, the uniform continuity of T results in $T(v_n) - T(u_n) \to 0$ (Theorem 7.4 (b)). From which, at the limit, $\widetilde{T}(u) = \overline{T}(u)$.

(b) *Uniform continuity of* $\overline{T}$. Let $\{\| \; \|_{E;\nu} : \nu \in \mathcal{N}_E\}$ and $\{\| \; \|_{F;\mu} : \mu \in \mathcal{N}_F\}$ be the families of semi-norms of E and F. The mapping T being uniformly continuous (Definition 7.1 (b)) on X, for all $\epsilon > 0$ and $\mu \in \mathcal{N}_F$, there exist a finite set $N \subset \mathcal{N}_E$ and $\eta > 0$ such that:

$$u \in X,\ v \in X,\ \sup_{\nu \in N} \|v - u\|_{E;\nu} \leq \eta \quad \Rightarrow \quad \|T(v) - T(u)\|_{F;\mu} \leq \epsilon.$$

Let $u \in X^{\text{lim seq}}$ and $v \in X^{\text{lim seq}}$ be such that $\sup_{\nu \in N} \|v - u\|_{E;\nu} \leq \eta/2$. Let us choose two sequences $(u_n)_{n \in \mathbb{N}}$ and $(v_n)_{n \in \mathbb{N}}$ of X such that $u_n \to u$ and $v_n \to v$. If n is large enough, $\sup_{\nu \in N} \|v_n - u_n\|_{E;\nu} \leq \eta$, and thus $\|T(v_n) - T(u_n)\|_{F;\mu} \leq \epsilon$. At the limit, $\|\overline{T}(v) - \overline{T}(u)\|_{F;\mu} \leq \epsilon$, which proves that $\overline{T}$ is uniformly continuous.

(c) *Extension property.* The mapping $\overline{T}$ indeed extends T; as a matter of fact, if $u \in X$, then $\overline{T}(u) = T(u)$ because (a) is verified with $u_n = u$.

Extension unicity. Let $\widetilde{T}$ be another sequentially continuous extension of T. Given $u \in X^{\text{lim seq}}$, let $(u_n)_{n \in \mathbb{N}}$ be a sequence of X such that $u_n \to u$. Then $\widetilde{T}(u_n) = T(u_n)$ and therefore, at the limit, $\widetilde{T}(u) = \overline{T}(u)$.

This is *a fortiori* true if $\widetilde{T}$ is continuous or uniformly continuous since it is then sequentially continuous (Theorem 7.2 (a)). □

Let us now show that such a mapping has, in fact, a **continuous extension** to the whole sequential closure $\overline{X}^{\text{seq}}$ of its domain of definition X.

THEOREM 10.4.– *Let E be a separated semi-normed space, $X \subset E$ and*

$$F \text{ a Neumann space,}$$

$$T \text{ a uniformly continuous mapping from } X \text{ into } F.$$

Then:

(a) T *has a uniformly continuous extension from* $\overline{X}^{\text{seq}}$ *into* F.

(b) $\begin{cases} T \text{ has no other uniformly continuous, continuous, or} \\ \text{sequentially continuous extension from } \overline{X}^{\text{seq}} \text{ into } F. \end{cases}$ ∎

Comparison to Theorem 10.3. The extension domain $\overline{X}^{\text{seq}}$ of Theorem 10.4 is larger (Theorem 2.16) than $\overline{X}^{\text{lim seq}}$, that of Theorem 10.3. It is strictly larger when, as in (2.5), p. 24, X is not sequentially dense in $\overline{X}^{\text{seq}}$ (which needs that E is non-metrizable, see Theorem 4.6 (f)).

On the other hand, the explicit expression of Theorem 10.3 (a) is no longer valid under the conditions of Theorem 10.4 since a point $u \in \overline{X}^{\text{seq}}$ is no longer necessarily the limit of a sequence of X. □

Proof of Theorem 10.4. (b) Two potential sequentially continuous extensions to $\overline{X}^{\text{seq}}$ are equal according to Theorem 10.1 (b). This is also true for continuous or uniformly continuous extensions since they are sequentially continuous (Theorem 7.2 (a)).

(a) *Extension to a set $\mathcal{A}$.* Let A be a set such that:

$$X \subset A \subset \overline{X} \text{ and } T \text{ has a continuous extension } \overline{T}^A \text{ from } A \text{ into } F.$$

Let $\mathcal{E}$ be the set of these sets A and let $\mathcal{A}$ be their union. If B is another set of $\mathcal{E}$, it follows that

$$\overline{T}^A = \overline{T}^B \text{ in } A \cap B.$$

Indeed, being equal in X, these mappings are so in $\overline{X} \cap A \cap B$ (Theorem 10.1 (a) with $X = A \cap B$), which is equal to $A \cap B$.

Therefore, there is a unique mapping $\overline{T}$ defined on $\mathcal{A}$ such that:

$$\overline{T} = \overline{T}^A \text{ in each } A \in \mathcal{E}.$$

Uniform continuity of $\overline{T}$. Let $\{\| \ \|_{E;\nu} : \nu \in \mathcal{N}_E\}$ and $\{\| \ \|_{F;\mu} : \mu \in \mathcal{N}_F\}$ be the families of semi-norms of E and F. The mapping T being uniformly continuous (Definition 7.1 (b)), for all $\mu \in \mathcal{N}_F$ and $\epsilon > 0$, there exist a finite set $N \subset \mathcal{N}_E$ and $\eta > 0$ such that:

$$u \in X,\ v \in X,\ \sup_{\nu \in N} \|v - u\|_{E;\nu} \leq \eta \quad \Rightarrow \quad \|T(v) - T(u)\|_{F;\mu} \leq \epsilon. \tag{10.1}$$

Let us show that, for all $\eta' < \eta$:

$$u' \in \mathcal{A},\ v' \in \mathcal{A},\ \sup_{\nu \in N} \|v' - u'\|_{E;\nu} \leq \eta' \quad \Rightarrow \quad \|\overline{T}(v') - \overline{T}(u')\|_{F;\mu} \leq \epsilon. \quad (10.2)$$

Otherwise, there would exist $u' \in A$ where $A \in \mathcal{E}$, $v' \in B$ where $B \in \mathcal{E}$ and $a > 0$ such that:

$$\sup_{\nu \in N} \|v' - u'\|_{E;\nu} \leq \eta', \quad \|\overline{T}^B(v') - \overline{T}^A(u')\|_{F;\mu} = \epsilon + a.$$

Since $\overline{T}^A$ is continuous at u', there would exist a finite set $N' \subset \mathcal{N}_E$ and $b' > 0$ such that:

$$u'' \in A,\ \sup_{\nu \in N'} \|u'' - u'\|_{E;\nu} \leq b' \quad \Rightarrow \quad \|\overline{T}^A(u'') - \overline{T}^A(u')\|_{F;\mu} \leq \frac{a}{3}. \quad (10.3)$$

Since $u' \in \overline{X}$, there would exist (according to the characterization of the closure of Theorem 2.17 (b)) $u \in X$ such that, denoting $d = (\eta - \eta')/2$,

$$\sup_{\nu \in N \cup N'} \|u - u'\|_{E;\nu} \leq \min\{b', d\}.$$

In addition, it would verify, according to (10.3) with $u'' = u$,

$$\|T(u) - \overline{T}^A(u')\|_{F;\mu} = \|\overline{T}^A(u) - \overline{T}^A(u')\|_{F;\mu} \leq \frac{a}{3}.$$

The continuity of $\overline{T}^B$ in v' would likewise result in the existence of $v \in X$ such that:

$$\sup_{\nu \in N} \|v - v'\|_{E;\nu} \leq d \quad \text{and} \quad \|T(v) - \overline{T}^B(v')\|_{F;\mu} \leq \frac{a}{3}.$$

We would then have $\sup_{\nu \in N} \|v - u\|_{E;\nu} \leq \eta' + 2d = \eta$ and

$$\|T(v) - T(u)\|_{F;\mu} \geq \|\overline{T}^B(v') - \overline{T}^A(u')\|_{F;\mu} - \\ - \|T(u) - \overline{T}^A(u')\|_{F;\mu} - \|T(v) - \overline{T}^B(v')\|_{F;\mu} \geq \epsilon + \frac{a}{3}.$$

This would contradict (10.1), which proves (10.2) and therefore the uniform continuity of $\overline{T}$ in $\mathcal{A}$.

Inclusion $\overline{X}^{\text{seq}} \subset \mathcal{A}$. Theorem 10.3 provides a uniformly continuous extension of $\overline{T}$, and thus of T, in the set $\mathcal{A}^{\text{lim seq}}$ of the limits of the sequences of $\mathcal{A}$. Then, $\mathcal{A}^{\text{lim seq}} \in \mathcal{E}$. As a result, $\mathcal{A}^{\text{lim seq}} \subset \mathcal{A}$ from which (Definition 2.9 (b)) $\mathcal{A}$ is sequentially closed. Since $\overline{X}^{\text{seq}}$ is (Definition 2.15 (c)) the smallest sequentially closed set containing X, it is included in $\mathcal{A}$.

Uniqueness of $\overline{T}$. Being continuous (Theorem 7.2 (a)), any uniformly continuous extension to $\overline{X}^{\text{seq}}$ is equal to $\overline{T}$ according to (b). □

Continuous extension to $\overline{X}$. If F is a Banach or a Fréchet space, both Theorems 10.3 and 10.4 provide a continuous extension to $\overline{X}$ since then the latter coincides with $X^{\text{lim seq}}$ and $\overline{X}^{\text{seq}}$ (Theorem 4.6 (f)).

More generally, any uniformly continuous mapping in X has a continuous extension to $\overline{X}$ if the space of values F is complete; this is a stronger notion than sequentially complete, which we shall not utilize for the reasons presented in the commentary "*Complete spaces*", p. 63. □

10.3. Equicontinuous families of mappings

Let us now define equicontinuous families of mappings[1], namely that verify Definition 7.1 of continuity with parameters N and η independent of the mapping.

DEFINITION 10.5.– *Let E and F be separated semi-normed spaces, $X \subset E$ and $\mathcal{T}$ a family of mappings from X into F.*

(a) *$\mathcal{T}$ is* ***equicontinuous*** *if, for all $u \in X$, $\mu \in \mathcal{N}_F$ and $\epsilon > 0$, there exist a finite set $N \subset \mathcal{N}_E$ and $\eta > 0$ such that: for all $T \in \mathcal{T}$,*

$$v \in X, \ \sup_{\nu \in N} \|v - u\|_{E;\nu} \le \eta \quad \Rightarrow \quad \|T(v) - T(u)\|_{F;\mu} \le \epsilon.$$

(b) *$\mathcal{T}$ is* ***uniformly equicontinuous*** *if, for all $\mu \in \mathcal{N}_F$ and $\epsilon > 0$, there exist a finite set $N \subset \mathcal{N}_E$ and $\eta > 0$ such that: for all $T \in \mathcal{T}$,*

$$u \in X, \ v \in X, \ \sup_{\nu \in N} \|v - u\|_{E;\nu} \le \eta \quad \Rightarrow \quad \|T(v) - T(u)\|_{F;\mu} \le \epsilon. \ \blacksquare$$

Motivation for introducing the notion of equicontinuity. The equicontinuity of a family of mappings is crucial in order to extract thereof a converging sequence with Ascoli's theorem 11.12; the latter will be extensively used, especially in Volume II. □

Let us show that equicontinuous families are preserved by topological equalities.

THEOREM 10.6.– *For any family of mappings from a subset of a separated semi-normed space E into a separated semi-normed space F:*

$$\begin{cases} \textit{Equicontinuity and uniform equicontinuity are preserved when the family} \\ \textit{of semi-norms } E \textit{ and/or that of } F \textit{ is replaced by an equivalent family.} \end{cases} \ \blacksquare$$

1 **History of the notion of equicontinuity.** Giulio ASCOLI defined in 1883 [5, p. 545] the *equicontinuity* of a family of mappings in $\mathcal{C}(I)$, I being a real interval. Cesare ARZELÀ used this concept at the same time [4].

Proof. This is a consequence of Theorems 10.7 and 10.8, below. □

Let us give the effect of changing the space of values on an equicontinuous family of mappings.

THEOREM 10.7.– *Let E, F_1 and F_2 be separated semi-normed spaces such that:*

$$F_1 \underset{\rightarrow}{\subsetneq} F_2$$

and let $X \subset E$. Then, for any family of mappings from X into F_1:

(a) *Equicontinuous from X into F_1 $\Rightarrow$ equicontinuous from X into F_2.*

(b) *Uniformly equicontinuous from X into F_1*
$\Rightarrow$ *uniformly equicontinuous from U into F_2.* ∎

Proof. These properties result from Definition 10.5 of equicontinuity and 3.6 (c) of topological inclusion. □

Let us now give the effect of changing the space of definition.

THEOREM 10.8.– *Let E_1, E_2 and F be separated semi-normed spaces such that:*

$$E_1 \underset{\rightarrow}{\subsetneq} E_2.$$

Let $X \subset E_1$ and denote by X_{E_1} the subset X of the semi-normed space E_1 and X_{E_2} that of E_2. Then, every family of mappings from X into F verifies:

(a) *Equicontinuous from X_{E_2} into F $\Rightarrow$ equicontinuous from X_{E_1} into F.*

(b) *Uniformly equicontinuous from X_{E_2} into F*
$\Rightarrow$ *uniformly equicontinuous from X_{E_1} into F.* ∎

Different formulation. In terms of topology, X_{E_1} is the set X endowed with the topology of E_1. □

Proof. These properties result from Definitions 10.5 of equicontinuity and Definition 3.6 (c) of topological inclusion. □

Let us characterize equicontinuous families of linear mappings.

THEOREM 10.9.– *Let E and F be separated semi-normed spaces, $\{\| \ \|_{E;\nu} : \nu \in \mathcal{N}_E\}$ and $\{\| \ \|_{F;\mu} : \mu \in \mathcal{N}_F\}$ their families of semi-norms, and*

$$\mathcal{T} \text{ a family of linear mappings from } E \text{ into } F.$$

Then:

(a) $\mathcal{T}$ *is equicontinuous* $\Leftrightarrow$ $\begin{cases} \textit{for all } \mu \in \mathcal{N}_F, \textit{ there exist a finite set } N \subset \mathcal{N}_E \\ \textit{and } c \geq 0 \textit{ such that: for all } L \in \mathcal{T} \textit{ and } u \in E, \\ \qquad \|Lu\|_{F;\mu} \leq c \sup_{\nu \in N} \|u\|_{E;\nu}. \end{cases}$

$\Leftrightarrow$ $\mathcal{T}$ *is uniformly equicontinuous.*

(b) *If the family of semi-norms* $\{\| \ \|_{E;\nu} : \nu \in \mathcal{N}_E\}$ *is filtering:*

$\mathcal{T}$ *is equicontinuous* $\Leftrightarrow$ $\begin{cases} \textit{for all } \mu \in \mathcal{N}_F, \textit{ there exist } \nu \in \mathcal{N}_E \textit{ and } c \geq 0 \\ \textit{such that: for all } L \in \mathcal{T} \textit{ and } u \in E, \\ \qquad \|Lu\|_{F;\mu} \leq c \|u\|_{E;\nu}. \end{cases}$ ∎

Proof. (a) *Equicontinuity* $\Rightarrow$ *majoration.* Since it is linear, every $L \in \mathcal{T}$ verifies $L(0_E) = 0_F$ thus the equicontinuity in 0_E yields (Definition 10.5 (a)), for all $\mu \in \mathcal{N}_F$, the existence of a finite set $N \subset \mathcal{N}_E$ and of $\eta > 0$ such that: for all $L \in \mathcal{T}$ and $u \in E$,

$$\sup_{\nu \in N} \|u\|_{E;\nu} \leq \eta \quad \Rightarrow \quad \|Lu\|_{F;\mu} \leq 1.$$

Hence (Theorem 7.12) the stated majoration with $c = 1/\eta$.

Majoration $\Rightarrow$ *uniform equicontinuity.* The stated majoration gives, with linearity,

$$\|Lv - Lu\|_{F;\mu} \leq c \sup_{\nu \in N} \|v - u\|_{E;\nu}.$$

Uniform equicontinuity $\Rightarrow$ *equicontinuity.* This is always true according to Definition 10.5.

(b) If the family of semi-norms E is filtering (Definition 3.14), the majoration in (a) results in that in (b) because then, for every finite set $N \subset \mathcal{N}_E$, there exists $\lambda \in \mathcal{N}_E$ such that $\sup_{\nu \in N} \| \ \|_{E;\nu} \leq \| \ \|_{E;\lambda}$.

Conversely, the majoration of (b) is that of (a) for $N = \{\nu\}$. □

10.4. Banach–Steinhaus equicontinuity theorem

A family $\mathcal{T}$ of mappings defined on the same subset U of a separated semi-normed space is called **simply bounded**, if $\mathcal{T}(u)$ is bounded for every $u \in U$.

Let us express the **Banach–Steinhaus theorem**[2] which gives the equicontinuity of simply bounded families of continuous linear mappings defined on a Fréchet space.

THEOREM 10.10.– *Any family of continuous linear mappings from a Fréchet space into a separated semi-normed space that is simply bounded is equicontinuous.* ▮

Proof. This property is equivalent to Theorem 10.11 hereafter due to Definitions 10.5 of equicontinuity and 2.9 (a) of bounded sets. □

Let us express the Banach–Steinhaus theorem in terms of semi-norms. It is also known as the **uniform boundedness principle**.

THEOREM 10.11.– *Let E and F be separated semi-normed spaces, $\{\| \ \|_{E;\nu} : \nu \in \mathcal{N}_E\}$ and $\{\| \ \|_{F;\mu} : \mu \in \mathcal{N}_F\}$ their families of semi-norms, and $\mathcal{T}$ a set of mappings from E into F. Assume that*

$$E \text{ is a Fréchet space,}$$

$$\text{every } L \in \mathcal{T} \text{ is continuous linear,}$$

and, for all $u \in E$ and $\mu \in \mathcal{N}_F$,

$$\sup_{L \in \mathcal{T}} \|L(u)\|_{F;\mu} < \infty. \tag{10.4}$$

Then, for every $\mu \in \mathcal{N}_F$, there exist a finite set $N \subset \mathcal{N}_E$ and $c \in \mathbb{R}$ such that: for all $L \in \mathcal{T}$ and $u \in E$,

$$\|L(u)\|_{F;\mu} \leq c \sup_{\nu \in N} \|u\|_{E;\nu}.$$ ▮

Proof. Let $\mu \in \mathcal{N}_F$ and $(|\!|\!| \ |\!|\!|_{E;k})_{k \in \mathbb{N}}$ be an increasing sequence (Theorem 4.4) of semi-norms equivalent to the family of E. For each $n \in \mathbb{N}$, let

$$V_n = \{v \in E : \forall L \in \mathcal{T}, \ \|L(v)\|_{F;\mu} \leq n\}.$$

This set is closed because its complement $A = E \setminus V_n$ is open. In effect, if $a \in A$, there exists $L \in \mathcal{T}$ such that $\|L(a)\|_{F;\mu} = n + \epsilon$ where $\epsilon > 0$; since L is continuous, there exist (Theorem 7.14 (b)) $\ell \in \mathbb{N}$ and $c \geq 0$ such that $\|L(u)\|_{F;\mu} \leq c\, |\!|\!|u|\!|\!|_{E;\ell}$ for

2 **History of Banach–Steinhaus theorem 10.11.** Theorem 10.10 was established for a Banach space E in 1923 by Theophil Henry HILDEBRANDT [58]. It was extended to a Fréchet space in 1933 by Stanislaw MAZUR and Wladyslaw ORLICZ [71]. The result had been established for various linear mappings into a Banach space by Hans HAHN in 1922 [50]. Subsequently, it should rather be called the **Hildebrandt–Mazur–Orlicz theorem**.

Banach and Steinhaus's contribution. Stefan BANACH and Hugo STEINHAUS proved in 1927 [11, p. 53] the equicontinuity of any family of continuous linear mappings that is simply bounded on a Baire second category set of a Banach space, which is another generalization of Hildebrandt's result.

all $u \in E$, and then $|||v-a|||_{E;\ell} \leq \epsilon/2c$ yields $\|L(v)-L(a)\|_{F;\mu} \leq \epsilon/2$ and therefore $\|L(v)\|_{F;\mu} \geq n+\epsilon/2$, hence $v \in A$.

On the other hand, $E = \bigcup_n V_n$ according to (10.4). Furthermore, according to Baire's theorem 9.7, at least one of the V_n contains a semi-ball, that is to say there exist $n_0 \in \mathbb{N}$, $k \in \mathbb{N}$, $z \in E$ and $r > 0$ such that:

$$\{v \in E : |||v-z|||_{E;k} \leq r\} \subset V_{n_0}.$$

Then, for all $L \in \mathcal{T}$ and $u \in E$,

$$|||u|||_{E;k} \leq r$$

results in $\|L(u+z)\|_{F;\mu} \leq n_0$ from which, since $L(u) = L(u+z) - L(z)$,

$$\|L(u)\|_{F;\mu} \leq n_0 + s$$

where $s = \sup_{L\in\mathcal{T}} \|L(z)\|_{F;\mu}$. It thereof results (Theorem 7.12) that, for all $u \in E$,

$$\|L(u)\|_{F;\mu} \leq \frac{n_0+s}{r} |||u|||_{E;k}.$$

This inequality results in the stated one since, the families $\{\| \;\|_{E;\nu} : \nu \in \mathcal{N}_E\}$ and $\{||| \;|||_{E;k} : k \in \mathbb{N}\}$ being equivalent (Definition 3.1), there exist a finite set $N \subset \mathcal{N}_E$ and $c \in \mathbb{R}$ such that $|||u|||_{E;k} \leq c \sup_{\nu\in N} \|u\|_{E;\nu}$. □

Let us give a result similar to the Banach–Steinhaus theorem, that is to the uniform boundedness principle, for multilinear mappings.

THEOREM 10.12.– *Let E_1, E_2 …, E_d and F be separated semi-normed spaces, let $\{\| \;\|_{E_i;\nu_i} : \nu_i \in \mathcal{N}_{E_i}\}$, $i \in [\![1,d]\!]$, and $\{\| \;\|_{F;\mu} : \mu \in \mathcal{N}_F\}$ be their families of semi-norms, and let $\mathcal{T}$ be a set of mappings from $E_1 \times \cdots \times E_d$ into F. Assume that*

E_i is a Fréchet space, for all $i \in [\![1,d]\!]$,

every $T \in \mathcal{T}$ is continuous multilinear,

and, for all $u_1 \in E_1, \ldots, u_d \in E_d$ and $\mu \in \mathcal{N}_F$,

$$\sup_{T\in\mathcal{T}} \|T(u_1,\ldots,u_d)\|_{F;\mu} < \infty. \qquad (10.5)$$

Then, for all $\mu \in \mathcal{N}_F$, there exists $c \in \mathbb{R}$ and, for all $i \in [\![1,d]\!]$, there exist a finite set $N_i \subset \mathcal{N}_{E_i}$ such that, for all $T \in \mathcal{T}$ and $u_1 \in E_1, \ldots, u_d \in E_d$,

$$\|T(u_1,\ldots,u_d)\|_{F;\mu} \leq c \sup_{\nu_1\in N_1} \|u_1\|_{E_1;\nu_1} \cdots \sup_{\nu_d\in N_d} \|u_d\|_{E_d;\nu_d}.$$ ■

Proof. We will establish this result by induction on d. It is verified for $d = 1$ (this is the Banach–Steinhaus theorem 10.11), and therefore it merely has to be proven for d by assuming that it is true for $d - 1$.

Choice of the semi-norms. For every $i \in [\![1, d]\!]$, let $(|\!|\!| \ |\!|\!|_{E_i;k})_{k\in\mathbb{N}}$ be an increasing sequence (Theorem 4.4) of semi-norms equivalent to the family of E_i. Let us endow $E_1 \times \cdots \times E_d$ with the sequence of semi-norms, indexed by $k \in \mathbb{N}$, defined by

$$|\!|\!|(u_1, \ldots, u_d)|\!|\!|_k = \sup_{1\leq i\leq d} |\!|\!|u_i|\!|\!|_{E_i;k}.$$

It is equivalent to the family of Definition 6.1 of $E_1 \times \cdots \times E_d$ because, according to Theorems 6.5 and 6.9 (b), the latter is equivalent to the family of $\sup_{1\leq i\leq d} |\!|\!|u_i|\!|\!|_{E_i;k_i}$ indexed by $k_1 \in \mathbb{N}$, ..., $k_d \in \mathbb{N}$, which itself is equivalent to the above family since all $|\!|\!| \ |\!|\!|_{E_i;k}$ grow with k.

An estimation. Let us show that there exist $(z_1, \ldots, z_d) \in E_1 \times \cdots \times E_d$, $n_0 \in \mathbb{N}$, $k \in \mathbb{N}$ and $r \in \mathbb{R}$ such that:

$$|\!|\!|u_i|\!|\!|_{E_i;k} \leq r, \ \forall i \in [\![1, d]\!] \quad \Rightarrow \quad \sup_{T\in\mathcal{T}} \|T(u_1+z_1, \ldots, u_d+z_d)\|_{F;\mu} \leq n_0. \quad (10.6)$$

Let $\mu \in \mathcal{N}_F$ and, for all $n \in \mathbb{N}$, let

$$V_n = \{(v_1, \ldots, v_d) \in E_1 \times \cdots \times E_d : \forall T \in \mathcal{T}, \ \|T(v_1, \ldots, v_d)\|_{F;\mu} \leq n\}.$$

Every T, being continuous, is sequentially continuous (Theorem 7.2 (a)) therefore, if $(v_{1,i}, \ldots, v_{d,i})$ is an element of V_n and converges to $(v_1, \ldots, v_d)$ in $E_1 \times \cdots \times E_d$ when $i \to \infty$, then $\|T(v_{1,i}, \ldots, v_{d,i})\|_{F;\mu} \to \|T(v_1, \ldots, v_d)\|_{F;\mu}$, and consequently $(v_1, \ldots, v_d) \in V_n$. This shows that V_n is sequentially closed (Definition 2.9 (d)) and thus (Theorem 4.6 (a)) closed since $E_1 \times \cdots \times E_d$ is metrizable (Theorem 6.6 (a)).

On the other hand, $E_1 \times \cdots \times E_d = \bigcup_n V_n$ according to Assumption (10.5).

Moreover, $E_1 \times \cdots \times E_d$ is a Fréchet space (Theorem 6.7 (b)). Thus, according to Baire's theorem 9.7, at least one of the V_n contains a semi-ball, that is to say there exist $n_0 \in \mathbb{N}$, $k \in \mathbb{N}$, some points $z_1 \in E_1 \ldots, z_d \in E_d$ of E and $r > 0$ such that:

$$\{(v_1, \ldots, v_d) \in E_1 \times \cdots \times E_2 : |\!|\!|v_i - z_i|\!|\!|_{E_i;k} \leq r, \ \forall i \in [\![1, d]\!]\} \subset V_{n_0}.$$

This proves (10.6).

Another estimation. Now with the induction assumption, let us show that there exist $m \in \mathbb{N}$, $t > 0$ and $a \in \mathbb{R}$ such that:

$$|\!|\!|u_i|\!|\!|_{E_i;m} \leq t, \ \forall i \in [\![1, d]\!] \quad \Rightarrow \quad \|T(u_1, \ldots, u_d)\|_{F;\mu} \leq a. \quad (10.7)$$

Since T is multilinear, we can split

$$T(u_1 + z_1, \dots, u_d + z_d) = \sum_{1 \le \alpha \le 2^d} P_\alpha$$

where

$$P_\alpha = T(v_1^\alpha, \dots, v_d^\alpha) \text{ with } v_i^\alpha = u_i \text{ or } z_i.$$

By denoting $P_1 = T(u_1, \dots, u_d)$, it thus follows that

$$T(u_1, \dots, u_d) = T(u_1 + z_1, \dots, u_d + z_d) - \sum_{2 \le \alpha \le 2^d} P_\alpha. \tag{10.8}$$

Let $j \in [\![1, d]\!]$. The mapping $(v_i : i \neq j) \mapsto T(v_1, \dots, z_j, \dots, v_d)$ is multilinear with $d-1$ variables thus, following the induction assumption, there exists $c_j \in \mathbb{R}$ and for all $i \neq j$, there exists $k_{i,j} \in \mathbb{N}$ such that:

$$\|T(v_1, \dots, z_j, \dots, v_d)\|_{F;\mu} \le c_j \prod_{i \neq j} |||v_i|||_{E_i;k_{i,j}}. \tag{10.9}$$

Let us denote $m = \sup_{1 \le i \le d,\, 1 \le j \le d} k_{i,j}$, $s = \sup_{1 \le i \le d} |||z_i|||_{E_i;m}$, $t = \sup\{r, s\}$ and $b = \sup_{1 \le j \le d} c_j$. For all $\alpha \ge 2$, at least one of the v_j^α is equal to z_j therefore, if $|||u_i|||_{E_i;m} \le t$ for all $i \in [\![1, d]\!]$, (10.9) yields, due to the growth of $|||\ |||_{E_i,k}$ with k,

$$\|P_\alpha\|_{F;\mu} \le bt^{d-1}.$$

With (10.6), the splitting (10.8) thus gives (10.7) with $a = n_0 + 2^d bt^{d-1}$.

Conclusion. Since T is multilinear, it follows from (10.7) that (Theorem 7.21), for all $(u_1, \dots, u_d) \in E_1 \times \cdots \times E_d$,

$$\|T(u_1, \dots, u_d)\|_{F;\mu} \le \frac{a}{t^d} |||u_1|||_{E_1;m} \cdots |||u_d|||_{E_d;m}.$$

Hence the stated inequality since, the families of semi-norms $\{\|\ \|_{E_i;\nu_i} : \nu_i \in \mathcal{N}_{E_i}\}$ and $\{|||\ |||_{E_i;k} : k \in \mathbb{N}\}$ being equivalent (Definition 3.1) for every $i \in [\![1, d]\!]$, there exist a finite set $N_i \subset \mathcal{N}_{E_i}$ and $\sigma_i \in \mathbb{R}$ such that $|||\ |||_{E_i;m} \le \sigma_i \sup_{\nu_i \in N_i} \|\ \|_{E_i;\nu_i}$. □

The case of spaces that are not Fréchet spaces. The Banach–Steinhaus theorem 10.10 and its generalization 10.12 to multilinear mappings do not hold for all semi-normed spaces but, according to a result by BOURBAKI [17, Chap. III, § 4, no. 2, Theorem 1, p. 25], they extend to barreled spaces which are more general than Fréchet spaces. □

Chapter 11

Compactness in Mapping Spaces

11.1. The spaces $\mathcal{F}(X;F)$ and $\mathcal{C}(X;F)$-pt

Let us define the space of mappings from any set into a semi-normed space, endowed with pointwise convergence.

DEFINITION 11.1.– *Let X be a set, F a separated semi-normed space and $\{\| \ \|_{F;\mu} : \mu \in \mathcal{N}_F\}$ its family of semi-norms.*

$\mathcal{F}(X;F)$ is the vector space of the mappings from X into F endowed with the semi-norms, indexed by $u \in X$ and $\mu \in \mathcal{N}_F$,

$$\|f\|_{\mathcal{F}(X;F);u,\mu} \stackrel{\text{def}}{=} \|f(u)\|_{F;\mu}. \ \blacksquare$$

The addition and multiplication by a real number t of mappings that are valued in the same vector space are defined by

$$(f+g)(u) = f(u) + g(u), \quad (tf)(u) = tf(u).$$

Let us show that this space is separated.

THEOREM 11.2.– *Let X be a set and F a separated semi-normed space. Then:*

$$\mathcal{F}(X;F) \textit{ is a separated semi-normed space. } \blacksquare$$

Proof. The space $\mathcal{F}(X;F)$ is semi-normed by construction. It is separated (Definition 2.3 (c)) because, using the notations of Definition 11.1, if all semi-norms of f are zero, that is, if, for all $u \in X$, we have $\|f(u)\|_{F;\mu} = 0$ for all $\mu \in \mathcal{N}_F$, then, F being separated, $f(u) = 0_F$ hence $f = 0_{\mathcal{F}}$. $\square$

Observe that $\mathcal{F}(X;F)$ is sequentially complete if F is.

THEOREM 11.3.– *Let X be a set. Then:*

$$F \text{ is a Neumann space} \quad \Rightarrow \quad \mathcal{F}(X;F) \text{ is a Neumann space.} \blacksquare$$

Proof. Let $(f_n)_{n\in\mathbb{N}}$ be a Cauchy sequence of $\mathcal{F}(X;F)$. For all $u \in X$, the sequence $(f_n(u))_{n\in\mathbb{N}}$ is a Cauchy sequence of F thus, if F is sequentially complete, it has a limit denoted by $f(u)$. Therefore, f is an element of $\mathcal{F}(X;F)$ and $f_n \to f$ in $\mathcal{F}(X;F)$ which proves that the latter is sequentially complete. □

Let us define the space of continuous mappings endowed with pointwise convergence.

DEFINITION 11.4.– *Let E and F be separated semi-normed spaces, $\{\| \ \|_{F;\mu} : \mu \in \mathcal{N}_F\}$ the family of semi-norms of F and $X \subset E$.*

$\mathcal{C}(X;F)$-pt is the vector space of continuous mappings from X into F endowed with the semi-norms, indexed by $u \in E$ and $\mu \in \mathcal{N}_F$,

$$\|f\|_{\mathcal{C}(X;F)\text{-pt};u,\mu} \stackrel{\text{def}}{=} \|f(u)\|_{F;\mu}. \blacksquare$$

Let us show that this space is separated.

THEOREM 11.5.– *Let X be a set and F a separated semi-normed space. Then:*

$$\begin{cases} \mathcal{C}(X;F)\text{-pt is a separated semi-normed space} \\ \text{and is a topological subspace of } \mathcal{F}(X;F). \end{cases} \blacksquare$$

Proof. The space $\mathcal{C}(X;F)$-pt is a topological subspace (Definition 3.9) of $\mathcal{F}(X;F)$ since its semi-norms are (Definition 11.4) the restrictions of those of the latter (Definition 11.1). It is thus separated (Theorem 3.10 (b)). □

Non-completeness of $\mathcal{C}(X;F)$-pt. Generally, this space is not sequentially complete, for example:

$$\mathcal{C}(\mathbb{R};\mathbb{R})\text{-pt } \textit{is not sequentially complete.} \tag{11.1}$$

Proof. The mappings f_n defined by $f_n(x) = \sup(1 - n|x|, 0)$ form a Cauchy sequence of $\mathcal{C}(\mathbb{R};\mathbb{R})$-pt that does not converge. □

11.2. Zorn's lemma

Let us start with a few definitions related to orders (these have been defined in section 1.1).

DEFINITION 11.6.– *Let U be a set ordered by $\preceq$ and $V \subset U$.*

(a) *U is **inductive** if any totally ordered subset of U has an upper bound.*

(b) *u is a **maximal element** of V if $u \in V$ and if there exists no $v \in V$ such that $u \prec v$.* ∎

Observe that, when there exists a *maximum*, it is the unique *maximal element*. On the other hand, a maximal element is not necessarily a maximum, or even an upper bound, and is not necessarily unique.

Let us state **Zorn's lemma**[1].

THEOREM 11.7.– *Every inductive ordered set has a maximal element.* ∎

The proof will make use of the following result[2].

LEMMA 11.8.– *Let U be a non-empty ordered set such that*

any totally ordered subset of U has a least upper bound in U,

and let f be a mapping from U into itself such that, for all $v \in U$,

$$f(v) \succeq v.$$

Then, there exists $u \in U$ such that:

$$f(u) = u. \blacksquare$$

Proof of Lemma 11.8. Let us choose an element $x \in U$. We will say that a set $V \subset U$ is admissible if it satisfies the following three conditions:

$$x \in V; \tag{11.2}$$

$$f(V) \subset V; \tag{11.3}$$

The least upper bound of any totally ordered subset of V is an element of V. (11.4)

Let A be the intersection of all admissible sets. This intersection is defined because there are such sets, in particular $\{v \in U : v \succeq x\}$. The latter therefore contains A, hence

$$v \in A \quad \Rightarrow \quad v \succeq x. \tag{11.5}$$

1 **History of Zorn's lemma.** Max ZORN established a result which is equivalent to Theorem 11.7 in 1935 [118].

2 **History of Lemma 11.8.** This result is due to Nicolas BOURBAKI, see [LANG, 66, p. 12].

The intersection of admissible sets is admissible and thus A is admissible.

Let

$$B = \{b \in A : a \in A,\ a \prec b \Rightarrow f(a) \preceq b\}.$$

We intend first to show that

$$b \in B,\ a \in A \quad \Rightarrow \quad a \preceq b \text{ or } a \succeq f(b). \tag{11.6}$$

To this end, let us set $b \in B$ and denote

$$C = \{c \in A :\ c \preceq b \text{ or } c \succeq f(b)\}.$$

Let us show that C is admissible:
— (11.2), that is $x \in C$, is realized because $x \preceq b$ according to (11.5).
— (11.3) means that, if $c \in C$, then $f(c) \in C$. This is realized because: either $c \succeq f(b)$, and then $f(c) \succeq c \succeq f(b)$; or $c = b$, and then $f(c) = f(b)$; or $c \prec b$, and then $f(c) \preceq b$ because $b \in B$ and $c \in A$.
— (11.4) means that, if z is the least upper bound of a totally ordered subset Z of C, then $z \in C$. This is realized because: either $c \preceq b$ for all $c \in Z$, and then $z \preceq b$; or there exists $c \in Z$ such that $c \succeq f(b)$, and then $z \succeq f(b)$.

This proves that C is an admissible subset of A. Therefore, $C = A$, which leads to (11.6).

Let us now show that B is admissible:
— (11.2), that is $x \in B$, is realized because the condition $a \in A$, $a \prec x$ is never satisfied due to (11.5).
— (11.3) means that, if $b \in B$, then $f(b) \in B$; that is, if $a \in A$ and $a \prec f(b)$, then $f(a) \preceq f(b)$. This is realized because, according to (11.6), $a \in A$ leads to $a \preceq b$ or $a \succeq f(b)$; since $a \prec f(b)$, it yields $a \preceq b$. Thus, either $a = b$ and then $f(a) = f(b)$, or $a \prec b$, and then $b \in B$ yields $f(a) \preceq b \preceq f(b)$.
— (11.4) means that, if z is the least upper bound of a totally ordered subset Z of B, and if $a \in A$ and $a \prec z$, then

$$f(a) \preceq z. \tag{11.7}$$

Let us verify this. According to (11.6), every $v \in Z$ verifies $a \preceq v$ or $a \succeq f(v) \succeq v$. The second possibility cannot be satisfied by all v of Z because it would thereof result that $a \succeq z$. Therefore, there exists $v \in Z$ such that $a \preceq v$. If $a = v$, since $a \prec z$, there exists $y \in Z$ such that $a \prec y$, and then $y \in B$ gives $f(a) \preceq y \preceq z$. Otherwise, $a \prec v$, and then $v \in B$ gives $f(a) \preceq v \preceq z$. This proves (11.7).

This proves that B is an admissible subset of A. As a result, $B = A$. Then, (11.6) shows that, for all v and a in A, it follows that $a \preceq v$ or $a \succeq f(v) \succeq v$. Thus, A is totally ordered.

Let u be the least upper bound of A in E. According to (11.4), $u \in A$ and so, according to (11.3), $f(u) \in A$ and therefore $f(u) \preceq u$ since u is an upper bound of A. Since $f(u) \succeq u$, it follows that $f(u) = u$. □

Proof of Theorem 11.7. Let U be a set ordered by $\preceq$ and let

$$\mathcal{Y} = \{Y \subset U : Y \text{ is totally ordered}\}.$$

This set $\mathcal{Y}$ is not empty because it contains $\emptyset$. Let us order it by $\subset$. Then, any totally ordered subset $\mathcal{Z}$ of $\mathcal{Y}$ has an upper bound that is the union of the $Y \in \mathcal{Z}$.

If an element $Y \in \mathcal{Y}$ is not maximal, there exists $V \in \mathcal{Y}$ such that $V \supset Y$ and $V \neq Y$. If no Y were maximal, the mapping f from $\mathcal{Y}$ into itself which associates V to Y would contradict Lemma 11.8. Therefore, $\mathcal{Y}$ has (at least) a maximal element W.

Now suppose that U is inductive (Definition 11.6 (a)). Then, W, being totally ordered (since it is an element of $\mathcal{Y}$), has an upper bound w. Let us verify that w is a maximal element of U.

Otherwise, there would exist $v \in U$ such that $v \succ w$. The set $X = W \cup \{w, v\}$ would be totally ordered and thus it would be an element of $\mathcal{Y}$. Since $X \supset W$, it would follow that $X = W$ (since W is maximal) hence $v \in W$ and therefore w would not be an upper bound of W. Therefore, w is indeed maximal. □

Axiom of choice. In the proof of Theorem 11.7, the axiom of choice has been used when considering "*the mapping f from $\mathcal{Y}$ into itself which to Y associates V*". This axiom may be stated as follows:

$$\begin{cases} \textit{For any non-empty family } (X_i)_{i \in \mathcal{I}} \textit{ of non-empty sets, there exists} \\ \textit{a mapping } f \textit{ defined on } \mathcal{I} \textit{ such that } f(i) \in X_i \textit{ for all } i \in \mathcal{I}. \end{cases} \qquad (11.8)$$

Resorting to this axiom is here inevitable because it is equivalent to Zorn's lemma [SCHWARTZ, 99, theorem 1.6.20, p. 95]. It is independent of the other axioms of set theory known as Zermelo–Fraenkel set theory [99, remark 4, p. 50 and chap. I, pp. 13–127]. □

Use of Zorn's lemma. We will use Zorn's lemma to prove the Hahn–Banach extension theorem 14.1 and Tychonoff's compactness theorem 11.9. □

11.3. Compactness in $\mathcal{F}(X; F)$

Let us show that the set of mappings that are valued in given compact sets is compact for pointwise convergence topology, which is a **variant of Tychonoff's theorem**[3].

---

3 **History of Theorem 11.9.** This theorem is the formulation in terms of mappings of a specific case of Tychonoff's theorem 11.10, see p. 160. The proof given here is an adaptation of a proof by BOURBAKI, see [LANG, 66, p. 35], of Tychonoff's theorem. □

THEOREM 11.9.– *Let X be a set, F a separated semi-normed space and, for all $u \in X$, a set $K_u \subset F$. Denote*

$$\mathcal{T} = \{f \in \mathcal{F}(X; F) : \forall u \in X,\ f(u) \in K_u\}.$$

Then:

$$\forall u \in X,\ K_u \text{ is compact in } F \quad \Rightarrow \quad \mathcal{T} \text{ is compact in } \mathcal{F}(X; F).\ \blacksquare$$

Proof. Let $A = \{A_i\}_{i \in \mathcal{I}}$ be a family of subsets of $\mathcal{F}(X; F)$ such that:

$$\begin{cases} \text{For all } i \in \mathcal{I}, \quad A_i \text{ is closed in } \mathcal{F}(X; F); \\ \text{For any finite set } I \subset \mathcal{I}, \quad \bigcap_{i \in I} A_i \cap \mathcal{T} \neq \emptyset. \end{cases}$$

According to the characterization of compact sets by the finite intersection property (Theorem 2.27), it suffices to prove that:

$$\text{there exists } f \in \textstyle\bigcap_{i \in \mathcal{I}} A_i \cap \mathcal{T}. \qquad (11.9)$$

Let B be a family of (non-necessarily closed) subsets of $\mathcal{T}$ such that:

$$\begin{cases} B \supset \{A_i \cap \mathcal{T}\}_{i \in \mathcal{I}}; \\ \text{Any set of } B \text{ is included in } \mathcal{T}; \\ \text{Any finite subfamily of } B \text{ has a non-empty intersection.} \end{cases} \qquad (11.10)$$

Let us order the set $\mathcal{Y}$ of these families B by means of inclusion. Then, $\mathcal{Y}$ is inductive (Definition 11.6 (a)) because any totally ordered subset $\mathcal{Z}$ of $\mathcal{Y}$ has an upper bound, namely the union C of the $B \in \mathcal{Z}$. In effect:

— C is an upper bound (that is contains) each $B \in \mathcal{Z}$.

— $C \in \mathcal{Y}$, that is verifies (11.10); indeed, its first two properties are immediate and, for the third one, any finite subfamily of C is included in a finite number of $B \in \mathcal{Z}$ and thereby is a finite subfamily of the largest of these B which implies that it has a non-empty intersection.

According to Zorn's lemma (Theorem 11.7), $\mathcal{Y}$ therefore has (at least) a maximal element $M = \{M_j\}_{j \in \mathcal{J}}$, that is (Definition 11.6 (b)) such that:

$$\begin{cases} M \text{ verifies (11.10)}; \\ \text{No family verifying (11.10) strictly contains } M. \end{cases} \qquad (11.11)$$

Let us set $u \in X$ and denote $M_j(u) = \{f(u) : f \in M_j\}$.

Any finite subfamily of $\{M_j\}_{j \in \mathcal{J}}$ has a non-empty intersection, so does any finite subfamily of $\{M_j(u)\}_{j \in \mathcal{J}}$ and *a fortiori* any finite subfamily of $\{\overline{M_j(u)}\}_{j \in \mathcal{J}}$. The set K_u being compact by assumption, it is closed (Theorem 2.26 (a)) and thus,

$\overline{M_j(u)} \subset \overline{K_u} = K_u$. The $\overline{M_j(u)}$ being closed, there exists, according again to the finite intersection property of Theorem 2.27,

$$f_u \in \bigcap_{j\in\mathcal{J}} \overline{M_j(u)}.$$

Let f be the mapping that associates f_u to u. Let us observe that $f \in \mathcal{T}$ and set $i \in \mathcal{I}$.

To establish (11.9), we still have to show that $f \in A_i$. Since A_i is closed, it suffices that $f \in \overline{A_i}$. According to the characterization of closure of Theorem 2.17 (b), it suffices that, for any finite family of semi-norms of $\mathcal{F}(X;F)$, that is (Definition 11.1) for any finite family $u_1, \ldots, u_k$ of X and $\mu_1, \ldots, \mu_k$ of $\mathcal{N}_F$ and every $\eta > 0$, there exists g such that:

$$g \in A_i, \qquad \sup_{1\le n\le k} \|g(u_n) - f(u_n)\|_{F;\mu_n} \le \eta. \tag{11.12}$$

Let

$$\mathcal{O}_n = \{g \in \mathcal{T} : \|g(u_n) - f(u_n)\|_{F;\mu_n} \le \eta\}$$

and $j \in \mathcal{J}$. Since $f(u_n) \in \overline{M_j(u_n)}$, there exists, once again thanks to the characterization of closure of Theorem 2.17 (b), $\gamma \in M_j(u_n)$ such that $\|\gamma - f(u_n)\|_{F;\mu_n} \le \eta$. Let $g \in M_j$ be such that $\gamma = g(u_n)$. Then, $g \in \mathcal{O}_n$ since $M_j \subset \mathcal{T}$, and therefore

$$M_j \cap \mathcal{O}_n \neq \emptyset. \tag{11.13}$$

In order to show that $\mathcal{O}_n \in M$, observe that:

$$\text{For any finite set } Q \subset \mathcal{J}, \quad \textstyle\bigcap_{j\in Q} M_j \in M. \tag{11.14}$$

Indeed, by denoting $N = \bigcap_{j\in Q} M_j$, the family $M \cup \{N\}$ verifies (11.10) thus, according to (11.11), it is not strictly greater than M, hence $N \in M$.

The intersection of any finite subfamily of $M \cup \{\mathcal{O}_n\}$ is thus of the form M_j or $M_j \cap \mathcal{O}_n$, and as a result it is not empty according to (11.13). Subsequently, $M \cup \{\mathcal{O}_n\}$ verifies (11.10) thus, according to (11.11), it is not strictly greater than M, hence

$$\mathcal{O}_n \in M.$$

Since $A_i \cap \mathcal{T}$ is also an element of M, its finite subfamily $\{\mathcal{O}_1, \ldots, \mathcal{O}_k, A_i \cap \mathcal{T}\}$ has a non-empty intersection. Any element g from this intersection verifies (11.12), which concludes the proof of Theorem 11.9. □

Tychonoff's theorem. Theorem 11.9 above is a variant of Tychonoff's theorem on compactness of (finite or not) product of compact sets.

More precisely, let $\{F_u\}_{u\in X}$ be a family of semi-normed spaces and $\{\| \|_{F_u;\mu_u} : \mu_u \in \mathcal{N}_{F_u}\}$ the family of semi-norms of F_u. The **product of the** F_u is to the vector space

$$\prod_{u\in X} F_u \stackrel{\text{def}}{=} \{\{z_u\}_{u\in X} : z_u \in F_u,\ \forall u \in X\} \tag{11.15}$$

endowed with the semi-norms, indexed by $v \in X$ and $\mu_v \in \mathcal{N}_{F_v}$,

$$\|z\|_{\prod_{u\in X} F_u;v,\mu_v} \stackrel{\text{def}}{=} \|z_v\|_{F_v,\mu_v}. \tag{11.16}$$

We can now state **Tychonoff's theorem**[4].

THEOREM 11.10.– *For every u in a set X, let F_u be a separated semi-normed space and $K_u \subset F_u$.*

Then:

$$\forall u \in X,\ K_u \textit{ is compact in } F_u \quad\Rightarrow\quad \prod_{u\in X} K_u \textit{ is compact in } \prod_{u\in X} F_u. \ \blacksquare$$

Proof. Theorem 11.9 gives this result when $F_u = F$ for all u because then the mapping that associates the mapping $u \mapsto z_u$ to $\{z_u\}_{u\in X}$ is an isomorphism from $\prod_{u\in X} F_u$ onto $\mathcal{F}(X;F)$. Its demonstration is easily extended to the case of distinct spaces F_u. □

Let us now characterize relatively compact sets of mappings for pointwise convergence.

THEOREM 11.11.– *Let X be a set, F a separated semi-normed space and $\mathcal{T} \subset \mathcal{F}(X;F)$. Then:*

$$\mathcal{T} \textit{ is relatively compact in } \mathcal{F}(X;F) \quad\Leftrightarrow$$
$$\Leftrightarrow\quad \forall u \in X,\ \mathcal{T}(u) \textit{ is relatively compact in } F. \ \blacksquare$$

Proof. Let $\{\| \|_{F;\mu} : \mu \in \mathcal{N}_F\}$ be the family of semi-norms of F and $u \in X$. For all $\mu \in \mathcal{N}_F$, Definition 11.1 gives

$$\|f(u)\|_{F;\mu} = \|f\|_{\mathcal{F}(X;F);u,\mu}$$

therefore the linear mapping $\Phi_u : f \mapsto f(u)$ is continuous from $\mathcal{F}(X;F)$ into F according to the characterization of the continuous linear mappings of Theorem 7.14 (a).

If $\mathcal{T}$ is relatively compact (Definition 2.23 (b)), its closure $\overline{\mathcal{T}}$ in $\mathcal{F}(X;F)$ is compact, so the image $\overline{\mathcal{T}}(u)$ of $\overline{\mathcal{T}}$ under Φ_u is compact in F according to Theorem 8.7 (a). Being included in this compact set, $\mathcal{T}(u)$ is relatively compact (Theorem 2.26 (c)).

Conversely, if, for every $u \in X$, the set $\mathcal{T}(u)$ is relatively compact in F, then $\overline{\mathcal{T}(u)}$ is therein compact therefore the set

$$\{f \in \mathcal{F}(X;F) : \forall u \in X,\ f(u) \in \overline{\mathcal{T}(u)}\}$$

4 **History of Tychonoff's theorem 11.10.** Andrei Nikolaevitch TYCHONOFF proved in 1930 [110, p. 548] that any product of compact topological spaces is compact.

is compact in $\mathcal{F}(X;F)$ according to the variant of Tychonoff's theorem given in Theorem 11.9. Being included in this compact set, $\mathcal{T}$ is relatively compact (Theorem 2.26 (c), again). □

Characterization of compact sets. Unlike relative compactness, the compactness of the $\mathcal{T}(u)$ does not characterize that of $\mathcal{T}$, because it does not guarantee that the latter be closed. □

11.4. An Ascoli compactness theorem in $\mathcal{C}(X;F)$-pt

Let us give a sufficient condition for relative compactness in $\mathcal{C}(X;F)$-pt, which is an **Ascoli's theorem**[5].

THEOREM 11.12.– *Let E and F be two separated semi-normed spaces, $X \subset E$ and a set of mappings $\mathcal{T} \subset \mathcal{C}(X;F)$ such that:*

$$\mathcal{T} \textit{ is equicontinuous,}$$

$$\forall u \in X, \mathcal{T}(u) \textit{ is relatively compact in } F.$$

Then:

$$\mathcal{T} \textit{ is relatively compact in } \mathcal{C}(X;F)\text{-pt.}$$ ▮

Proof. The $\mathcal{T}(u)$ being relatively compact in F, $\mathcal{T}$ is relatively compact in $\mathcal{F}(X;F)$ according to the characterization of Theorem 11.11. Thus:

$$\text{the closure } \overline{\mathcal{T}} \text{ of } \mathcal{T} \text{ in } \mathcal{F}(X;F) \text{ is compact.} \tag{11.17}$$

Let us show that, on the other hand,

$$\overline{\mathcal{T}} \subset \mathcal{C}(X;F)\text{-pt.} \tag{11.18}$$

According to Definition 10.5 (a) of equicontinuity, for all $u \in X$, $\mu \in \mathcal{N}_F$ and $\epsilon > 0$, there exist a finite set $N \subset \mathcal{N}_E$ and $\eta > 0$ such that, for all $v \in X$ such that $\sup_{\nu \in N} \|v-u\|_{E;\nu} \leq \eta$ and all $T \in \mathcal{T}$,

$$\|T(v) - T(u)\|_{F;\mu} \leq \epsilon.$$

Every $S \in \overline{\mathcal{T}}$ verifies this property because, according to the characterization of closure of Theorem 2.17 (b), for all $\alpha > 0$, there exists $T \in \mathcal{T}$ such that:

$$\sup\{\|S-T\|_{\mathcal{F}(X;F);u,\mu}, \|S-T\|_{\mathcal{F}(X;F);v,\mu}\} \leq \alpha$$

that is $\sup\{\|(S-T)(u)\|_{F;\mu}, \|(S-T)(v)_{F;\mu}\} \leq \alpha$. Then, $\|S(v) - S(u)\|_{F;\mu} \leq \epsilon + 2\alpha$, this for all $\alpha > 0$ and thus for $\alpha = 0$. This proves that S is continuous (and even that $\overline{\mathcal{T}}$ is equicontinuous) and thereby (11.18).

5 **History of Ascoli's theorem 11.12.** Giulio ASCOLI characterized in 1883 [5] compactness in $\mathcal{C}(I)$, I being a real interval.

The space $\mathcal{C}(X;F)$-pt being (Theorem 11.5) a topological subspace of $\mathcal{F}(X;F)$, the properties (11.17) and (11.18) lead to (Theorem 3.11 (b)) that $\overline{\mathcal{T}}$ is compact in $\mathcal{C}(X;F)$-pt. Its subset $\mathcal{T}$ is thus relatively compact (Theorem 2.26 (c)) in the space $\mathcal{C}(X;F)$-pt. □

Chapter 12

Spaces of Linear or Multilinear Mappings

12.1. The space $\mathcal{L}(E;F)$

Let us define the space $\mathcal{L}(E;F)$ of continuous linear mappings from E into F and endow it with two families of semi-norms.

DEFINITION 12.1.– *Let E and F be two separated semi-normed spaces and denote $\{\| \ \|_{F;\mu} : \mu \in \mathcal{N}_F\}$ the family of semi-norms of F.*

(a) *$\mathcal{L}(E;F)$ is the vector space of continuous linear mappings from E into F endowed with the semi-norms, indexed by the bounded sets B of E and by $\mu \in \mathcal{N}_F$,*

$$\|L\|_{\mathcal{L}(E;F);B,\mu} \stackrel{\text{def}}{=} \sup_{u\in B} \|Lu\|_{F;\mu}.$$

(b) *$\mathcal{L}(E;F)$-pt is the same vector space endowed with the semi-norms indexed by $u \in E$ and $\mu \in \mathcal{N}_F$,*

$$\|L\|_{\mathcal{L}(E;F)\text{-pt};u,\mu} \stackrel{\text{def}}{=} \|Lu\|_{F;\mu}. \ \blacksquare$$

Let us recall that the addition and the multiplication by a real number t of, linear or not, mappings that are valued in the same vector space are defined by

$$(L+M)(u) = L(u) + M(u), \quad (tL)(u) = tL(u).$$

Justification of Definition 12.1 (a). The mapping $L \mapsto \sup_{u\in B} \|Lu\|_{F;\mu}$ indeed defines a semi-norm on $\mathcal{L}(E;F)$ according to Theorem 12.2 hereafter because the mapping $L \mapsto \|Lu\|_{F;\mu}$ is a semi-norm for each u and $\sup_{u\in B} \|Lu\|_{F;\mu} < \infty$ for each L (since any continuous linear mapping transforms bounded sets into bounded sets, see Theorem 7.15). $\square$

It remains to verify the following property of the upper envelope of semi-norms.

THEOREM 12.2.– *Let $\mathcal{P}$ be a family of semi-norms over a vector space E such that:*

$$\forall u \in E, \ \sup_{p \in \mathcal{P}} p(u) < \infty.$$

Then:

$$u \mapsto \sup_{p \in \mathcal{P}} p(u) \textit{ is a semi-norm on } E. \ \blacksquare$$

Proof. For all $u \in E$, $v \in E$ and $t \in \mathbb{R}$, it follows that $\sup_{p \in \mathcal{P}} p(u) \geq 0$ and

$$\sup_{p \in \mathcal{P}} p(tu) = \sup_{p \in \mathcal{P}} tp(u) = t \sup_{p \in \mathcal{P}} p(u),$$

$$\sup_{p \in \mathcal{P}} p(u+v) \leq \sup_{p \in \mathcal{P}} (p(u) + p(v)) \leq \sup_{p \in \mathcal{P}} (p(u) + \sup_{q \in \mathcal{P}} q(v)) = \sup_{p \in \mathcal{P}} p(u) + \sup_{p \in \mathcal{P}} p(v).$$

□

Let us show that the spaces of linear mapping are separated and let us compare them.

THEOREM 12.3.– *Let E and F be two separated semi-normed spaces. Then:*

(a) $\mathcal{L}(E;F)$ *and* $\mathcal{L}(E;F)$-pt *are separated semi-normed spaces.*

(b) $$\mathcal{L}(E;F) \subsetneq \mathcal{L}(E;F)\text{-pt}.$$

(c) $$\begin{cases} \mathcal{L}(E;F)\text{-pt } \textit{is a topological subspace of } \mathcal{F}(E;F) \\ \textit{and is a closed topological subspace of } \mathcal{C}(E;F)\text{-pt}. \ \blacksquare \end{cases}$$

Proof. Let $\{\| \ \|_{F;\mu} : \mu \in \mathcal{N}_F\}$ be the family of semi-norms of F.

(a) $\mathcal{L}(E;F)$ and $\mathcal{L}(E;F)$-pt spaces are semi-normed by construction (Definition 12.1). They are separated (Definition 2.3 (c)) because, if all semi-norms of L in $\mathcal{L}(E;F)$ or in $\mathcal{L}(E;F)$-pt are zero, that is, if, for all $u \in E$, we have $\|Lu\|_{F;\mu} = 0$ for all $\mu \in \mathcal{N}_F$, then, F being separated, $Lu = 0_F$ hence $L = 0_{\mathcal{L}}$.

(b) The semi-normed space $\mathcal{L}(E;F)$ is topologically included (Definition 3.6 (c)) in $\mathcal{L}(E;F)$-pt, because any semi-norm of $\mathcal{L}(E;F)$-pt is a semi-norm of $\mathcal{L}(E;F)$. More precisely, $\| \ \|_{\mathcal{L}(E;F)\text{-pt};u,\mu} = \| \ \|_{\mathcal{L}(E;F);\{u\},\mu}$ for all $u \in E$ and $\mu \in \mathcal{N}_F$.

(c) The space $\mathcal{L}(E;F)$-pt is a topological subspace (Definition 3.9) of $\mathcal{F}(E;F)$ and of $\mathcal{C}(E;F)$-pt because, according to their Definitions 11.1 and 11.4, it is endowed with the restrictions of their semi-norms.

Let us verify that $\mathcal{L}(E;F)$ is closed in $\mathcal{C}(E;F)$-pt, that is $V = \mathcal{C}(E;F)\backslash\mathcal{L}(E;F)$ is open therein. Let $f \in V$. It is not linear, so there exist $u \in E$ and $v \in E$ such that $f(u+v) - f(u) - f(v) = b \neq 0_F$. There exists then $\mu \in \mathcal{N}_F$ such that $\|b\|_{F;\mu} = a > 0$. If a mapping $g \in \mathcal{C}(E;F)$ verifies (by abbreviating $\mathcal{C}(E;F)$ into $\mathcal{C}$)

$$\sup\{\|g-f\|_{\mathcal{C}\text{-pt};u+v,\mu}, \|g-f\|_{\mathcal{C}\text{-pt};u,\mu}, \|g-f\|_{\mathcal{C}\text{-pt};v,\mu}\} \leq \frac{a}{4}$$

that is if $\sup\{\|(f-g)(u+v)\|_{F;\mu}, \|(f-g)(u)\|_{F;\mu}, \|(f-g)(v)\|_{F;\mu}\} \leq a/4$, then $\|g(u+v) - g(u) - g(v)\|_{F;\mu} \geq a/4$, therefore $g(u+v) - g(u) - g(v) \neq 0_F$ hence $g \in V$. This proves that V is open (Definition 2.9 (b)). □

Let us observe that linear mapping spaces are preserved by topological equalities.

THEOREM 12.4.– *Let E and F be two separated semi-normed spaces. Then:*

(a) *$\mathcal{L}(E;F)$ and $\mathcal{L}(E;F)$-pt and their topologies are preserved when the family of semi-norms of E and/or that of F is replaced by an equivalent family.*

(b) *The family of semi-norms of $\mathcal{L}(E;F)$ is filtering.* ▮

Proof. (a) When the family of semi-norms of E and/or that of F is replaced by an equivalent family (Definition 3.1):
— The set $\mathcal{L}(E;F)$ is preserved because continuous mappings are (Theorem 7.5).
— The semi-norms of $\mathcal{L}(E;F)$ are preserved because they are indexed (Definition 12.1 (a)) by the bounded subsets of E, which are preserved (Theorem 3.4 (b)).
— The semi-norms of $\mathcal{L}(E;F)$-pt are preserved because they are indexed (Definition 12.1 (b)) by E, which is preserved.

(b) The family of semi-norms of $\mathcal{L}(E;F)$ is filtering (Definition 3.14) because, given a finite family $\{B_\sigma : \sigma \in M\}$ of bounded subsets of E, their union B is bounded (Theorem 2.12 (f)) and for all $L \in \mathcal{L}(E;F)$ and $\mu \in \mathcal{N}_F$,

$$\sup_{\sigma\in M} \|L\|_{\mathcal{L}(E;F);B_\sigma,\mu} = \sup_{\sigma\in M}\sup_{u\in B_\sigma} \|Lu\|_{F;\mu} = \sup_{u\in B} \|Lu\|_{F;\mu} = \|L\|_{\mathcal{L}(E;F);B,\mu}.$$ □

12.2. Bounded sets in $\mathcal{L}(E;F)$

Let us characterize the bounded subsets of $\mathcal{L}(E;F)$.

THEOREM 12.5.– *Let E and F be two separated semi-normed spaces, $\{\| \|_{F;\mu} : \mu \in \mathcal{N}_F\}$ the family of semi-norms of F and $\mathcal{T} \subset \mathcal{L}(E;F)$. Then:*

(a) *$\mathcal{T}$ is bounded in $\mathcal{L}(E;F)$* $\Leftrightarrow$ $\begin{cases} \textit{For every bounded set } B \textit{ of } E \textit{ and } \mu \in \mathcal{N}_F\textit{:} \\ \quad \sup_{L\in\mathcal{T}} \sup_{u\in B} \|Lu\|_{F;\mu} < \infty. \end{cases}$

$\Leftrightarrow$ $\begin{cases} \textit{For every bounded set } B \textit{ of } E\textit{, the set} \\ \{L(u) : L \in \mathcal{T},\ u \in B\} \textit{ is bounded in } F. \end{cases}$

(b) *$\mathcal{T}$ is bounded in $\mathcal{L}(E;F)$-pt* $\Leftrightarrow$ $\begin{cases} \text{For every } u \in E \text{ and } \mu \in \mathcal{N}_F: \\ \quad \sup_{L\in\mathcal{T}} \|Lu\|_{F;\mu} < \infty. \end{cases}$ ∎

Proof. (a) These two equivalences are the simple writing of Definition 2.9 (a) of a bounded set with respectively the semi-norms of $\mathcal{L}(E;F)$ (Definition 12.1 (a)) and of F.

(b) This equivalence is the simple writing of the definition of a bounded set with the semi-norms of $\mathcal{L}(E;F)$-pt (Definition 12.1 (b)). □

Let us show that $\mathcal{L}(E;F)$ and $\mathcal{L}(E;F)$-pt have the same bounded sets if E is a Fréchet space.

THEOREM 12.6.– *Let F be a separated semi-normed space and*

E a Fréchet space.

Then, for any subset of $\mathcal{L}(E;F)$:

Bounded in $\mathcal{L}(E;F)$-pt $\Leftrightarrow$ *bounded in $\mathcal{L}(E;F)$*

$\Leftrightarrow$ *equicontinuous from E into F.* ∎

Proof. Let us denote by $\{\|\ \|_{E;\nu} : \nu \in \mathcal{N}_E\}$ and $\{\|\ \|_{F;\mu} : \mu \in \mathcal{N}_F\}$ the families of semi-norms of E and F and let $\mathcal{T} \subset \mathcal{L}(E;F)$.

If $\mathcal{T}$ is bounded in $\mathcal{L}(E;F)$-pt, then, for all $u \in E$ and $\mu \in \mathcal{N}_F$,

$$\sup_{L\in\mathcal{T}} \|Lu\|_{F;\mu} = \sup_{L\in\mathcal{T}} \|L\|_{\mathcal{L}(E;F)\text{-pt};u,\mu} < \infty$$

therefore the Banach–Steinhaus theorem 10.11 shows that, for all $\mu \in \mathcal{N}_F$, there exist a finite set $N \subset \mathcal{N}_E$ and $c \in \mathbb{R}$ such that, for all $L \in \mathcal{T}$ and $u \in E$,

$$\|L(u)\|_{F;\mu} \leq c \sup_{\nu\in N} \|u\|_{E;\nu}. \tag{12.1}$$

That is to say, according to Theorem 10.9 (a), that $\mathcal{T}$ is equicontinuous.

Furthermore, this inequality, namely the equicontinuity of $\mathcal{T}$, implies that the latter is bounded in $\mathcal{L}(E;F)$ because it gives, for any bounded set B of E,

$$\sup_{L\in\mathcal{T}} \|L\|_{\mathcal{L}(E;F);B,\mu} = \sup_{L\in\mathcal{T}} \sup_{u\in B} \|L(u)\|_{F;\mu} \leq c \sup_{\nu\in N} \sup_{u\in B} \|u\|_{E;\nu} < \infty.$$

Finally, if $\mathcal{T}$ is bounded in $\mathcal{L}(E;F)$, it is bounded in $\mathcal{L}(E;F)$-pt since, then, $\|L\|_{\mathcal{L}(E;F)\text{-pt};u,\mu} = \|Lu\|_{F;\mu} = \|L\|_{\mathcal{L}(E;F);\{u\},\mu} < \infty$ for all $u \in E$ and $\mu \in \mathcal{N}_F$. □

Equivalent formulation of Theorem 12.6. When E is a Fréchet space, according to Theorem 12.6, the inequality (12.1) above characterizes the bounded subsets of $\mathcal{L}(E; F)$ since it characterizes the equicontinuous subsets of linear mappings (Theorem 10.9).

When E is normed, this inequality is reduced to the existence, for $\mu \in \mathcal{N}_F$, of $c \in \mathbb{R}$ such that: $\forall u \in E$,

$$\sup_{L \in \mathcal{T}} \|Lu\|_{F;\mu} \leq c\,\|u\|_E. \ \square$$

12.3. Sequential completeness of $\mathcal{L}(E; F)$ when E is metrizable

Let us show that $\mathcal{L}(E; F)$ is sequentially complete if E is metrizable and F is sequentially complete which is **Hausdorff's theorem**[1].

THEOREM 12.7.– *Let*

E be a semi-normed metrizable space and F a Neumann space.

Then:

$\mathcal{L}(E; F)$ is a Neumann space. ∎

Proof. Let $\{\| \ \|_{F;\mu} : \mu \in \mathcal{N}_F\}$ be the family of semi-norms of F.

Let $(L_n)_{n\in\mathbb{N}}$ be a Cauchy sequence (Definition 2.6 (b)) of $\mathcal{L}(E; F)$. Following Definition 12.1 (a) of its semi-norms, for any bounded set B of E and any $\mu \in \mathcal{N}_F$ and $\epsilon > 0$, there exists $n_0 \in \mathbb{N}$ such that, for all $n' \geq n \geq n_0$,

$$\|L_n - L_{n'}\|_{\mathcal{L}(E;F);B,\mu} = \sup_{v\in B} \|(L_n - L_{n'})(v)\|_{F;\mu} \leq \epsilon. \tag{12.2}$$

For all $u \in E$, this inequality relative to $B = \{u\}$ shows that $(L_n(u))_{n\in\mathbb{N}}$ is a Cauchy sequence in the space F which is sequentially complete and subsequently it has a limit that we denote by $L(u)$. As a result, a mapping L from E into F is obtained. We show that it is linear by passing to the limit in the equalities $L_n(u + v) = L_n(u) + L_n(v)$ and $L_n(tu) = tL_n(u)$.

To establish its continuity, let us evaluate the limit when $n' \to \infty$ in the majoration of (12.2). It follows that, for all $u \in B$,

$$\|L_n(u) - L(u)\|_{F;\mu} \leq \epsilon \tag{12.3}$$

from which

$$\sup_{u\in B} \|L(u)\|_{F;\mu} \leq \sup_{u\in B} \|L_n(u)\|_{F;\mu} + \epsilon = \|L_n\|_{\mathcal{L}(E;F);B,\mu} + \epsilon < \infty.$$

1 **History of Hausdorff's theorem 12.7.** Felix HAUSDORFF proved this result in 1932 [54, p. 299], for normed spaces E and F.

Therefore, the image $\{L(u) : u \in B\}$ of any bounded set B of E is bounded in F. Since E is metrizable and L is linear, the Banach–Mazur–Orlicz–Mackey theorem 9.2 thus shows that L is continuous.

We now obtain $\|L_n - L\|_{\mathcal{L}(E;F);B,\mu} \leq \epsilon$ by taking in (12.3) the least upper bound for all $u \in B$. Whence $L_n \to L$ in $\mathcal{L}(E;F)$ which proves that $\mathcal{L}(E;F)$ is sequentially complete. □

Let us show that $\mathcal{L}(E;F)$-pt is sequentially complete if E is a Fréchet space and F is sequentially complete.

THEOREM 12.8.– *Let*

E be a Fréchet space and F a Neumann space.

Then:

$\mathcal{L}(E;F)$-pt *is a Neumann space.* ▮

Proof. Let us denote by $\{\| \ \|_{E;\nu} : \nu \in \mathcal{N}_E\}$ and $\{\| \ \|_{F;\mu} : \mu \in \mathcal{N}_F\}$ the families of semi-norms of E and F.

Let $(L_n)_{n\in\mathbb{N}}$ be a Cauchy sequence (Definition 2.6 (b)) of $\mathcal{L}(E;F)$-pt. According to Definition 12.1 (b) of its semi-norms, for every $u \in E$, $\mu \in \mathcal{N}_F$ and $\epsilon > 0$, there exists $n_0 \in \mathbb{N}$ such that, for all $n' \geq n \geq n_0$,

$$\|L_n - L_{n'}\|_{\mathcal{L}(E;F)\text{-pt};u,\mu} = \|(L_n - L_{n'})(u)\|_{F;\mu} \leq \epsilon.$$

Therefore, $(L_n u)_{n\in\mathbb{N}}$ is a Cauchy sequence in F which is sequentially complete, thereby it has a limit that we denote $L(u)$. We show that L is linear from E into F by passing to the limit in the equalities $L_n(u + v) = L_n(u) + L_n(v)$ and $L_n(tu) = tL_n(u)$.

In order to establish its continuity, we observe that since any Cauchy sequence is bounded (Theorem 2.7 (b)), it follows that, for all $u \in E$ and $\mu \in \mathcal{N}_F$,

$$\sup_{n\in\mathbb{N}} \|L_n u\|_{F;\mu} < \infty.$$

Since E is a Fréchet space, the Banach–Steinhaus theorem 10.11 therefore shows that, for all $\mu \in \mathcal{N}_F$, there exist a finite set $N \subset \mathcal{N}_E$ and $c \geq 0$ such that, for all $n \in \mathbb{N}$ and $u \in E$,

$$\|L_n u\|_{F;\mu} \leq c \sup_{\nu\in N} \|u\|_{E;\nu}.$$

At the limit,

$$\|Lu\|_{F;\mu} \leq c \sup_{\nu\in N} \|u\|_{E;\nu}$$

hence (Theorem 7.14 (a)) L is continuous.

The convergence $L_n u \to Lu$ in F for all $u \in E$ then results in $L_n \to L$ in $\mathcal{L}(E;F)$-pt. This proves that $\mathcal{L}(E;F)$-pt is sequentially complete. □

Non-optimality of Theorems 12.7 and 12.8. It is not necessary that E be metrizable in order for $\mathcal{L}(E;F)$ and $\mathcal{L}(E;F)$-pt to be Neumann spaces.

For example, $\mathcal{D}(\mathbb{R})$ is not metrizable and yet $\mathcal{L}(\mathcal{D}(\mathbb{R});F)$-pt is a Neumann space because it is equal to $\mathcal{D}'(\mathbb{R};F)$ endowed with pointwise convergence on $\mathcal{D}(\mathbb{R})$ which is a Neumann space, see Volume II. The space $\mathcal{L}(\mathcal{D}(\mathbb{R});F)$ is also a Neumann space because it has the same converging sequences and the same Cauchy sequences as $\mathcal{L}(\mathcal{D}(\mathbb{R});F)$-pt (which is not immediate to verify). □

12.4. Semi-norms and norm on $\mathcal{L}(E;F)$ when E is normed

Let us build, when E is normed, another family of semi-norms defining the topology of $\mathcal{L}(E;F)$.

THEOREM 12.9.– *Let F be a separated semi-normed space, $\{\| \ \|_{F;\mu} : \mu \in \mathcal{N}_F\}$ its family of semi-norms and*

E a normed space.

For all $L \in \mathcal{L}(E;F)$, $u \in E$ and $\mu \in \mathcal{N}_F$, by denoting $B_E = \{u \in E : \|u\|_E \leq 1\}$:

(a) *The topology of $\mathcal{L}(E;F)$ is generated by the family of semi-norms, indexed by $\mu \in \mathcal{N}_F$,*

$$|||L|||_{\mathcal{L}(E;F);\mu} \stackrel{\text{def}}{=} \sup_{u \in E,\, u \neq 0} \frac{\|Lu\|_{F;\mu}}{\|u\|_E}.$$

(b)
$$|||L|||_{\mathcal{L}(E;F);\mu} = \sup\nolimits_{u \in B_E} \|Lu\|_{F;\mu}.$$

(c)
$$\|Lu\|_{F;\mu} \leq |||L|||_{\mathcal{L}(E;F);\mu} \|u\|_E.$$

(d) $(L,u) \mapsto Lu$ *is continuous bilinear from $\mathcal{L}(E;F) \times E$ into F.* ▮

Proof. (a) Let B be a bounded subset of E and $\mu \in \mathcal{N}_F$. According to Definition 12.1 (a) of semi-norms of $\mathcal{L}(E;F)$, there is, by denoting $r = \sup_{u \in B} \|u\|_E$,

$$\|L\|_{\mathcal{L}(E;F);B,\mu} = \sup_{u \in B} \|Lu\|_{F;\mu} \leq \sup_{\|u\|_E \leq r,\, u \neq 0} r \frac{\|Lu\|_{F;\mu}}{\|u\|_E} \leq r\, |||L|||_{\mathcal{L}(E;F);\mu}. \tag{12.4}$$

Conversely,

$$|||L|||_{\mathcal{L}(E;F);\mu} = \sup_{u \in E,\, u \neq 0} \left\| L\Big(\frac{u}{\|u\|_E}\Big) \right\|_{F;\mu} \leq \sup_{v \in B_E} \|Lv\|_{F;\mu} = \|L\|_{\mathcal{L}(E;F);B_E,\mu}. \tag{12.5}$$

The family of the semi-norms $|||\ \ |||_{\mathcal{L}(E;F);\mu}$ is thus equivalent (Definition 3.1) to the family of the semi-norms of $\mathcal{L}(E;F)$.

(b) This equality results from the inequalities (12.4) relative to B_E and (12.5).

(c) This inequality results from Definition (a).

(d) Bilinearity is elementary. Continuity results from the inequality (c) according to the characterization of the continuous multilinear mappings of Theorem 7.20 (a). □

Let us show that $\mathcal{L}(E;F)$ is normable if E and F are normed.

THEOREM 12.10.– *Let*

E and F be normed spaces.

Then, for all $L \in \mathcal{L}(E;F)$ and $u \in E$, by denoting $B_E = \{u \in E : \|u\|_E \leq 1\}$:

(a) *The topology of $\mathcal{L}(E;F)$ is generated by the norm defined by*

$$|||L|||_{\mathcal{L}(E;F)} \stackrel{\text{def}}{=} \sup_{u\in E,\, u\neq 0} \frac{\|Lu\|_F}{\|u\|_E}.$$

(b) $$|||L|||_{\mathcal{L}(E;F)} = \sup_{u\in B_E} \|Lu\|_F.$$

(c) $$\|Lu\|_F \leq |||L|||_{\mathcal{L}(E;F)} \|u\|_E.$$

(d) *If F is a Banach space, $\mathcal{L}(E;F)$ is a Banach space for the norm of* (a). ▮

Proof. (a) The family of semi-norms given by Theorem 12.9 (a) is here reduced to this norm.

(b) and (c). These are particular cases of parts (b) and (c) of Theorem 12.9.

(d) If F is sequentially complete, $\mathcal{L}(E;F)$ also is (Hausdorff's theorem 12.7). It is therefore a Banach space (Definition 4.10 (c)). □

Relation between the semi-norms and the norm of $\mathcal{L}(E;F)$ when E and F are normed. The norm of Theorem 12.10 (a) is connected to the semi-norms of Definition 12.1 (a), following Theorem 12.10 (b), by

$$|||L|||_{\mathcal{L}(E;F)} = \|L\|_{\mathcal{L}(E;F);B_E}. \tag{12.6}$$

□

Metrizability of $\mathcal{L}(E;F)$. Theorems 12.9 (a) and 12.7 lead to:

$$E \text{ is normed and } F \text{ is metrizable} \quad \Rightarrow \quad \mathcal{L}(E;F) \text{ is metrizable.} \tag{12.7}$$

$$E \text{ is normed and } F \text{ is a Fréchet space} \quad \Rightarrow \quad \mathcal{L}(E;F) \text{ is a Fréchet space.} \tag{12.8}$$

□

Normability of $\mathcal{L}(E;F)$. Theorem 12.10 (a) leads to:

$$E \text{ and } F \text{ are normable} \quad \Rightarrow \quad \mathcal{L}(E;F) \text{ is normable.} \tag{12.9}$$

□

12.5. Continuity of the composition of linear mappings

Observe first that any composite of linear mappings is linear.

THEOREM 12.11.– (a) *Any mapping obtained by composing linear mappings is linear.*

(b) *The composition of linear mappings is a bilinear mapping.* ▮

Proof. Let E, F and G be three vector spaces, L a linear mapping from E into F and M and N linear mappings from F into G.

(a) The mapping $M \circ L$ is linear because, for all $u \in E$, $v \in E$ and $t > 0$

$$(M \circ L)(tu) = M(L(tu)) = M(tL(u)) = tM(L(u)) = t(M \circ L)(u),$$

$$(M \circ L)(u+v) = M(L(u+v)) = M(L(u) + L(v)) =$$
$$= M(L(u)) + M(L(v)) = (M \circ L)(u) + (M \circ L)(v).$$

(b) The mapping $M \mapsto M \circ L$ is linear because, for all $M, N, t > 0$ and $u \in E$,

$$((tM) \circ L)(u) = (tM)(L(u)) = t(M(L(u))) = t(M \circ L)(u),$$

$$((M+N) \circ L)(u) = (M+N)(L(u)) = M(L(u)) + N(L(u)) =$$
$$= (M \circ L)(u) + (N \circ L)(u).$$

We show in a similar manner that the mapping $L \mapsto M \circ L$ is linear. The mapping $(L, M) \mapsto M \circ L$ is therefore bilinear. □

Let us give continuity properties of the composition of continuous linear mappings[2].

THEOREM 12.12.– *Let E, F and G be three separated semi-normed spaces and*

$$L \in \mathcal{L}(E; F), \quad M \in \mathcal{L}(F; G).$$

Then:

(a) $$M \circ L \in \mathcal{L}(E; G).$$

(b) $M \mapsto M \circ L$ *is continuous linear from* $\mathcal{L}(F; G)$ *into* $\mathcal{L}(E; G)$.

2 **History of Theorem 12.12.** Nicolas BOURBAKI has showed [17, chap. IV, § 4, no. 2, proposition 9, p. 42] that the mapping $(L, M) \mapsto M \circ L$ is hypocontinuous, that is property (e) of Theorem 12.12.

(c) $L \mapsto M \circ L$ *is continuous linear from* $\mathcal{L}(E;F)$ *into* $\mathcal{L}(E;G)$.

(d) $\begin{cases} (L,M) \mapsto M \circ L \text{ is bilinear sequentially continuous} \\ \text{from } \mathcal{L}(E;F) \times \mathcal{L}(F;G) \text{ into } \mathcal{L}(E;G). \end{cases}$

(e) $\begin{cases} \text{For every bounded subset } \mathcal{T} \text{ of } \mathcal{L}(E;F), \\ (L,M) \mapsto M \circ L \text{ is continuous from } \mathcal{T} \times \mathcal{L}(F;G) \text{ into } \mathcal{L}(E;G). \end{cases}$

(f) F *is normed* $\Rightarrow$ $\begin{cases} (L,M) \mapsto M \circ L \text{ is continuous bilinear} \\ \text{from } \mathcal{L}(E;F) \times \mathcal{L}(F;G) \text{ into } \mathcal{L}(E;G). \end{cases}$ ∎

Proof. Let $\{\| \ \|_{E;\nu} : \nu \in \mathcal{N}_E\}$, $\{\| \ \|_{F;\mu} : \mu \in \mathcal{N}_F\}$ and $\{\| \ \|_{G;\lambda} : \lambda \in \mathcal{N}_G\}$ be the families of semi-norms of E, F and G.

(a) The mapping $M \circ L$ is linear (Theorem 12.11 (a)) and it is continuous as any composite of continuous mappings (Theorem 7.9 (a)). Therefore, $M \circ L \in \mathcal{L}(E;G)$.

(b) The mapping $M \mapsto M \circ L$ is linear (Theorem 12.11 (b)). Let us verify that it is continuous. Let $\lambda \in \mathcal{N}_G$ and B be a bounded subset of E. According to Definition 12.1 (a) of the semi-norms of $\mathcal{L}(E;F)$,

$$\begin{aligned} \|M \circ L\|_{\mathcal{L}(E;G);B,\lambda} &= \sup_{u \in B} \|M(Lu)\|_{G;\lambda} = \\ &= \sup_{v \in L(B)} \|M(v)\|_{G;\lambda} = \|M\|_{\mathcal{L}(F;G);L(B),\lambda} \end{aligned} \tag{12.10}$$

because $L(B)$ is bounded in F (Theorem 7.15). According to the characterization of continuous linear mappings of Theorem 7.14 (a), this shows that $M \mapsto M \circ L$ is continuous from $\mathcal{L}(F;G)$ into $\mathcal{L}(E;G)$.

(c) The mapping $L \mapsto M \circ L$ is linear (Theorem 12.11 (b)). Let us verify that it is continuous. Let $\lambda \in \mathcal{N}_G$. Since M is continuous linear, there exist (again by Theorem 7.14 (a)) a finite set $P \subset \mathcal{N}_F$ and $c \in \mathbb{R}$ such that:

$$\|Mv\|_{G;\lambda} \le c \sup_{\mu \in P} \|v\|_{F;\mu}. \tag{12.11}$$

In addition, let B be a bounded subset of E. Then:

$$\begin{aligned} \|M \circ L\|_{\mathcal{L}(E;G);B,\lambda} &= \sup_{u \in B} \|M(Lu)\|_{G;\lambda} \le \\ &\le c \sup_{\mu \in P} \sup_{u \in B} \|Lu\|_{F;\mu} = c \sup_{\mu \in P} \|L\|_{\mathcal{L}(E;F);B,\mu}. \end{aligned} \tag{12.12}$$

This shows (again by Theorem 7.14 (a)) that $L \mapsto M \circ L$ is continuous from $\mathcal{L}(E;F)$ into $\mathcal{L}(E;G)$.

(e) Let $\mathcal{T}$ be a bounded subset of $\mathcal{L}(E;F)$, L and L_0 in $\mathcal{T}$ and M and M_0 in $\mathcal{L}(F;G)$. Let us split

$$M \circ L - M_0 \circ L_0 = (M - M_0) \circ L + M_0 \circ (L - L_0). \tag{12.13}$$

Let B be a bounded subset of E and $\lambda \in \mathcal{N}_G$. Since $\mathcal{T}$ is bounded in $\mathcal{L}(E;F)$, the set $V = \{L(u) : L \in \mathcal{T},\ u \in B\}$ is bounded in F (Theorem 12.5 (a)). Therefore, analogously to (12.10), we bound

$$\begin{aligned} \|(M - M_0) \circ L\|_{\mathcal{L}(E;G);B,\lambda} &= \sup_{u \in B} \|(M - M_0)(Lu)\|_{G;\lambda} \le \\ &\le \sup_{v \in V} \|(M - M_0)(v)\|_{G;\lambda} = \|(M - M_0)\|_{\mathcal{L}(F;G);V,\lambda}. \end{aligned} \tag{12.14}$$

On the other hand, P being a finite subset of $\mathcal{N}_F$ and c a real number satisfying the inequality (12.11) with respect to M_0, the majoration (12.12) gives

$$\|M_0 \circ (L - L_0)\|_{\mathcal{L}(E;G);B,\lambda} \le c \sup_{\mu \in P} \|L - L_0\|_{\mathcal{L}(E;F);B,\mu}. \tag{12.15}$$

These two majorations jointly with the splitting (12.13) result in the continuity at the point (L_0, M_0) of the mapping $(L, M) \mapsto M \circ L$, from $\mathcal{T} \times \mathcal{L}(F;G)$ into $\mathcal{L}(E;G)$. In effect, given $\epsilon > 0$ and $\eta = \epsilon/(2\sup\{1, c\})$, if

$$\sup_{\mu \in P} \|(L - L_0, M - M_0)\|_{\mathcal{L}(E;F)\times\mathcal{L}(F;G);B,\mu,V,\lambda} \le \eta$$

that is, following Definition 6.1 of the semi-norms of the product, if

$$\sup_{\mu \in P} (\|L - L_0\|^2_{\mathcal{L}(E;F);B,\mu} + \|M - M_0\|^2_{\mathcal{L}(F;G);V,\lambda})^{1/2} \le \eta$$

then $\|(M - M_0) \circ L\|_{\mathcal{L}(E;G);B,\lambda} \le \epsilon/2$ and $\|M_0 \circ (L - L_0)\|_{\mathcal{L}(E;G);B,\lambda} \le \epsilon/2$ according to (12.14) and (12.15) from which, with (12.13),

$$\|M \circ L - M_0 \circ L_0\|_{\mathcal{L}(E;G);B,\lambda} \le \epsilon.$$

(d) The mapping $(L, M) \mapsto M \circ L$ is bilinear (Theorem 12.11 (b)). Let us verify that it is sequentially continuous.

Let $(L_n)_{n\in\mathbb{N}}$ and $(M_n)_{n\in\mathbb{N}}$ be two sequences converging respectively to L in $\mathcal{L}(E;F)$ and M in $\mathcal{L}(F;G)$. The set $\mathcal{T} = \{L_n : n \in \mathbb{N}\}$ is bounded (Theorem 2.7 (a)) in $\mathcal{L}(E;F)$ thus, according to (e), the mapping $(L, M) \mapsto M \circ L$ is continuous, and *a fortiori* sequentially continuous (Theorem 7.2 (a)) from $\mathcal{T} \times \mathcal{L}(F;G)$ into $\mathcal{L}(E;G)$. As a result, $M_n \circ L_n \to M \circ L$ in $\mathcal{L}(E;G)$.

(f) Assume that F is normed and let us verify that the mapping $(L, M) \mapsto M \circ L$ is continuous. Remember that (Theorem 12.9 (a)) the topology of $\mathcal{L}(F; G)$ is then generated by the family of the semi-norms, indexed by $\lambda \in \mathcal{N}_G$,

$$|||M|||_{\mathcal{L}(F;G);\lambda} = \sup_{v \in F,\, v \neq 0} \frac{\|Mv\|_{G;\lambda}}{\|v\|_F}.$$

For every bounded set B of E and every $\lambda \in \mathcal{N}_G$, there is (Definition 12.1 (a) and Theorem 12.9 (c))

$$\|M \circ L\|_{\mathcal{L}(E;G);B,\lambda} = \sup_{u \in B} \|M(Lu)\|_{G;\lambda} \leq$$

$$\leq \sup_{u \in B} |||M|||_{\mathcal{L}(F;G);\lambda} \|L(u)\|_F = |||M|||_{\mathcal{L}(F;G);\lambda} \|L\|_{\mathcal{L}(E;F);B}.$$

According to the characterization of continuous bilinear mappings of Theorem 7.20 (a), this shows that $(L, M) \mapsto M \circ L$ is continuous from $\mathcal{L}(E; F) \times \mathcal{L}(F; G)$ into $\mathcal{L}(E; G)$. □

QUESTION: Continuity when F is not normed. The author does not know if the assumption "F is normed" is necessary in Theorem 12.12 (f). In other words:

$$\begin{cases} \textit{Is there any non-normed space } F \textit{ for which the mapping} \\ (L, M) \mapsto M \circ L \textit{ be continuous from } \mathcal{L}(E; F) \times \mathcal{L}(F; G) \textit{ into } \mathcal{L}(E; G)? \end{cases}$$

A positive response would make it possible to obtain (in Theorem 19.18) the continuity of the differential of the composite of continuously differentiable mappings without having to assume that the intermediate space F is normed (or that E is metrizable). Likewise, it would also allow obtaining (in Theorem 21.6) that the composite of m times differentiable mappings is m times differentiable without these assumptions. □

Let us bound the norm of the composite of linear mappings in normed spaces.

THEOREM 12.13.– *Let E, F and G be three normed spaces, $L \in \mathcal{L}(E; F)$ and $M \in \mathcal{L}(F; G)$. Then, for the norm of Theorem 12.10 (a):*

$$|||M \circ L|||_{\mathcal{L}(E;G)} \leq |||M|||_{\mathcal{L}(F;G)} |||L|||_{\mathcal{L}(E;F)}. \blacksquare$$

Proof. We have (Theorem 12.10 (a) and (c)):

$$|||M \circ L|||_{\mathcal{L}(E;G)} = \sup_{u \in E,\, u \neq 0} \frac{\|M(Lu)\|_G}{\|u\|_E} \leq$$

$$\leq \sup_{u \in E,\, u \neq 0} |||M|||_{\mathcal{L}(F;G)} \frac{\|Lu\|_F}{\|u\|_E} = |||M|||_{\mathcal{L}(F;G)} |||L|||_{\mathcal{L}(E;F)}. \square$$

12.6. Inversibility in the neighborhood of an isomorphism

Let us first show that any perturbation of the identity of a Banach space by a strictly contracting linear mapping is invertible.

THEOREM 12.14.– *Let*

$$E \text{ be a Banach space}$$

and $L \in \mathcal{L}(E; E)$ *such that:*

$$|||L|||_{\mathcal{L}(E;E)} < 1.$$

where $|||L|||_{\mathcal{L}(E;E)} = \sup_{u \in E,\, u \neq 0} \|Lu\|_E / \|u\|_E$.

Then, by denoting $L^n = L \circ \cdots \circ L$ *(n times),* $L^0 = I_E$ *and* I_E *the identity of* E*:*

(a) $$\text{The series } \sum_{n \geq 0} L^n \text{ converges in } \mathcal{L}(E; E).$$

(b) $$\sum_{n \geq 0} L^n \text{ is the inverse mapping of } I_E - L.$$

(c) $$\Big|\Big|\Big| \sum_{n \geq 0} L^n \Big|\Big|\Big|_{\mathcal{L}(E;E)} \leq \frac{1}{1 - |||L|||_{\mathcal{L}(E;E)}}.$$

(d) $$\Big|\Big|\Big| \sum_{n \geq 1} L^n \Big|\Big|\Big|_{\mathcal{L}(E;E)} \leq \frac{|||L|||_{\mathcal{L}(E;E)}}{1 - |||L|||_{\mathcal{L}(E;E)}}. \blacksquare$$

Proof. (a) The $\mathcal{L}(E; E)$ space thus normed is a Banach space (Theorem 12.10 (d)) since E is a Banach space. To ensure that the series $\sum_{n \geq 0} L^n$ converges, it thus suffices (Theorem 4.22) that it be absolutely converging. Now,

$$|||L^n|||_{\mathcal{L}(E;E)} \leq (|||L|||_{\mathcal{L}(E;E)})^n. \tag{12.16}$$

In effect, this property holds for $n = 0$ and 1 and, if it holds for $n \geq 1$, it does for $n + 1$ according to Theorem 12.13.

Therefore, by denoting $k = |||L|||_{\mathcal{L}(E;E)}$, it follows that, see (1.10), p. 7, for all $i \in \mathbb{N}$,

$$\sum_{0 \leq n \leq i} |||L^n|||_{\mathcal{L}(E;E)} \leq 1 + k + \cdots + k^i \leq \frac{1}{1-k}.$$

This proves that $\sum_{n \geq 0} L^n$ is absolutely converging (Definition 4.21 (b)) and therefore that it converges.

(b) For all $i \in \mathbb{N}$, we have

$$\Big(\sum_{0 \leq n \leq i} L^n \Big) \circ (I_E - L) = I_E - L^{i+1} \tag{12.17}$$

because the left-hand side is equal to $\sum_{0\le n\le i} L^n - \sum_{1\le n\le i+1} L^n = L^0 - L^{i+1}$ since $L^0 = I_E$. By passing to the limit in $\mathcal{L}(E;E)$ when $i \to \infty$, (12.17) gives

$$\Big(\sum_{n\ge 0} L^n\Big) \circ (I_E - L) = I_E. \tag{12.18}$$

As a matter of fact, the left-hand side converges because $\sum_{0\le n\le i} L^n \to \sum_{n\ge 0} L^n$ (Definition 4.21 (a)) and because the composition of mappings is continuous (Theorem 12.12 (f)) since E is a Banach space. In addition, in the right-hand side, $L^{i+1} \to 0$ because, according to the inequality (12.16), $|||L^{i+1}|||_{\mathcal{L}(E;E)} \le k^{i+1}$ which tends to 0 (Theorem 1.4) since it is assumed that $k < 1$.

This establishes (12.18) which proves that $I_E - L$ is invertible and $\sum_{n\ge 0} L^n$ is its inverse mapping.

(c) According to the inequalities (12.16) and, again, (1.10),

$$\Big|\Big|\Big|\sum_{0\le n\le i} L^n\Big|\Big|\Big|_{\mathcal{L}(E;E)} \le \sum_{0\le n\le i} |||L^n|||_{\mathcal{L}(E;E)} \le$$
$$\le \sum_{0\le n\le i} (|||L|||_{\mathcal{L}(E;E)})^n \le \frac{1}{1 - |||L|||_{\mathcal{L}(E;E)}}.$$

Hence, at the limit, the stated inequality.

(d) Similarly,

$$\Big|\Big|\Big|\sum_{1\le n\le i} L^n\Big|\Big|\Big|_{\mathcal{L}(E;E)} \le \sum_{1\le n\le i} (|||L|||_{\mathcal{L}(E;E)})^n \le \frac{|||L|||_{\mathcal{L}(E;E)}}{1 - |||L|||_{\mathcal{L}(E;E)}}. \quad \square$$

Let us now show that any continuous linear mapping in Banach spaces that is close enough to an isomorphism is also invertible. We denote:

$\mathcal{I}(E;F)$ is the set of isomorphisms from E onto F.

(It is not a vector space, because it does not contain the zero mapping.)

THEOREM 12.15.– *Let*

E and F be Banach spaces.

Then:

(a) *$\mathcal{I}(E;F)$ is open in $\mathcal{L}(E;F)$.*

(b) $T \mapsto T^{-1}$ *is continuous from (the subset)* $\mathcal{I}(E;F)$ *(of* $\mathcal{L}(E;F)$*) into* $\mathcal{L}(F;E)$. ∎

Proof. (a) Let $S \in \mathcal{I}(E;F)$ and $T \in \mathcal{L}(E;F)$ be such that:

$$|||T - S|||_{\mathcal{L}(E;F)} < \frac{1}{|||S^{-1}|||_{\mathcal{L}(F;E)}}. \tag{12.19}$$

Let us show that T is invertible. Let

$$R = (S - T) \circ S^{-1} = I_F - T \circ S^{-1}$$

where I_F is the identity of F. Since (Theorem 12.13)

$$|||R|||_{\mathcal{L}(F;F)} \leq |||S - T|||_{\mathcal{L}(E;F)}\, |||S^{-1}|||_{\mathcal{L}(F;E)} \tag{12.20}$$

we have $|||R|||_{\mathcal{L}(F;F)} < 1$. Hence (Theorem 12.14 (a) and (b)) $\sum_{n \geq 0} R^n \in \mathcal{L}(F;F)$ and

$$T \circ S^{-1} \circ \sum_{n \geq 0} R^n = (I_F - R) \circ \sum_{n \geq 0} R^n = I_F.$$

Therefore, T is invertible and

$$T^{-1} = S^{-1} \circ \sum_{n \geq 0} R^n. \tag{12.21}$$

Then, (12.19) yields $T \in \mathcal{I}(E;F)$ which proves that $\mathcal{I}(E;F)$ is open.

(b) Let us again assume (12.19). Since S^{-1} is linear (Theorem 7.17 (b)) and R^0 is equal to I_F (according to the notations of Theorem 12.14), Equality (12.21) leads to

$$T^{-1} - S^{-1} = S^{-1} \circ \Big(\sum_{n \geq 0} R^n - I_F \Big) = S^{-1} \circ \sum_{n \geq 1} R^n.$$

Thus (Theorems 12.13 and 12.14 (d))

$$|||T^{-1} - S^{-1}|||_{\mathcal{L}(F;E)} \leq |||S^{-1}|||_{\mathcal{L}(F;E)} \Big|\Big|\Big| \sum_{n \geq 1} R^n \Big|\Big|\Big|_{\mathcal{L}(F;F)} \leq$$

$$\leq \frac{|||S^{-1}|||_{\mathcal{L}(F;E)}\, |||R|||_{\mathcal{L}(F;F)}}{1 - |||R|||_{\mathcal{L}(F;F)}}.$$

With (12.20), we get, by denoting $r = 1/(1 - |||R|||_{\mathcal{L}(F;F)})$,

$$|||T^{-1} - S^{-1}|||_{\mathcal{L}(F;E)} \leq r \left(|||S^{-1}|||_{\mathcal{L}(F;E)} \right)^2 |||T - S|||_{\mathcal{L}(E;F)}.$$

Now, if $|||T - S|||_{\mathcal{L}(E;F)} \leq 1/(2|||S^{-1}|||_{\mathcal{L}(F;E)})$, then $|||R|||_{\mathcal{L}(F;F)} \leq 1/2$ again according to (12.20) and thus $r \leq 2$. Therefore, the above inequality implies that the ma pping $T \mapsto T^{-1}$ is continuous at the point S. □

Assumptions on E and F. The proof of Theorem 12.15 only uses the sequential completeness of E, not that of F. However, if there exists an isomorphism from E into F, the sequential completeness of E implies that of F. Consequently, there is no gain in omitting the latter. □

12.7. The space $\mathcal{L}^d(E_1 \times \cdots \times E_d; F)$

Let us define spaces of continuous multilinear mappings.

DEFINITION 12.16.– *Let $E_1, E_2 \ldots, E_d$ and F be separated semi-normed spaces and $\{\| \ \|_{F;\mu} : \mu \in \mathcal{N}_F\}$ be the family of semi-norms of F.*

$\mathcal{L}^d(E_1 \times \cdots \times E_d; F)$ is the vector space of continuous multilinear mappings from $E_1 \times \cdots \times E_d$ into F endowed with the semi-norms, indexed by the bounded sets B_1 of $E_1 \ldots, B_d$ of E_d and by $\mu \in \mathcal{N}_F$,

$$\|T\|_{\mathcal{L}^d(E_1 \times \cdots \times E_d; F); B_1, \ldots, B_d, \mu} \stackrel{\text{def}}{=} \sup_{u_1 \in B_1, \ldots, u_d \in B_d} \|T(u_1, \ldots, u_d)\|_{F;\mu}. \ \blacksquare$$

Justification.– The mapping $T \mapsto \sup_{u_1 \in B_1, \ldots, u_d \in B_d} \|T(u_1, \ldots, u_d)\|_{F;\mu}$ is indeed a semi-norm on $\mathcal{L}^d(E_1 \times \cdots \times E_d; F)$ according to Theorem 12.2 because, for each $(u_1, \ldots, u_d)$, the mapping $T \mapsto \|T(u_1, \ldots, u_d)\|_{F;\mu}$ is a semi-norm and, for each T,

$$\sup_{u_1 \in B_1, \ldots, u_d \in B_d} \|T(u_1, \ldots, u_d)\|_{F;\mu} < \infty$$

(this quantity is finite since every continuous multilinear mapping transforms any product of bounded sets into a bounded set according to Theorem 9.3). □

Let us show that this space is separated.

THEOREM 12.17.– *Let $E_1, E_2 \ldots, E_d$ and F be separated semi-normed spaces. Then:*

$$\mathcal{L}^d(E_1 \times \cdots \times E_d; F) \text{ is a separated semi-normed space. } \blacksquare$$

Proof. This space is semi-normed by construction. It is separated (Definition 2.3 (c)) because, using the notations of Definition 12.16, if all semi-norms of T are zero, that is, if, for all $u_1 \in E_1 \ldots, u_d \in E_d$, we have $\|T(u_1, \ldots, u_d)\|_{F;\mu} = 0$ for all $\mu \in \mathcal{N}_F$, then, F being separated, $T(u_1, \ldots, u_d) = 0_F$ hence $T = 0_{\mathcal{L}^d}$. □

Let us show that $\mathcal{L}^d(E_1 \times \cdots \times E_d; F)$ is sequentially complete if the E_i are metrizable and F is sequentially complete.

THEOREM 12.18.– *Let*

$E_1, E_2 \ldots, E_d$ be metrizable semi-normed spaces and F a Neumann space.

Then:

$$\mathcal{L}^d(E_1 \times \cdots \times E_d; F) \text{ is a Neumann space. } \blacksquare$$

Proof. Let $\{\| \ \|_{F;\mu} : \mu \in \mathcal{N}_F\}$ be the family semi-norms of F.

Let $(T_n)_{n\in\mathbb{N}}$ be a Cauchy sequence (Definition 2.6 (b)) of $\mathcal{L}^d(E_1 \times \cdots \times E_d; F)$ (Definition 12.16). For every bounded set B_1 of $E_1 \dots, B_d$ of E_d and every $\mu \in \mathcal{N}_F$ and $\epsilon > 0$, there thus exists $n_0 \in \mathbb{N}$ such that, for all $n' \geq n \geq n_0$,

$$\|T_n - T_{n'}\|_{\mathcal{L}^d(E_1\times\cdots\times E_d;F);B_1,\dots,B_d,\mu} = \\ = \sup_{u_1\in B_1,\dots,u_d\in B_d} \|(T_n - T_{n'})(u_1,\dots,u_d)\|_{F;\mu} \leq \epsilon. \tag{12.22}$$

For each $u_1 \in B_1 \dots, u_d \in B_d$, the sequence $(T_n(u_1,\dots,u_d))_{n\in\mathbb{N}}$ is therefore a Cauchy sequence in F that is sequentially complete and thereby it has a limit that we denote by $T(u_1,\dots,u_d)$. As a result, we obtain a mapping T from $E_1 \times \cdots \times E_d$ into F. We show that it is multilinear by passing to the limit in the equalities characterizing the linearity (Definition 7.13) of T_n with respect to each u_i.

In order to establish its continuity, let us pass to the limit in (12.22) when $n' \to \infty$. It follows, for all $u_1 \in B_1 \dots, u_d \in B_d$,

$$\|T_n(u_1,\dots,u_d) - T(u_1,\dots,u_d)\|_{F;\mu} \leq \epsilon \tag{12.23}$$

from which

$$\sup_{u_1\in B_1,\dots,u_d\in B_d} \|T(u_1,\dots,u_d)\|_{F;\mu} \leq \sup_{u_1\in B_1,\dots,u_d\in B_d} \|T_n(u_1,\dots,u_d)\|_{F;\mu} + \epsilon = \\ = \|T_n\|_{\mathcal{L}^d(E_1\times\cdots\times E_d;F);B_1,\dots,B_d,\mu} + \epsilon < \infty.$$

Thus, the image $T(B_1 \times \cdots \times B_d)$ of any product of bounded sets is bounded in F. Since the E_i are metrizable and T is multilinear, the generalization of the Banach–Mazur–Orlicz–Mackey theorem given in Theorem 9.3 then shows that T is continuous. Therefore, $T \in \mathcal{L}^d(E_1 \times \cdots \times E_d; F)$.

By taking in (12.23) the least upper bound for all $u_1 \in B_1 \dots, u_d \in B_d$, we obtain $\|T_n - T\|_{\mathcal{L}^d(E_1\times\cdots\times E_d;F);B_1,\dots,B_d,\mu} \leq \epsilon$, hence $T_n \to T$ in $\mathcal{L}^d(E_1 \times \cdots \times E_d; F)$. This proves that $\mathcal{L}^d(E_1 \times \cdots \times E_d; F)$ is sequentially complete. □

Let us build, when the E_i are normed, another family of semi-norms generating the topology of $\mathcal{L}^d(E_1 \times \cdots \times E_d; F)$.

THEOREM 12.19.– *Let F be a separated semi-normed space, $\{\| \ \|_{F;\mu} : \mu \in \mathcal{N}_F\}$ its family of semi-norms and*

$E_1, E_2 \dots, E_d$ be normed spaces.

Then, for all $T \in \mathcal{L}^d(E_1 \times \cdots \times E_d; F)$, $u_1 \in E_1, \ldots, u_d \in E_d$ *and* $\mu \in \mathcal{N}_F$*:*

(a) *The topology of* $\mathcal{L}^d(E_1 \times \cdots \times E_d; F)$ *is generated by the family of semi-norms, indexed by* $\mu \in \mathcal{N}_F$*,*

$$|||T|||_{\mathcal{L}^d(E_1\times\cdots\times E_d;F);\mu} \stackrel{\text{def}}{=} \sup_{u_i \in E_i,\, u_i \neq 0,\, 1\le i\le d} \frac{\|T(u_1,\ldots,u_d)\|_{F;\mu}}{\|u_1\|_{E_1}\cdots\|u_d\|_{E_d}}.$$

(b) $$\|T(u_1,\ldots,u_d)\|_{F;\mu} \le |||T|||_{\mathcal{L}^d(E_1\times\cdots\times E_d;F);\mu}\|u_1\|_{E_1}\cdots\|u_d\|_{E_d}.$$

(c) $$\begin{cases}(T, u_1, \ldots, u_d) \mapsto T(u_1,\ldots,u_d) \text{ is continuous multilinear}\\ \text{from } \mathcal{L}^d(E_1\times\cdots\times E_d;F)\times E_1\times\cdots\times E_d \text{ into } F.\end{cases}$$ ∎

Proof. (a) Let $B_1\ldots, B_d$ be respectively bounded subsets of $E_1\ldots, E_d$ and let $\mu \in \mathcal{N}_F$. According to Definition 12.16 of semi-norms of $\mathcal{L}^d(E_1 \times \cdots \times E_d; F)$, we have, by denoting $r_i = \sup_{u_i\in B_i} \|u_i\|_{E_i}$,

$$\|T\|_{\mathcal{L}^d(E_1\times\cdots\times E_d;F);B_1,\ldots,B_d,\mu} = \sup_{u_1\in B_1,\ldots,\, u_d\in B_d} \|T(u_1,\ldots,u_d)\|_{F;\mu} \le$$

$$\le \sup_{\|u_i\|_{E_i}\le r_i,\, u_i\neq 0,\, \forall i} r_1\cdots r_d \frac{\|T(u_1,\ldots,u_d)\|_{F;\mu}}{\|u_1\|_{E_1}\cdots\|u_d\|_{E_d}} \le$$

$$\le r_1\cdots r_d\, |||T|||_{\mathcal{L}^d(E_1\times\cdots\times E_d;F);\mu}.$$

Conversely, by denoting B_{E_i} the unit ball of E_i, we have

$$|||T|||_{\mathcal{L}^d(E_1\times\cdots\times E_d;F);\mu} = \sup_{u_i\in E_i,\, u_i\neq 0,\, \forall i} \left\|T\Big(\frac{u_1}{\|u_1\|_{E_1}}, \ldots, \frac{u_d}{\|u_d\|_{E_d}}\Big)\right\|_{F;\mu} \le$$

$$\le \sup_{v_1\in B_{E_1},\ldots,\, v_d\in B_{E_d}} \|T(v_1,\ldots,v_d)\|_{F;\mu} = \|T\|_{\mathcal{L}^d(E_1\times\cdots\times E_d;F);B_{E_1},\ldots,B_{E_d},\mu}.$$

The family of the semi-norms $|||\ |||_{\mathcal{L}^d(E_1\times\cdots\times E_d;F);\mu}$ is therefore equivalent (Definition 3.1) to the family of $\mathcal{L}^d(E_1 \times \cdots \times E_d; F)$ (Definition 12.16).

(b) This inequality results from Definition (a).

(c) Multilinearity is elementary. Continuity results from the inequality (b) according to the characterization of the continuous multilinear mappings of Theorem 7.20 (a). □

Let us show that $\mathcal{L}^d(E_1 \times \cdots \times E_d; F)$ is normable if the E_i and F are normed.

THEOREM 12.20.– *Let*

$E_1, E_2\ldots, E_d$ and F be normed spaces.

Then, for all $T \in \mathcal{L}^d(E_1 \times \cdots \times E_d; F)$ *and* $u_1 \in E_1, \ldots, u_d \in E_d$*:*

(a) *The topology of* $\mathcal{L}^d(E_1 \times \cdots \times E_d; F)$ *is generated by the norm*

$$|||T|||_{\mathcal{L}^d(E_1 \times \cdots \times E_d; F)} \stackrel{\text{def}}{=} \sup_{u_i \in E_i,\, u_i \neq 0,\, 1 \leq i \leq d} \frac{\|T(u_1, \ldots, u_d)\|_F}{\|u_1\|_{E_1} \cdots \|u_d\|_{E_d}}.$$

(b) $$\|T(u_1, \ldots, u_d)\|_F \leq |||T|||_{\mathcal{L}^d(E_1 \times \cdots \times E_d; F)} \|u_1\|_{E_1} \cdots \|u_d\|_{E_d}.$$

(c) *If* F *is a Banach space,* $\mathcal{L}^d(E_1 \times \cdots \times E_d; F)$ *is a Banach space for the norm of* (a). ▮

Proof. (a) The family of semi-norms given by Theorem 12.19 (a) is here reduced to this norm.

(b) This is a particular case of Theorem 12.19 (b).

(c) If F is sequentially complete, $\mathcal{L}^d(E_1 \times \cdots \times E_d; F)$ also is (Theorem 12.18) since the E_i are metrizable (Theorem 4.2). It is thus a Banach space (Definition 4.10 (c)). □

Relation between the semi-norms and the norm of $\mathcal{L}^d(E_1 \times \cdots \times E_d; F)$ **when** E_i **and** F **are normed.** The norm of Theorem 12.20 (a) is related to the semi-norms of Definition 12.16, by denoting $B_{E_i} = \{v \in E_i : \|v\|_{E_i} \leq 1\}$, by

$$|||T|||_{\mathcal{L}^d(E_1 \times \cdots \times E_d; F)} = \|T\|_{\mathcal{L}^d(E_1 \times \cdots \times E_d; F); B_{E_1}, \ldots, B_{E_d}}. \ \square$$

12.8. Separation of the variables of a multilinear mapping

Let us separate the variables in a d-linear mapping, which gives a $d-1$-linear mapping valued in a linear mappings space.

THEOREM 12.21.– *Let* $E_1, E_2 \ldots, E_d$ *and* F *be separated semi-normed spaces. For any*

$$T \in \mathcal{L}^d(E_1 \times \cdots \times E_d; F),$$

and $i \in [\![1, d]\!]$*, let us denote* T_i *the mapping obtained by separating the* i*-th variable, that is*

$$(T_i(u_1, \ldots, u_{i-1}, u_{i+1}, \ldots, u_d))(u_i) \stackrel{\text{def}}{=} T(u_1, \ldots, u_d).$$

Then:

(a) $$T_i \in \mathcal{L}^{d-1}(E_1 \times \cdots \times E_{i-1} \times E_{i+1} \times \cdots \times E_d; \mathcal{L}(E_i; F)).$$

(b) $$\begin{cases} T \mapsto T_i \text{ is an isomorphism from } \mathcal{L}^d(E_1 \times \cdots \times E_d; F) \text{ onto its} \\ \text{image in } \mathcal{L}^{d-1}(E_1 \times \cdots \times E_{i-1} \times E_{i+1} \times \cdots \times E_d; \mathcal{L}(E_i; F)). \end{cases}$$

(c) *If the E_j are metrizable:*

$$\begin{cases} T \mapsto T_i \text{ is an isomorphism from } \mathcal{L}^d(E_1 \times \cdots \times E_d; F) \\ \text{onto } \mathcal{L}^{d-1}(E_1 \times \cdots \times E_{i-1} \times E_{i+1} \times \cdots \times E_d; \mathcal{L}(E_i; F)). \end{cases}$$

(d) *If the E_j and F are normed, then for the norms defined in Theorems 12.10 (a) and 12.20 (a):*

$$|||T_i|||_{\mathcal{L}^{d-1}(E_1\times\cdots\times E_{i-1}\times E_{i+1}\times\cdots\times E_d;\mathcal{L}(E_i;F))} = |||T|||_{\mathcal{L}^d(E_1\times\cdots\times E_d;F)}. \blacksquare$$

Proof. To simplify the writing, let us write the proof for $i = d$. The general case results thereof by permuting variables.

(a) Let $\{\| \ \|_{E_j;\nu_j} : \nu_j \in \mathcal{N}_{E_j}\}$ and $\{\| \ \|_{F;\mu} : \mu \in \mathcal{N}_F\}$ be the families of semi-norms of E_j and F. According to the characterization of continuous multilinear mappings of Theorem 7.20 (a), for all $\mu \in \mathcal{N}_F$, there exist finite sets $N_j \subset \mathcal{N}_{E_j}$ and $c \geq 0$ such that $\|T(u_1, \ldots, u_d)\|_{F;\mu} \leq c \sup_{\nu_1 \in N_1} \|u_1\|_{E_1;\nu_1} \cdots \sup_{\nu_d \in N_d} \|u_d\|_{E_d;\nu_d}$ for all $u_1 \in E_1, \ldots, u_d \in E_d$. That is to say

$$\|T_d(u_1, \ldots, u_{d-1})(u_d)\|_{F;\mu} \leq c \sup_{\nu_1 \in N_1} \|u_1\|_{E_1;\nu_1} \cdots \sup_{\nu_d \in N_d} \|u_d\|_{E_d;\nu_d}. \qquad (12.24)$$

The mapping $T_d(u_1, \ldots, u_{d-1})$ being linear from E_d into F (it is the partial mapping $u_d \mapsto T(u_1, \ldots, u_d)$ of Definition 7.19 of a multilinear mapping), this inequality gives, according to the characterization of continuous linear mappings (Theorem 7.14 (a)),

$$T_d(u_1, \ldots, u_{d-1}) \in \mathcal{L}(E_d; F).$$

According to Definition 12.1 (a) of the semi-norms of $\mathcal{L}$, (12.24) gives, for any bounded set B_d of E_d

$$\|T_d(u_1, \ldots, u_{d-1})\|_{\mathcal{L}(E_d;F);B_d,\mu} = \sup_{u_d \in B_d} \|T_d(u_1, \ldots, u_{d-1})(u_d)\|_{F;\mu} \leq$$

$$\leq c' \sup_{\nu_1 \in N_1} \|u_1\|_{E_1;\nu_1} \cdots \sup_{\nu_{d-1} \in N_{d-1}} \|u_{d-1}\|_{E_{d-1};\nu_{d-1}}$$

where $c' = c \sup_{\nu_d \in N_d} \sup_{u_d \in B_d} \|u_d\|_{E_d;\nu_d}$.

The mapping T_d being multilinear from $E_1 \times \cdots \times E_{d-1}$ into $\mathcal{L}(E_d; F)$ (it is immediate), this inequality shows (Theorem 7.20 (a), again) that it is continuous from $E_1 \times \cdots \times E_{d-1}$ into $\mathcal{L}(E_d; F)$, that is $T_d \in \mathcal{L}^{d-1}(E_1 \times \cdots \times E_{d-1}; \mathcal{L}(E_d; F))$.

(b) According to Definitions 12.16 and 12.1 (a) of the semi-norms of $\mathcal{L}^d$ and $\mathcal{L}$, for all bounded set B_j of E_j and all $\mu \in \mathcal{N}_F$, there follows that

$$\begin{aligned}
&\|T_d\|_{\mathcal{L}^{d-1}(E_1\times\cdots\times E_{d-1};\mathcal{L}(E_d;F));B_1,\ldots,B_{d-1},B_d,\mu} = \\
&\qquad = \sup_{u_1\in B_1,\ldots,u_{d-1}\in B_{d-1}} \|T_d(u_1,\ldots,u_{d-1})\|_{\mathcal{L}(E_d;F);B_d,\mu} \\
&\qquad = \sup_{u_1\in B_1,\ldots,u_d\in B_d} \|T(u_1,\ldots,u_d)\|_{F;\mu} \\
&\qquad = \|T\|_{\mathcal{L}^d(E_1\times\cdots\times E_d;F);B_1,\ldots,B_d,\mu}.
\end{aligned}$$

The families of semi-norms of $\mathcal{L}^d(E_1 \times \cdots \times E_d; F)$ and of $\mathcal{L}^{d-1}(E_1 \times \cdots \times E_{d-1}; \mathcal{L}(E_d; F))$ being both indexed by the bounded sets B_1 of E_1, ..., B_d of E_d, this equality shows that $T \mapsto T_d$ is bijective and bicontinuous from $\mathcal{L}^d(E_1 \times \cdots \times E_d; F)$ onto its image.

(c) The point here is to verify that the image of $\mathcal{L}^d(E_1 \times \cdots \times E_d; F)$ by the mapping $T \mapsto T_d$ is the whole space $\mathcal{L}^{d-1}(E_1 \times \cdots \times E_{d-1}; \mathcal{L}(E_d; F))$.

Let thus $S \in \mathcal{L}^{d-1}(E_1 \times \cdots \times E_{d-1}; \mathcal{L}(E_d; F))$. We define a multilinear mapping T in $E_1 \times \cdots \times E_d$ by

$$T(u_1,\ldots,u_d) = (S(u_1,\ldots,u_{d-1}))(u_d). \tag{12.25}$$

To obtain $T_d = S$ and thus surjectivity, it has yet to be shown that T is continuous. If the E_j are metrizable, it suffices (Theorem 9.3) that T transforms any product of bounded sets into a bounded set. Let thus $B_1 \ldots, B_d$ be bounded sets, respectively of $E_1 \ldots, E_d$. Again according to Definitions 12.16 and 12.1 (a),

$$\begin{aligned}
&\sup_{u_1\in B_1,\ldots,u_d\in B_d} \|T(u_1,\ldots,u_d)\|_{F;\mu} = \\
&\qquad = \sup_{u_1\in B_1,\ldots,u_{d-1}\in B_{d-1}} \|S(u_1,\ldots,u_{d-1})\|_{\mathcal{L}(E_d;F);B_d,\mu} \\
&\qquad = \|S\|_{\mathcal{L}^{d-1}(E_1\times\cdots\times E_{d-1};\mathcal{L}(E_d;F));B_1,\ldots,B_d,\mu} < \infty.
\end{aligned}$$

Therefore, $T(B_1 \times \cdots \times B_d)$ is bounded, hence T is continuous and $T_d = S$. This proves that the mapping $T \mapsto T_d$ is surjective and subsequently is an isomorphism.

(d) If all the E_j and F are normed, following the definition of norms of $\mathcal{L}^d$ and $\mathcal{L}$ (Theorems 12.20 (a) and 12.10 (a)),

$$\begin{aligned}
&|||T_d|||_{\mathcal{L}^{d-1}(E_1\times\cdots\times E_{d-1};\mathcal{L}(E_d;F))} = \\
&= \sup_{u_1\in E_1,\, u_1\neq 0,\dots,u_{d-1}\in E_{d-1},\, u_{d-1}\neq 0} \frac{|||T_d(u_1,\dots,u_{d-1})|||_{\mathcal{L}(E_d;F)}}{\|u_1\|_{E_1}\cdots\|u_{d-1}\|_{E_{d-1}}} \\
&= \sup_{u_1\in E_1,\, u_1\neq 0,\dots,u_d\in E_d,\, u_d\neq 0} \frac{\|T(u_1,\dots,u_d)\|_F}{\|u_1\|_{E_1}\cdots\|u_d\|_{E_d}} \\
&= |||T|||_{\mathcal{L}^d(E_1\times\cdots\times E_d;F)}. \ \square
\end{aligned}$$

Caution with the identification of $\mathcal{L}(E;\mathcal{L}(E;F))$ **to** $\mathcal{L}^2(E\times E;F)$. The space $\mathcal{L}(E;\mathcal{L}(E;F))$ is frequently identified to $\mathcal{L}^2(E\times E;F)$ when E is normed. This identification is also possible when E is metrizable according to Theorem 12.21 (c), but **is it valid for every separated semi-normed space** E? The author of these lines does not know.

More precisely, given $S\in\mathcal{L}(E;\mathcal{L}(E;F))$, it is not guaranteed that the bilinear mapping T defined by $T(u,v) = (S(u))(v)$ be in $\mathcal{L}^2(E\times E;F)$. As a point of fact, the continuity of S is equivalent (Theorem 7.14 (a) and Definition 12.1 (a)) to the existence, for any bounded set B of E and any $\mu\in\mathcal{N}_F$, of a finite set $N\in\mathcal{N}_E$ and of $c\geq 0$ such that, for all $u\in E$,

$$\sup_{v\in B}\|T(u,v)\|_{F;\mu} \leq c\sup_{\nu\in N}\|u\|_{E;\nu}. \tag{12.26}$$

This does not seem to ensure the continuity of T, that is (Theorem 7.20 (a)) the existence, for all $\mu\in\mathcal{N}_F$, of finite sets N and M of $\mathcal{N}_E$ and of $c\geq 0$ such that, for all $u\in E$ and $v\in E$,

$$\|T(u,v)\|_{F;\mu} \leq c\sup_{\nu\in N}\|u\|_{E;\nu}\sup_{\mu\in M}\|v\|_{E;\nu}. \tag{12.27}$$

It would be interesting to establish that (12.26) does not lead to (12.27).

In addition, we should always be cautious with identifications, as we explain in section 14.6 *Dangerous identifications*, p. 216. □

Let us end with a continuity property of the composite of a linear mapping by a multilinear mapping that will be used for the study of differentiability.

THEOREM 12.22.– *Let E, $F_1\dots$, F_d and G be separated semi-normed spaces and*

$$S\in\mathcal{L}^d(F_1\times\cdots\times F_d;G),\quad L\in\mathcal{L}(E;F_i).$$

For all $i\in[\![1,d]\!]$, $v_1\in F_1,\dots,v_d\in F_d$ and $z\in E$, let us denote

$$T(v_1,\dots,v_{i-1},L,v_{i+1},\dots,v_d)z \overset{\text{def}}{=} S(v_1,\dots,v_{i-1},Lz,v_{i+1},\dots,v_d).$$

Then:

$$T\in\mathcal{L}^d(F_1\times\cdots\times F_{i-1}\times\mathcal{L}(E;F_i)\times F_{i+1}\times\cdots\times F_d;\mathcal{L}(E;G)).\ \blacksquare$$

Proof. The fact that T is multilinear with values into $\mathcal{L}(E;G)$ is immediate, it remains to verify that it is continuous.

Let $\{\| \ \|_{E;\nu} : \nu \in \mathcal{N}_E\}$, $\{\| \ \|_{F_j;\mu_j} : \mu_j \in \mathcal{N}_{F_j}\}$ and $\{\| \ \|_{G;\sigma} : \sigma \in \mathcal{N}_G\}$ be the families of semi-norms of E, F_j and G. According to the characterization of continuous multilinear mappings of Theorem 7.20 (a), for all $\sigma \in \mathcal{N}_G$, there exist $c \geq 0$ and finite sets $M_j \subset \mathcal{N}_{F_j}$ such that, for all $v_j \in F_j$,

$$\|S(v_1, \dots, v_d)\|_{G;\sigma} \leq c \sup_{\mu_1 \in M_1} \|v_1\|_{F_1;\mu_1} \cdots \sup_{\mu_d \in M_d} \|v_d\|_{F_d;\mu_d}.$$

Let B be a bounded subset of E. According to Definition 12.1 of the semi-norms of $\mathcal{L}(E;G)$,

$$\|T(v_1, \dots, L, \dots, v_d)\|_{\mathcal{L}(E;G);B,\sigma} = \sup_{z \in B} \|T(v_1, \dots, L, \dots, v_d)z\|_{G;\sigma} =$$

$$= \sup_{z \in B} \|S(v_1, \dots, Lz, \dots, v_d)\|_{G;\sigma} \leq$$

$$\leq c \sup_{\mu_1 \in M_1} \|v_1\|_{F_1;\mu_1} \cdots \sup_{\mu_i \in M_i} \sup_{z \in B} \|Lz\|_{F_i;\mu_i} \cdots \sup_{\mu_d \in N_d} \|v_d\|_{F_d;\mu_d} =$$

$$= c \sup_{\mu_1 \in M_1} \|v_1\|_{F_1;\mu_1} \cdots \sup_{\mu_i \in M_i} \|L\|_{\mathcal{L}(E;F_i);B,\mu_i} \cdots \sup_{\mu_d \in N_d} \|v_d\|_{F_d;\mu_d}.$$

Thus (Theorem 7.20, again), T is continuous from $F_1 \times \cdots \times \mathcal{L}(E; F_i) \times \cdots \times F_d$ into $\mathcal{L}(E;G)$. □

PART 3

WEAK TOPOLOGIES

Chapter 13

Duality

13.1. Dual

Let us define the dual[1] of a separated semi-normed space.

DEFINITION 13.1.– *Let E be a separated semi-normed space.*

(a) *A **linear form** on E is a linear mapping from E into $\mathbb{R}$.*

(b) *The **dual** of E, denoted E', is the vector space of continuous linear forms on E endowed with the semi-norms, indexed by the bounded sets B of E,*

$$\|e'\|_{E';B} \stackrel{\text{def}}{=} \sup_{e \in B} |\langle e', e\rangle|.$$

We denote $\langle e', e\rangle \stackrel{\text{def}}{=} e'(e)$ for $e' \in E'$ and $e \in E$. If there is any ambiguity on the spaces, we denote it $\langle e', e\rangle_{E' \times E}$.

(c) *The **bidual** of E is the dual $E'' = (E')'$ of its dual.* ∎

Let us characterize the elements of E'.

THEOREM 13.2.– *Let E be a separated semi-normed space, $\{\| \; \|_{E;\nu} : \nu \in \mathcal{N}_E\}$ its family of semi-norms and e' a linear form on E. Then:*

---

1 **History of the notion of dual space.** The dual of a general normed space was studied and endowed with a normed space structure by Hans HAHN, in 1927, in his fundamental essay [51]. This was done independently by Stefan BANACH [8] in 1929.
The first dual studied, that of $\mathcal{C}(I)$, I being a real compact interval, was by Jacques HADAMARD in 1903 [49]. In 1907, Maurice FRÉCHET [44] and Frigyes RIESZ [90] studied the dual of $L^2(I)$.

(a) $e' \in E' \quad \Leftrightarrow \quad \begin{cases} \textit{there exist a finite set } N \subset \mathcal{N}_E \textit{ and } c \geq 0 \textit{ such that, for all } e \in E, \\ \qquad |\langle e', e \rangle| \leq c \sup_{\nu \in N} \|e\|_{E;\nu}. \end{cases}$

(b) *If the family of semi-norms* $\{\| \ \|_{E;\nu} : \nu \in \mathcal{N}_E\}$ *is filtering:*

$$e' \in E' \quad \Leftrightarrow \quad \begin{cases} \textit{there exist } \nu \in \mathcal{N}_E \textit{ and } c \geq 0 \textit{ such that, for all } e \in E, \\ \qquad |\langle e', e \rangle| \leq c \|e\|_{E;\nu}. \end{cases}$$ ∎

Proof. This is a particular case of the characterizations of continuous linear mappings of Theorem 7.14. □

Let us characterize the elements of E''.

THEOREM 13.3.– *Let E be a separated semi-normed space. Then:*

(a) *E'' is endowed with the semi-norms, indexed by the bounded sets B' of E',*

$$\|e''\|_{E'';B'} \stackrel{\text{def}}{=} \sup_{e' \in B'} |\langle e'', e' \rangle|.$$

(b) *If e'' is a linear form on E':*

$$e'' \in E'' \quad \Leftrightarrow \quad \begin{cases} \textit{there exist a bounded set } B \textit{ of } E \textit{ and } c \geq 0 \textit{ such that, for all } e' \in E', \\ \qquad |\langle e'', e' \rangle| \leq c \sup_{e \in B} |\langle e', e \rangle|. \end{cases}$$ ∎

Proof. (a) It is the Definition 13.1 (a) of the semi-norms of a dual, here that of E'.

(b) *Preliminary.* First, let us observe that:

$$\textit{The family of semi-norms of } E' \textit{ is filtering} \text{ (Definition 3.14).} \qquad (13.1)$$

Indeed, given a finite family $\{B_i : i \in I\}$ of bounded subsets of E, their union B is bounded (Theorem 2.12 (f)) and, for all $e' \in E'$,

$$\sup_{i \in I} \|e'\|_{E';B_i} = \sup_{i \in I} \sup_{e \in B_i} |\langle e', e \rangle| = \sup_{e \in B} |\langle e', e \rangle| = \|e'\|_{E';B}.$$

Characterization of e''. This is the characterization of Theorem 13.2 (b) relative to the dual of E', according to Definition 13.1 (b) of the semi-norms of E'. □

Let us define two other topologies on the dual.

DEFINITION 13.4.– *Let E be a separated semi-normed space.*

(a) E'-weak *is the vector space E' endowed with the semi-norms, indexed by $e'' \in E''$,*

$$\|e'\|_{E'\text{-weak};e''} \stackrel{\text{def}}{=} |\langle e'', e'\rangle|.$$

(b) E'-$*$weak *is the vector space E' endowed with the semi-norms, indexed by $e \in E$,*

$$\|e'\|_{E'\text{-}*\text{weak};e} \stackrel{\text{def}}{=} |\langle e', e\rangle|. \blacksquare$$

The $*$weak topology is specific to a dual. Rather, the weak topology is defined for any separated semi-normed space (Definition 15.1).

With Definition 12.1 of spaces of continuous linear mappings, Definitions 13.1 (b) and 13.4 (b) give

$$E' \underset{\leftrightarrows}{\equiv} \mathcal{L}(E;\mathbb{R}), \quad E'\text{-}*\text{weak} \underset{\leftrightarrows}{\equiv} \mathcal{L}(E;\mathbb{R})\text{-pt}. \tag{13.2}$$

Let us compare the three topologies that we have defined on a dual.

THEOREM 13.5.– *For every separated semi-normed space E:*

(a) E', E'-weak *and* E-$*$weak *are separated semi-normed spaces.*

(b) $$E' \underset{\rightarrow}{\subsetneq} E'\text{-weak} \underset{\rightarrow}{\subsetneq} E'\text{-}*\text{weak}. \blacksquare$$

Proof. (a) *Separation of E' and E'-$*$weak.* These are particular cases of the fact that the spaces $\mathcal{L}(E;F)$ and $\mathcal{L}(E;F)$-pt are separated (Theorem 12.3 (a)) since, according to (13.2), $E' \underset{\leftrightarrows}{\equiv} \mathcal{L}(E;\mathbb{R})$ and E'-$*$weak $\underset{\leftrightarrows}{\equiv} \mathcal{L}(E;\mathbb{R})$-pt.

Separation of E'-weak. Let us first observe that, given $e \in E$, we define $e'' \in E''$ by: for all $e' \in E'$,

$$\langle e'', e'\rangle = \langle e', e\rangle. \tag{13.3}$$

As a matter of fact, e'' is linear and verifies the characterization of Theorem 13.3 (b) with $B = \{e\}$.

The space E'-weak is separated (Definition 2.3 (c)) because, if an element e' of E' verifies $\|e'\|_{E'\text{-weak};u''} = 0$ for all $u'' \in E''$, then for all $e \in E$, by defining $e'' \in E''$ by (13.3), it follows that

$$|\langle e', e\rangle| = |\langle e'', e'\rangle| = \|e'\|_{E'\text{-weak};e''} = 0$$

and thus $e' = 0_{E'}$.

(b) *Inclusion* $E' \underset{\rightarrow}{\subsetneq} E'$-weak. Let $e'' \in E''$. The characterization of the bidual of Theorem 13.3 (b) shows that there exist a bounded set B of E and $c \in \mathbb{R}$ such that $|\langle e'', e' \rangle| \leq c \sup_{e \in B} |\langle e', e \rangle|$ for all $e' \in E'$. That is, according to Definitions 13.1 (b) and 13.4 (a) of the semi-norms of E' and E'-weak,

$$\|e'\|_{E'\text{-weak};e''} \leq c \|e'\|_{E';B}.$$

This proves (Definition 3.6 (c)) that $E' \underset{\rightarrow}{\subsetneq} E'$-weak.

Inclusion E'-weak $\underset{\rightarrow}{\subsetneq} E'$-$*$weak. Let $e' \in E'$ and $e \in E$. By defining $e'' \in E''$ by (13.3), it follows that

$$\|e'\|_{E'\text{-}*\text{weak};e} = |\langle e', e \rangle| = |\langle e'', e' \rangle| = \|e'\|_{E'\text{-weak};e''}$$

which proves that E'-weak $\underset{\rightarrow}{\subsetneq} E'$-$*$weak. □

Another proof of the separation of E'**-weak.** We may also, as in the proof of the separation of a general space E-weak (Theorem 15.2), conclude by using the Hahn–Banach theorem (via Theorem 14.4) that is itself based on the difficult Zorn's lemma (Theorem 11.7). □

Another proof of the inclusion $E' \underset{\rightarrow}{\subsetneq} E'$**-weak.** This is a particular case of the proof of the inclusion $E \underset{\rightarrow}{\subsetneq} E$-weak of Theorem 15.2. □

Equality of the weak and $*$weak topologies. The topological equality E'-weak $\underset{\leftrightarrows}{=} E'$-$*$weak holds if E is semi-reflexive (Theorem 17.21), and thus in particular (Theorem 17.12) if E is a Hilbert space. □

Let us observe that dual spaces are preserved by topological equalities and let us give a filtration property.

THEOREM 13.6.– *For all semi-normed space E:*

(a) E', E'-weak, E'-$*$weak *and their semi-norms are preserved when the family of the semi-norms of E is replaced by an equivalent family.*

(b) *The family of semi-norms of E' is filtering.* ▮

Proof. (a) If the family of the semi-norms of E is replaced by an equivalent family:
— The set E' is preserved because continuous mappings on E are (Theorem 7.5).
— The semi-norms of E' are preserved because they are indexed (Definition 13.1 (b)) by the bounded subsets of E, which are preserved (Theorem 3.4 (b)).
— The semi-norms of E'-weak are preserved because they are indexed (Definition 13.4 (a)) by the elements of $E'' = (E')'$, which is preserved since E' is.
— The semi-norms of E'-$*$weak are preserved because they are indexed (Definition 13.4 (b)) by the elements of E, which is preserved.

(b) This was established in the proof of Theorem 13.3, see property (13.1). □

Let us characterize the converging sequences for the weak topologies.

THEOREM 13.7.– *Let E be a separated semi-normed space, $(u'_n)_{n\in\mathbb{N}} \subset E'$ and $u' \in E'$. Then:*

(a) $$u'_n \to u' \textit{ in } E'\text{-weak} \quad\Leftrightarrow\quad \langle e'', u'_n\rangle \to \langle e'', u'\rangle,\ \forall e'' \in E''.$$

(b) $$u'_n \to u' \textit{ in } E'\text{-}{*}\text{weak} \quad\Leftrightarrow\quad \langle u'_n, e\rangle \to \langle u', e\rangle,\ \forall e \in E.$$ ▮

Proof. (a) According to Definitions 2.6 (a) of the convergence of a sequence and 13.4 (a) of the semi-norms of E'-weak, we have $u'_n \to u'$ in E'-weak if and only if, for all $e'' \in E''$, we have $\|u'_n - u'\|_{E'\text{-weak};e''} = |\langle e'', u'_n - u'\rangle| \to 0$, that is $\langle e'', u'_n - u'\rangle \to 0$.

(b) Likewise, we conclude here with $\|u'_n - u'\|_{E'\text{-}*\text{weak};e} = |\langle u'_n - u', e\rangle|$ for all $e \in E$. □

Terminology. The topology of E' (Definition 13.1), that is the *topology of uniform convergence on bounded subsets of* E, is sometimes called $*$**strong topology** of E' [SCHWARTZ, 99, p. 288] and is denoted by $\beta(E', E)$. This is so by analogy with the $*$*weak topology* (Definition 13.4 (b)), that is the *topology of pointwise convergence on* E, denoted by $\sigma(E', E)$.

The **strong topology** of E' designates here (and in others books!) the *topology of uniform convergence on the bounded subsets of* E'' (that is Property (17.9), p. 251, applied to E') which is denoted by $\beta(E', E'')$. This is by analogy with the *weak topology* (Definition 13.4 (a)), that is the *topology of pointwise convergence on* E'', denoted by $\sigma(E', E'')$. □

CAUTION. The theory of *spaces in duality* [BOURBAKI, 17, chap. IV, § 1.1] attributes the same role to two semi-normed spaces F and G, via a bilinear form $\langle\ ,\ \rangle$ on $F \times G$. In this case, the *weak topology* of one designates the topology of the pointwise convergence on the other and $\sigma(F, G)$ denotes the one on F and $\sigma(G, F)$ that on G.

A pair (E, E') is a particular case of spaces in duality. In this case, the notation $\sigma(E', E)$ has no ambiguity but, in contrast, the designation **weak topology of** E' **is dangerous** if it is not supplemented by "*for the duality* (E', E)" since it refers to our $*$*weak topology* and it does not always coincide with the *weak topology of* E' *for the duality* (E', E''), that is $\sigma(E', E'')$, which is our *weak* topology. □

13.2. Dual of a metrizable or normed space

Let us show that the dual of a metrizable space is sequentially complete.

THEOREM 13.8.– *For every separated semi-normed space E:*

(a) *E is metrizable $\Rightarrow$ E' is a Neumann space.*

(b) *E is a Fréchet space $\Rightarrow$ E'-$*$weak is a Neumann space.* ▮

Proof. These are particular cases of the sequential completeness properties of spaces $\mathcal{L}(E;F)$ and $\mathcal{L}(E;F)$-pt given in Theorems 12.7 and 12.8 since $E' \underset{\leftrightarrows}{\equiv} \mathcal{L}(E;\mathbb{R})$ and E'-$*$weak $\underset{\leftrightarrows}{\equiv} \mathcal{L}(E;\mathbb{R})$-pt, see (13.2), p. 191, and since $\mathbb{R}$ is a Banach space (Theorem 4.15) and therefore a Neumann space (Theorem 4.11). □

Non-necessity of metrizability. There exist non-metrizable spaces whose dual is a Neumann space. This is the case, among others, of every non-metrizable semi-reflexive space (Theorem 17.7), such as $\mathcal{D}$. □

Let us show that the dual of a normed space is a Banach space[2].

THEOREM 13.9.– *Let*

E be a normed space.

Then, for all e ∈ E and e′ ∈ E′, by denoting $B_E = \{e \in E : \|e\|_E \le 1\}$*:*

(a) *The topology of E′ is generated by the* ***dual norm***

$$|||e'|||_{E'} \overset{\text{def}}{=} \sup_{e \in E,\, e \neq 0} \frac{|\langle e', e\rangle|}{\|e\|_E}.$$

(b)
$$|||e'|||_{E'} = \sup\nolimits_{e \in B_E} |\langle e', e\rangle|.$$

(c)
$$|\langle e', e\rangle| \le |||e'|||_{E'} \|e\|_E.$$

(d) *E′ is a Banach space for the dual norm.* ■

Proof. This is a particular case of the normability properties of $\mathcal{L}(E;F)$ given in Theorem 12.10 since $E' \underset{\leftrightarrows}{\equiv} \mathcal{L}(E;\mathbb{R})$, see (13.2), p. 191, and, for (d), since $\mathbb{R}$ is a Banach space (Theorem 4.15). □

Dual equality. The roles of E and E' can be exchanged in the equality of Theorem 13.9 (a) defining the dual norm, see Theorem 15.12. □

Relation between the semi-norms and the dual norm of E' when E is normed. With Definition 13.1 (b) of the semi-norms of E', the equality of Theorem 13.9 (b) may be written as

$$|||e'|||_{E'} = \|e'\|_{E';B_E}. \tag{13.4}$$

□

Let us show that the bidual of a normed space is a Banach space.

THEOREM 13.10.– *Let*

E be a normed space.

Then, for all e′ ∈ E′ and e″ ∈ E″:

2 **History of Theorem 13.9.** Hans HAHN defined in 1927 the dual norm and showed that it makes the dual sequentially complete [51]. This result was independently found again, in 1929, by Stefan BANACH [8].

(a) *The topology of E'' is generated by the dual norm of the dual norm $|||\ |||_{E'}$ defined in Theorem 13.9 (a), known as* **bidual norm**, *that is*

$$|||e''|||_{E''} \stackrel{\text{def}}{=} \sup_{e' \in E',\, e' \neq 0} \frac{|\langle e'', e' \rangle|}{|||e'|||_{E'}}.$$

(b) *By denoting $B_{E'} = \{e' \in E' : |||e'|||_{E'} \leq 1\}$, it follows that:*

$$|||e''|||_{E''} = \sup_{e' \in B_{E'}} |\langle e'', e' \rangle|.$$

(c) $$|\langle e'', e' \rangle_{E'' \times E'}| \leq |||e''|||_{E''} |||e'|||_{E'}.$$

(d) *E'' is a Banach space for the bidual norm.* ▮

Proof. Let us here denote E^* the vector space E' endowed with the dual norm. We denote, as we always do, by E' this same vector space endowed with the semi-norms of Definition 13.1 (b).

(a) Since $E^* \underset{\leftrightarrows}{=} E'$ (Theorem 13.9 (a)), their duals coincide (Theorem 13.6 (a)), that is $E^{*\prime} \underset{\leftrightarrows}{=} E''$. Since E^* is normed, $E^{**} \underset{\leftrightarrows}{=} E^{*\prime}$ (Theorem 13.9 (a), again). Therefore, $E^{**} \underset{\leftrightarrows}{=} E''$. Thus, the bidual norm indeed generates the topology of E''.

(b), (c) *and* (d). These are properties (b), (c) and (d) of Theorem 13.9 relatively to E^* since the latter is normed according to (a). □

Let us show that $\mathbb{R}'$ is isomorphic to $\mathbb{R}$.

THEOREM 13.11.– (a) *The mapping $x' \mapsto \langle x', 1 \rangle$ is an isomorphism from $\mathbb{R}'$ onto $\mathbb{R}$.*

(b) *The dual norm of $\mathbb{R}'$ defined in Theorem 13.9 (a) is equal to, for all $x' \in \mathbb{R}'$,*

$$|||x'|||_{\mathbb{R}'} = |\langle x', 1 \rangle|. \ ▮$$

Proof. (b) Any $x' \in \mathbb{R}'$ being a linear form, for all $x \in \mathbb{R}$, we have

$$\langle x', x \rangle = \langle x', x1 \rangle = \langle x', 1 \rangle x. \tag{13.5}$$

Therefore, $|||x'|||_{\mathbb{R}'} = \sup_{x \neq 0} |\langle x', x \rangle| / |x| = |\langle x', 1 \rangle|$, according to the definition of the dual norm.

(a) The mapping $x' \mapsto \langle x', 1\rangle$ is surjective because, given $x \in \mathbb{R}$, we define $x' \in \mathbb{R}'$ such that $\langle x', 1\rangle = x$ by $\langle x', y\rangle = xy$ for all $y \in \mathbb{R}$. The equality (b) then implies that $x' \mapsto \langle x', 1\rangle$ is bijective and bicontinuous. □

Henceforth:

$$\textbf{Each } x' \in \mathbb{R}' \textbf{ is identified to the real number } \langle x', 1\rangle. \tag{13.6}$$

THEOREM 13.12.– *The identification (13.6) implies:*

(a)
$$\mathbb{R}' \underset{\leftrightarrows}{\equiv} \mathbb{R}.$$

(b)
$$\langle x', x\rangle_{\mathbb{R}'\times\mathbb{R}} = x'x. \ \blacksquare$$

Proof. Let us denote by I the mapping $x' \mapsto \langle x', 1\rangle$.

(a) Theorem 13.11 gives $I\mathbb{R}' = \mathbb{R}$ and $\|Ix'\|_{\mathbb{R}} = |||x'|||_{\mathbb{R}'}$, and thus $I\mathbb{R}' \underset{\leftrightarrows}{\equiv} \mathbb{R}$ following Definition 3.6 (a) of the identity between semi-normed spaces. From which $\mathbb{R}' \underset{\leftrightarrows}{\equiv} \mathbb{R}$ with the identification (13.6) of $I\mathbb{R}'$ to $\mathbb{R}'$.

(b) The equality (13.5) is written $\langle x', x\rangle = Ix'x$. From which $\langle x', x\rangle = x'x$ with the identification (13.6) of Ix' to x'. □

13.3. Dual of a Hilbert space

Let us show that any continuous linear form on a Hilbert space E can be "represented" by an element of E which is the **Riesz–Fréchet representation theorem**[3].

THEOREM 13.13.– *Let E be a Hilbert space, $(\ ,\ )_E$ its scalar product and $f \in E'$.*

Then, there exists a unique $u \in E$ such that, for all $v \in E$,

$$(u, v)_E = \langle f, v\rangle_{E'\times E}. \ \blacksquare$$

Proof. This results from Theorem 13.14 (c) hereafter. □

More specifically, let us show that every Hilbert space is isomorphic to its dual.

3 **History of the Riesz–Fréchet theorem 13.13.** The fact that $L^2(I)$, I being a real bounded interval, be isomorphic to its dual was discovered, independently by Maurice FRÉCHET [44] and Frigyes RIESZ [90] and published in the same issue of the *Comptes rendus de l'Académie des Sciences de Paris*, in 1907.

Another Riesz–Fréchet representation theorem. A similar theorem for the representation of the dual of L^p bears the same name, see Volume III.

THEOREM 13.14.– *Let E be a Hilbert space, $(\ ,\ )_E$ its scalar product and $u \in E$.*

(a) *We define $Ju \in E'$ by: for all $v \in E$,*

$$\langle Ju, v\rangle_{E'\times E} \stackrel{\text{def}}{=} (u,v)_E.$$

(b) *For the dual norm $|||\ |||_{E'}$ defined in Theorem 13.9 (a),*

$$|||Ju|||_{E'} = \|u\|_E.$$

(c) *J is an isomorphism, known as* **canonical,** *from E onto E'.* ▮

Proof. (a) *Membership of Ju to E'.* The linear mapping Ju is continuous, and is thus indeed an element of E', because according to the Cauchy–Schwarz inequality (Theorem 5.3 (a)),

$$|\langle Ju, v\rangle_{E'\times E}| = |(u,v)_E| \leq \|u\|_E\,\|v\|_E.$$

(b) *Norms equality.* The definition of $|||\ |||_{E'}$ (Theorem 13.9 (a)) and the above inequality give

$$|||Ju|||_{E'} = \sup_{v\in E,\ v\neq 0} \frac{|\langle Ju, v\rangle_{E'\times E}|}{\|v\|_E} \leq \|u\|_E.$$

Conversely, if $u \neq 0_E$, it follows that (with Definition 5.1 (b) of the Hilbertian norm $\|\ \|_E$)

$$|||Ju|||_{E'} \geq \frac{|\langle Ju, u\rangle_{E'\times E}|}{\|u\|_E} = \frac{|(u,u)_E|}{\|u\|_E} = \|u\|_E.$$

If $u = 0_E$, this inequality is obvious (its two members are equal to zero).

(c) *Isomorphism.* Let us show that J is surjective which, with (b), will imply that it is an isomorphism. Let thus $f \in E'$. The point is to find $u \in E$ such that $Ju = f$. We will do this by minimizing the mapping g defined, for all $v \in E$, by

$$g(v) = \frac{1}{2}\|v\|_E^2 - \langle f, v\rangle_{E'\times E}.$$

The inequality of Theorem 13.9 (c) and $(x-y)^2 = x^2 + y^2 - 2xy \geq 0$ lead to

$$|\langle f, v\rangle_{E'\times E}| \leq |||f'|||_{E'}\|v\|_E \leq \frac{1}{2}|||f'|||_{E'}^2 + \frac{1}{2}\|v\|_E^2,$$

therefore $g(v)$ is bounded on below by $-|||f'|||_{E'}^2/2$. Thus (Theorem 1.2 (b)), g has a greatest lower bound $\gamma = \inf_{v\in E} g(v)$.

Let $(v_n)_{n\in\mathbb{N}}$ be a minimizing sequence, that is such that $g(v_n)\to\gamma$. For all m and n in $\mathbb{N}$, we have

$$\begin{aligned} g(v_m)+g(v_n)-2g\Big(\frac{v_m+v_n}{2}\Big) &= \frac{1}{2}\|v_m\|_E^2+\frac{1}{2}\|v_n\|_E^2-\frac{1}{4}\|v_m+v_n\|_E^2 \\ &= \frac{1}{4}\|v_m-v_n\|_E^2 \end{aligned}$$

(according to the parallelogram law, see Theorem 5.2 (b)). Consequently,

$$\|v_m-v_n\|_E^2 \le 4(g(v_m)+g(v_n)-2\gamma).$$

Given $\epsilon>0$, there exists $n_0\in\mathbb{N}$ such that $\gamma\le g(v_n)\le\gamma+\epsilon^2/8$ for all $n\ge n_0$. For all $m\ge n\ge n_0$, it follows that $\|v_m-v_n\|_E^2\le\epsilon^2$. Therefore, $(v_n)_{n\in\mathbb{N}}$ is a Cauchy sequence of E, which is sequentially complete, subsequently it has a limit u. Since g is sequentially continuous (according to Theorems 7.10, 7.26 (a), 7.9 (a) and 7.2 (a)), $g(v_n)\to g(u)$ from which

$$g(u)=\gamma.$$

For all $v\in E$ and $t>0$,

$$0\le\frac{1}{t}(g(u+tv)-g(u))=\frac{t}{2}\|v\|_E^2+(u,v)_E-\langle f,v\rangle_{E'\times E}.$$

Thus, $(u,v)_E-\langle f,v\rangle_{E'\times E}\ge 0$. This inequality relative to $-v$ gives the inverse inequality, which is therefore an equality. Hence

$$\langle Ju,v\rangle_{E'\times E}=(u,v)_E=\langle f,v\rangle_{E'\times E}.$$

This proves that $Ju=f$ and thus J is surjective.

The equality (b) implies that J is injective and therefore bijective as well as that it is bicontinuous. It is an isomorphism. □

Let us show that the dual of a Hilbert space is also a Hilbert space.

THEOREM 13.15.– *Let E be a Hilbert space with scalar product $(\ ,\ )_E$. Then:*

(a) *A scalar product on E', known as the* **dual scalar product**, *is defined by: for all $u'\in E'$ and $v'\in E'$,*

$$(u',v')_{E'}\overset{\text{def}}{=}(J^{-1}u',J^{-1}v')_E \tag{13.7}$$

where J is the isometry from E onto E' given by the Riesz–Fréchet theorem 13.14.

(b) *The dual scalar product generates the dual norm of E' defined in Theorem 13.9 (a) which, itself, generates the topology of E'.*

(c) *Once endowed with the dual scalar product, E' is a Hilbert space.* ▮

Proof. (a) The form $(\ ,\ )_{E'}$ has indeed the properties (symmetry and positivity) required in Definition 5.1 of a scalar product.

(b) This scalar product generates the dual norm $|||\ \ |||_{E'}$ because

$$(u', u')_{E'}^{1/2} = (J^{-1}u', J^{-1}u')_E^{1/2} = \|J^{-1}u'\|_E = |||u'|||_{E'}$$

(the last equality is given in the Riesz–Fréchet theorem 13.14 (b)). The dual norm generates the topology of E' according to Theorem 13.9 (a).

(c) The space E' is sequentially complete as is the dual of any normed space (Theorem 13.9 (d)). Consequently, it is indeed a Hilbert space. □

Another proof of the completeness of the dual E' of a Hilbert space. Rather than using, as above, Theorem 13.9 (d) (which itself is based on Haussdorff's Theorem 12.7 via Theorem 12.10), we can observe that, if $(u_n)_{n\in\mathbb{N}}$ is a Cauchy sequence of E', then $(J^{-1}u_n)_{n\in\mathbb{N}}$ is a Cauchy sequence of E and therefore it has a limit w (since E is sequentially complete) which implies that $u_n \to Jw$ in E'. □

13.4. Extraction of $*$ weakly converging subsequences

Let us show that, if E is a separable Fréchet space, the bounded subsets of E'-$*$weak are relatively sequentially compact, which is the **Banach–Bourbaki theorem**[4].

THEOREM 13.16.– *Let*

E be a separable Fréchet space.

Then:

$$\begin{cases} \textit{Any bounded sequence of } E', E'\text{-weak } \textit{or } E'\text{-}{*}\text{weak}, \\ \textit{has a subsequence that converges in } E'\text{-}{*}\text{weak}. \end{cases}$$ ▮

Proof. Let us denote by $\{\|\ \ \|_{E;\nu} : \nu \in \mathcal{N}_E\}$ the family of semi-norms of E.

Sequence of E'-$*$weak. Let $(e'_n)_{n\in\mathbb{N}}$ be a bounded sequence of E'-$*$weak. For all $e \in E$,

$$\sup_{n\in\mathbb{N}} |\langle e'_n, e\rangle| = \sup_{n\in\mathbb{N}} \|e'_n\|_{E'\text{-}*\text{weak};e} < \infty$$

---

4 **History of the Banach–Bourbaki theorem (Theorem 13.16).** Stefan BANACH established this result in 1932 [10, Theorem 3, p. 123] for a separable Banach space E. Nicolas BOURBAKI generalized this result to a separable barreled space (which is the case of a Fréchet space).

Comparison to Alaoglu's theorem. Leonidas ALAOGLU established the relative compactness (which does not imply the sequential compactness, see (2.6) p. 27) of bounded sets in the dual $*$weak of a Banach space [2, p. 255].

thus, E being a Fréchet space and each e'_n being continuous, the Banach–Steinhaus theorem 10.11 shows that the family $\{e'_n\}_{n\in\mathbb{N}}$ is equicontinuous, that is that there exist a finite set $N \subset \mathcal{N}_E$ and $c \in \mathbb{R}$ such that, for all $n \in \mathbb{N}$ and $e \in E$,

$$|\langle e'_n, e\rangle| \leq c \sup_{\nu\in N} \|e\|_{E;\nu}. \tag{13.8}$$

Since E is separable (Definition 2.21 (a)), there exists a sequence $(x_k)_{k\in\mathbb{N}}$ that is dense in E. For a fixed k, the set $\{\langle e'_n, x_k\rangle\}_{n\in\mathbb{N}}$ is bounded in $\mathbb{R}$ therefore (Bolzano-Weierstrass theorem 4.17) there exists a subsequence $(e'_{\sigma_k(n)})_{n\in\mathbb{N}}$ of $(e'_n)_{n\in\mathbb{N}}$ such that $(\langle e'_{\sigma_k(n)}, x_k\rangle)_{n\in\mathbb{N}}$ converges to a limit $\ell(x_k)$. By extracting the subsequence relating to x_{k+1} from that relating to x_k, we obtain successive subsequences whose diagonal subsequence[5] $(e'_{\sigma_n(n)})_{n\in\mathbb{N}}$ verifies, for each $k \in \mathbb{N}$, when $n \to \infty$,

$$\langle e'_{\sigma_n(n)}, x_k\rangle \to \ell(x_k) \tag{13.9}$$

(Theorem 2.8 (a), since $(\langle e'_{\sigma_n(n)}, x_k\rangle)_{n\in\mathbb{N}}$ is a subsequence of $(\langle e'_{\sigma_k(n)}, x_k\rangle)_{n\in\mathbb{N}}$). Henceforth, let us denote

$$e'_n = e'_{\sigma_n(n)} \tag{13.10}$$

and let $u \in E$. We will now show that

$$(\langle e'_n, u\rangle)_{n\in\mathbb{N}} \text{ is a Cauchy sequence of } \mathbb{R}. \tag{13.11}$$

Let $\epsilon > 0$. According to Definition 2.19 (a) of density, $\overline{(x_k)_{k\in\mathbb{N}}} = E$ thus, following the characterization of closure of Theorem 2.17 (b), there exists $k \in \mathbb{N}$ such that:

$$\sup_{\nu\in N} \|x_k - u\|_{E;\nu} \leq c/\epsilon.$$

On the other hand, according to (13.9), there exists $n_0 \in \mathbb{N}$ such that, for all $n \geq n_0$, there follow $|\langle e'_n, x_k\rangle - \langle e', x_k\rangle| \leq \epsilon$ and so, for $m \geq n \geq n_0$,

$$|\langle e'_n - e'_m, x_k\rangle| \leq 2\epsilon.$$

Then, with (13.8),

$$|\langle e'_n - e'_m, u\rangle| \leq |\langle e'_n - e'_m, x_k\rangle| + |\langle e'_n, u - x_k\rangle| + |\langle e'_m, u - x_k\rangle| \leq$$
$$\leq 2\epsilon + 2c \sup_{\nu\in N} \|u - x_k\|_{E;\nu} \leq 4\epsilon.$$

Thus, $(\langle e'_n, u\rangle)_{n\in\mathbb{N}}$ is a Cauchy sequence, which ends the proof of (13.11).

The space $\mathbb{R}$ being sequentially complete (Theorem 4.15), this sequence converges to a limit $\ell(u)$, that is

$$\langle e'_n, u\rangle \to \ell(u). \tag{13.12}$$

5 **History of the extraction of a diagonal subsequence.** The diagonal process for extracting a subsequence was introduced by Georg CANTOR, see his works [24].

The linearity of e'_n implies that of ℓ and (13.8) gives, at the limit,

$$|\ell(u)| \leq c \sup_{\nu \in N} \|u\|_{E;\nu}$$

hence (Theorem 13.2 (a)) $\ell \in E'$. Then, (13.12) gives (Theorem 13.7 (b)) $e'_n \to \ell$ in E'-$*$weak. According to notation (13.10), it is indeed a converging subsequence of the initial sequence.

Sequence of E' or E'-weak. This is a particular case of the previous one because (Theorem 13.5 (b)) $E' \underset{\rightarrow}{\subsetneq} E'$-weak $\underset{\rightarrow}{\subsetneq} E'$-$*$weak, thus (Theorem 3.8 (a)) any sequence that is bounded in E' or E'-weak is bounded in E'-$*$weak. □

Simplification of the proof when E is normed. It is not necessary to resort to the Banach–Steinhaus theorem for extracting a $*$weak converging subsequence from a sequence $(e'_n)_{n\in\mathbb{N}}$ that is bounded in the dual of a (sequentially) separable Banach space E. In effect, E' is then normable (Theorem 13.9 (a)) and $\sup_n |||e'_n|||_{E'} = c < \infty$ from which (Theorem 13.9 (c)), for all $n \in \mathbb{N}$ and $e \in E$,

$$|\langle e'_n, e\rangle| \leq |||e'_n|||_{E'} \|e\|_E \leq c\|e\|_E.$$

which is a particular case of (13.8). □

Let us complete this result with a majoration of the strong semi-norms of a $*$weak limit.

THEOREM 13.17.– *Let E be a separated semi-normed space, $(e'_n)_{n\in\mathbb{N}} \subset E'$, $e' \in E'$, B a bounded subset of E and $c \in \mathbb{R}$ such that:*

$$\|e'_n\|_{E';B} \leq c, \quad \forall n \in \mathbb{N},$$

$$e'_n \to e' \text{ in } E'\text{-}*\text{weak}.$$

Then:

$$\|e'\|_{E';B} \leq c. \blacksquare$$

Proof. For all $n \in \mathbb{N}$ and $e \in B$, according to Definition 13.1 (b) of the semi-norms of E',

$$|\langle e'_n, e\rangle| \leq \sup_{u\in B} |\langle e'_n, u\rangle| = \|e'_n\|_{E';B} \leq c.$$

At the limit (Theorem 13.7 (b)), $|\langle e', e\rangle| \leq c$. Thus, $\|e'\|_{E';B} = \sup_{e\in B} |\langle e', e\rangle| \leq c$. □

When E is normed, we get the following majoration of the dual norm of a $*$weak limit.

THEOREM 13.18.– *Let*

E be a normed space,

$(e'_n)_{n\in\mathbb{N}} \subset E'$ and $c \in \mathbb{R}$ such that, for the dual norm defined in Theorem 13.9 (a),

$$|||e'_n|||_{E'} \leq c, \quad \forall n \in \mathbb{N},$$

$$e'_n \to e' \text{ in } E'\text{-}*\text{weak}.$$

Then:

$$|||e'|||_{E'} \leq c. \ \blacksquare$$

Proof. This results from Theorem 13.17 because $|||e'|||_{E'} = \|e'\|_{E';B_E}$, see (13.4) p. 194. □

Another proof. For all $n \in \mathbb{N}$ and $e \in E$, we have $|\langle e'_n, e\rangle| \leq |||e'_n|||_{E'}\|e\|_E \leq c\,\|e\|_E$ (Theorem 13.9 (c)), and thus, at the limit, $|\langle e', e\rangle| \leq c\,\|e\|_E$ hence (Theorem 13.9 (a)) $|||e'|||_{E'} \leq c$. □

Let us show that when E is normed, the unit ball of E' is closed in E'-$*$weak, which is stronger than sequential closure established in Theorem 13.18.

THEOREM 13.19.– *Let*

E be a normed space,

and $B_{E'} = \{e' \in E' : |||e'|||_{E'} \leq 1\}$ be the unit ball of E' for the dual norm defined in Theorem 13.9 (a). Then:

$$B_{E'} \text{ is closed in } E'\text{-}*\text{weak}. \ \blacksquare$$

Proof. Denote $\overline{B_{E'}}^{\,*}$ the closure of $B_{E'}$ in E'-$*$weak.

Let $e' \in \overline{B_{E'}}^{\,*}$. The characterization of closure of Theorem 2.17 (b), with Definition 13.4 (b) of the semi-norms of E'-$*$weak, shows that, for all $e \in E$ and $\epsilon > 0$, there exists $f' \in B_{E'}$ such that:

$$\|e' - f'\|_{E'\text{-}*\text{weak};e} = |\langle e' - f', e\rangle| \leq \epsilon.$$

Since $|\langle f', e\rangle| \leq |||f'|||_{E'}\|e\|_E$ (Theorem 13.9 (c)) and $|||f'|||_{E'} \leq 1$, it follows that

$$|\langle e', e\rangle| \leq |\langle f', e\rangle| + \epsilon \leq \|e\|_E + \epsilon.$$

This holds for all $\epsilon > 0$, from which $|\langle e', e\rangle| \leq \|e\|_E$.

The definition of the dual norm (Theorem 13.9 (a)) then gives

$$|||e'|||_{E'} = \sup_{e\in E,\, e\neq 0} \frac{|\langle e', e\rangle|}{\|e\|_E} \leq 1.$$

Thus, $e' \in B_{E'}$. So, $B_{E'} = \overline{B_{E'}}^{*}$ which, by definition, is closed in E'-$*$weak. □

Supplement to Theorem 13.19. Despite that this will not be of any use to us, we observe that:

$$E \text{ is a normed space} \quad \Rightarrow \quad B_{E'} \text{ is compact in } E'\text{-}*\text{weak}.$$

Proof. Besides being closed in E'-$*$weak, $B_{E'}$ is therein relatively compact according to Theorem 13.20 hereafter since $|\langle e', u\rangle_{E'\times E}| \leq |||e'|||_{E'} \|u\|_E \leq \|u\|_E$ for all $e' \in B_{E'}$ and $u \in E$ according to Theorem 13.9 (c). It is therefore compact (Theorem 2.26 (a)). □

Let us give a sufficient condition for compactness in E'-$*$weak.

THEOREM 13.20.– *Let E be a separated semi-normed space, $\{\| \|_{E;\nu} : \nu \in \mathcal{N}_E\}$ its family of semi-norms and $V \subset E'$. If there exist a finite set $N \subset \mathcal{N}_E$ and $c \in \mathbb{R}$ such that, for all $e' \in V$ and $u \in E$,*

$$|\langle e', u\rangle_{E'\times E}| \leq c \sup_{\nu \in N} \|u\|_{E;\nu}$$

then

$$V \text{ is relatively compact in } E'\text{-}*\text{weak.} \ \blacksquare$$

Proof. Let

$$W = \{e' \in E' : \forall u \in E,\ |\langle e', u\rangle| \leq c \sup_{\nu \in N} \|u\|_{E;\nu}\}.$$

This set is equicontinuous (Definition 10.5 (a)) from E into $\mathbb{R}$ and, for all $u \in E$, the set $W(u) = \{\langle e', u\rangle : e' \in W\}$ is bounded in $\mathbb{R}$ and therefore is relatively compact according to the Borel–Lebesgue theorem 4.16 (b). Ascoli's theorem 11.12 then shows that W is relatively compact in $\mathcal{C}(E; \mathbb{R})$-pt.

Since W is included in $\mathcal{L}(E; \mathbb{R})$-pt which is a closed topological vector subspace of $\mathcal{C}(E; \mathbb{R})$-pt (Theorem 12.3 (c)), it thereof results (Theorem 3.13 (a)) that W is relatively compact in $\mathcal{L}(E; \mathbb{R})$-pt.

Being included in W, the set V is thus also relatively compact (Theorem 2.26 (c)) in $\mathcal{L}(E; \mathbb{R})$-pt, which proves the theorem since this space is topologically identical to E'-$*$weak, see (13.2) p. 191. □

13.5. Continuity of the bilinear form of duality

Let us show that the **bilinear form of duality**, that is the mapping $\langle \ , \ \rangle_{E'\times E}$, is sequentially continuous.

THEOREM 13.21.– *Let E be a separated semi-normed space. Then:*

$$\langle \ , \ \rangle_{E'\times E} \text{ is sequentially continuous from } E' \times E \text{ into } \mathbb{R}. \ \blacksquare$$

Proof. Let $(e_n)_{n\in\mathbb{N}}$ and $(e'_n)_{n\in\mathbb{N}}$ be two sequences such that $e_n \to e$ in E and $e'_n \to e'$ in E'. We split

$$\langle e'_n, e_n\rangle - \langle e', e\rangle = \langle e'_n - e', e_n\rangle + \langle e', e_n - e\rangle.$$

On the one hand, $\langle e', e_n - e\rangle \to 0$ because e', being continuous, is sequentially continuous (Theorem 7.2). On the other hand, $\langle e'_n - e', e_n\rangle \to 0$ because the set $B = \{e_n : n \in \mathbb{N}\}$ being bounded (Theorem 2.7 (a)), we have, by Definition 13.1 (b) of the semi-norms of E',

$$|\langle e'_n - e', e_n\rangle| \le \sup_{v\in B} |\langle e'_n - e', v\rangle| = \|e'_n - e'\|_{E';B}$$

which tends to 0. Subsequently, $\langle e'_n, e_n\rangle - \langle e', e\rangle \to 0$. □

The case of a normed space. Let us show that the continuity of the bilinear form of duality characterizes normable spaces.

THEOREM 13.22.– *Let E be a separated semi-normed space. Then:*

$$\langle\ ,\ \rangle_{E'\times E} \textit{ is continuous} \quad\Leftrightarrow\quad E \textit{ is normable.}\ \square$$

Proof. E is normable $\Rightarrow \langle\ ,\ \rangle$ is continuous. If the family of semi-norms of E is equivalent to a norm $\|\ \|_E$, we do not modify the dual E' and its topology (Theorem 13.6 (a)) by endowing E with this norm. Then, the topology of E' is defined by its dual norm (Theorem 13.9 (a)) and (Theorem 13.9 (c)) $|\langle e', e\rangle| \le |||e'|||_{E'}\|e\|_E$, which implies (Theorem 7.20 (a)) the continuity of the bilinear map $\langle\ ,\ \rangle$.

Converse. Assume that $\langle\ ,\ \rangle$ is continuous on $E' \times E$. Then (Theorem 7.20 (a)), there exist a finite set $N \subset \mathcal{N}_E$ and a finite number of bounded sets B_i of E and $c \in \mathbb{R}$ such that, for all $u \in E$ and $u' \in E'$,

$$|\langle u', u\rangle| \le c \sup_{\mu\in N} \|u\|_{E;\mu} \sup_i \|u'\|_{E';B_i}. \tag{13.13}$$

Let $\nu \in \mathcal{N}_E$ and $e \in E$. There exists (Theorem 14.3, corollary of the Hahn–Banach theorem that we will see in the next chapter, with respect to $p = \|\ \|_{E;\nu}$) $e' \in E'$ such that:

$$\langle e', e\rangle = \|e\|_{E;\nu}, \qquad |\langle e', u\rangle| \le \|u\|_{E;\nu},\ \forall u \in E. \tag{13.14}$$

By denoting B' the set spanned by these e' when e spans E, it results thereof that, for all $e \in E$,

$$\|e\|_{E;\nu} = \sup_{e'\in B'} |\langle e', e\rangle|.$$

With (13.13), it follows that

$$\|e\|_{E;\nu} \le c \sup_{\mu\in N} \|e\|_{E;\mu} \sup_i \sup_{e'\in B'} \|e'\|_{E';B_i}.$$

On the right-hand side, we bound, by means of the inequality in (13.14),

$$\sup_{e'\in B'} \|e'\|_{E';B_i} = \sup_{e'\in B'} \sup_{u\in B_i} |\langle e', u\rangle| \le \sup_{u\in B_i} \|u\|_{E;\nu}.$$

By denoting c_i the last term, which is finite since B_i is bounded, and $c_\nu = c \sup_i c_i$, it thus results that, for all $e \in E$,

$$\|e\|_{E;\nu} \le c_\nu \sup_{\mu\in N} \|e\|_{E;\mu}.$$

The semi-norm $|||\ ||| = \sup_{\mu\in N} \|\ \|_{E;\mu}$ thus dominates the family of semi-norms of E. It is therefore a norm. Since the family of semi-norms of E dominates this norm, it is thereto equivalent. Subsequently, E is normable. □

Terminology. The designation "*bilinear form of duality*" is used by L. SCHWARTZ [99, chap. XIX, § 7, p. 285–286]. Other authors refer to it as **canonical bilinear form**, see [BOURBAKI, 17, chap. IV, § 1, no 1, example 1, p. 48] or [SCHAEFER, 96, chap. IV, § 1, p. 123]. It is also sometimes called *scalar product on* $E' \times E$. □

13.6. Dual of a product

Let us show that the dual of a product of semi-normed spaces is isomorphic to the product of the duals.

THEOREM 13.23.– *Let $E_1, E_2 \ldots, E_d$ be separated semi-normed spaces, $E'_1, \ldots, E'_d$ their duals and $g \in (E_1 \times \cdots \times E_d)'$. Then:*

(a) *There exists a unique $h \in E'_1 \times \cdots \times E'_d$ such that, for all $e \in E_1 \times \cdots \times E_d$,*

$$\langle g, e\rangle_{(E_1\times\cdots\times E_d)'\times(E_1\times\cdots\times E_d)} = \sum_{i=1}^{d} \langle h_i, e_i\rangle_{E'_i\times E_i}.$$

(b) *The mapping $g \mapsto h$ is an isomorphism from $(E_1 \times \cdots \times E_d)'$ onto $E'_1 \times \cdots \times E'_d$.* ∎

Proof. (a) *Existence.* For every $i \in [\![1, d]\!]$, we define $h_i \in E'_i$ by: for all $e_i \in E_i$,

$$\langle h_i, e_i\rangle_{E'_i\times E_i} \stackrel{\text{def}}{=} \langle g, \widetilde{e_i}\rangle_{(E_1\times\cdots\times E_d)'\times(E_1\times\cdots\times E_d)} \tag{13.15}$$

where $\widetilde{e_i} \in E_1 \times \cdots \times E_d$ is defined by $(\widetilde{e_i})_i = e_i$ and $(\widetilde{e_i})_j = 0_{E_j}$ if $j \neq i$. In effect, since h_i is the composite of the continuous linear mappings $e_i \mapsto \widetilde{e_i}$ and g, it is continuous linear (Theorem 12.12 (a)) and is thus indeed an element of E'_i.

Unicity. Any solution h verifies (13.15), therefore the difference h' of two possible solutions verifies $\langle h'_i, e_i\rangle = 0$ for all $e_i \in E_i$ and therefore is equal to zero.

(b) *Linearity.* The mapping $g \mapsto h$ is linear according to its Definition (a).

Injectivity. It results from unicity in (a).

Surjectivity. Given $h \in E'_1 \times \cdots \times E'_d$, the equality $\langle g, e\rangle = \sum_{1\leq i\leq d}\langle h_i, e_i\rangle$ indeed defines $g \in (E_1 \times \cdots \times E_d)'$. As a matter of fact, each projection $e \mapsto e_i$ being continuous linear (Theorem 7.28), its composite by h_i also is (Theorem 12.12 (a)) as well as their sum g. The mapping $g \mapsto h$ is therefore surjective.

Continuity. For each i, let B_i be a bounded subset of E_i. Then:

$$\sup_{e_i\in B_i} |\langle h_i, e_i\rangle| = \sup_{\widetilde{e_i}\in\widetilde{B_i}} |\langle g, \widetilde{e_i}\rangle| \leq \sup_{a\in A} |\langle g, a\rangle|$$

where $\widetilde{B_i} = \{\widetilde{e_i} : e_i \in B_i\}$ and $A = \bigcup_i \widetilde{B_i}$. Each $\widetilde{B_i}$ is bounded in $E_1 \times \cdots \times E_d$ so (Theorem 2.12 (f)) their union A is also. Then, by Definition 6.1 of the semi-norms of the product and 13.1 (b) of the semi-norms of the dual,

$$\|h\|_{E'_1 \times \cdots \times E'_d; B_1, \ldots, B_d} = \Big(\sum_{1 \le i \le d} \|h_i\|^2_{E'_i; B_i} \Big)^{1/2} = \Big(\sum_{1 \le i \le d} \sup_{e_i \in B_i} |\langle h_i, e_i \rangle|^2 \Big)^{1/2} \le$$

$$\le \sqrt{d} \sup_{a \in A} |\langle g, a \rangle| = \sqrt{d}\, \|g\|_{(E_1 \times \cdots \times E_d)'; A}.$$

Hence (Theorem 7.14 (a)) the continuity of the bilinear mapping $g \mapsto h$.

Continuity of the inverse mapping. For every bounded set B of $E_1 \times \cdots \times E_d$, the set $B_i = \{e_i : e \in B\}$ is bounded in E_i thus (with the second inequality of Theorem 5.16 (a))

$$\|g\|_{(E_1 \times \cdots \times E_d)'; B} = \sup_{e \in B} |\langle g, e \rangle| = \sup_{e \in B} \sum_{1 \le i \le d} |\langle h_i, e_i \rangle| \le \sum_{1 \le i \le d} \sup_{e_i \in B_i} |\langle h_i, e_i \rangle| =$$

$$= \sum_{1 \le i \le d} \|h_i\|_{E'_i; B_i} \le \sqrt{d} \Big(\sum_{1 \le i \le d} \|h_i\|^2_{E'_i; B_i} \Big)^{1/2} = \sqrt{d}\, \|h\|_{E'_1 \times \cdots \times E'_d; B_1, \ldots, B_d}.$$

Hence (Theorem 7.14 (a), again) the continuity of the mapping $h \mapsto g$ since it is linear (Theorem 7.17 (b)). □

13.7. Dual of a direct sum

Let us show that the dual of a direct sum of semi-normed spaces is isomorphic to the product of their duals.

THEOREM 13.24.– *Let $E_1 \oplus \cdots \oplus E_d$ be a direct sum of separated semi-normed spaces and $g \in (E_1 \oplus \cdots \oplus E_d)'$. Then, by denoting E'_i the dual of E_i:*

(a) *For all $e_1 + \cdots + e_d \in E_1 \oplus \cdots \oplus E_d$,*

$$\langle g, e_1 + \cdots + e_d \rangle_{(E_1 \oplus \cdots \oplus E_d)' \times (E_1 \oplus \cdots \oplus E_d)} = \sum_{i=1}^{d} \langle g|_{E_i}, e_i \rangle_{E'_i \times E_i}.$$

(b) *The mapping $g \mapsto (g|_{E_1}, \ldots, g|_{E_d})$ is an isomorphism from $(E_1 \oplus \cdots \oplus E_d)'$ onto $E'_1 \times \cdots \times E'_d$.* ▮

Proof. (a) Since g is continuous linear and $e_i \in E_i$ (by Definition 6.28 of a direct sum),

$$\langle g, e_1 + \cdots + e_d \rangle = \sum_{1 \le i \le d} \langle g, e_i \rangle = \sum_{1 \le i \le d} \langle g|_{E_i}, e_i \rangle.$$

(b) We have indeed $(g|_{E_1},\dots,g|_{E_d}) \in E'_1 \times \cdots \times E'_d$ because each restriction $g|_{E_i}$ is continuous linear (Theorem 7.8 (a)) on E_i and is thus an element of E'_i.

Injectivity. The mapping $g \mapsto (g|_{E_1},\dots,g|_{E_d})$ is injective since any $e \in E_1 \oplus \cdots \oplus E_d$ has (Definition 6.28) a unique decomposition $e = e_1 + \cdots + e_d$ and

$$\langle g, e\rangle = \langle g, \sum_{1\le i\le d} e_i\rangle = \sum_{1\le i\le d} \langle g, e_i\rangle = \sum_{1\le i\le d} \langle g|_{E_i}, e_i\rangle,$$

and thus g is zero if each $g|_{E_i}$ is.

Surjectivity. The mapping $g \mapsto (g|_{E_1},\dots,g|_{E_d})$ is surjective since, $k \in E'_1 \times \cdots \times E'_d$ being given, we define $g \in (E_1 \oplus \cdots \oplus E_d)'$ such that $k_i = g|_{E_i}$ by

$$\langle g, e_1 + \cdots + e_d\rangle = \sum_{1\le i\le d} \langle k_i, e_i\rangle.$$

Continuity. For all $i \in [\![1, d]\!]$, let B_i be a bounded subset of E_i. Each B_i being bounded in $E_1 \oplus \cdots \oplus E_d$ (Theorem 3.11 (b), since E_i is thereof a topological subspace, see Theorem 6.32 (b)), their union A also is (Theorem 2.12 (f)). Then, following Definitions 6.1 of the semi-norms of the product and 13.1 (b) of the semi-norms of the dual,

$$\|(g|_{E_1},\dots,g|_{E_d})\|_{E'_1\times\cdots\times E'_d;B_1,\dots,B_d} = \Big(\sum_{1\le i\le d} \|g|_{E_i}\|^2_{E'_i;B_i}\Big)^{1/2} =$$

$$= \Big(\sum_{1\le i\le d} \sup_{e_i\in B_i} |\langle g, e_i\rangle|^2\Big)^{1/2} \le \sqrt{d}\,\sup_{a\in A} |\langle g, a\rangle| = \sqrt{d}\,\|g\|_{(E_1\oplus\cdots\oplus E_d)';A}.$$

Hence (Theorem 7.14 (a)) the continuity of the mapping $g \mapsto (g|_{E_1},\dots,g|_{E_d})$.

Continuity of the inverse mapping. Let B be a bounded subset of $E_1 \oplus \cdots \oplus E_d$ and for all $i \in [\![1, d]\!]$, let $B_i = \{e_i : e = e_1 + \cdots + e_d,\ e \in B\}$. Then, B_i is bounded in E_i (this follows from Definition 6.28 of the semi-norms of a direct sum) and

$$\|g\|_{(E_1\oplus\cdots\oplus E_d)';B} = \sup_{e\in B} |\langle g, e\rangle| = \sup_{e\in B} |\langle g, \sum_{1\le i\le d} e_i\rangle| \le$$

$$\le \sum_{1\le i\le d} \sup_{e_i\in B_i} |\langle g, e_i\rangle| = \sum_{1\le i\le d} \|g|_{E_i}\|_{E'_i;B_i} \le \sqrt{d}\Big(\sum_{1\le i\le d} \|g|_{E_i}\|^2_{E'_i;B_i}\Big)^{1/2} =$$

$$= \sqrt{d}\,\|(g|_{E_1},\dots,g|_{E_d})\|_{E'_1\times\cdots\times E'_d;B_1,\dots,B_d}.$$

Hence (Theorem 7.14 (a), again) the continuity of $(g|_{E_1},\dots,g|_{E_d}) \mapsto g$. □

Chapter 14

Dual of a Subspace

14.1. Hahn–Banach theorem

Let us state the **Hahn–Banach theorem**[1] on the extension of linear forms.

THEOREM 14.1.– *Let E be a vector space, F a vector subspace of E and p a real mapping on E such that, for all $u \in E$, $v \in E$ and $t \geq 0$,*

$$p(u+v) \leq p(u) + p(v), \quad p(tu) = tp(u). \tag{14.1}$$

Let ℓ be a linear form on F such that:

$$\ell(u) \leq p(u), \quad \forall u \in F.$$

Then, there exists a linear form L on E such that:

$$L(u) = \ell(u), \quad \forall u \in F;$$
$$L(u) \leq p(u), \quad \forall u \in E. \ \blacksquare$$

Properties of p. Properties (14.1) are in particular satisfied if p is a semi-norm. However, unlike a semi-norm, it is not necessary that $p \geq 0$ nor that $p(-u) = p(u)$. $\square$

Proof. Let X be the set of pairs (D, h) such that:

$$\begin{cases} D \text{ is a vector subspace of } E \text{ containing } F, \\ h \text{ is a linear form on } D,\ h = \ell \text{ in } F,\ h \leq p \text{ in } D. \end{cases} \tag{14.2}$$

Let us endow X with the order relation $\preceq$ defined by:

$$(D_1, h_1) \preceq (D_2, h_2) \text{ if } D_1 \subset D_2 \text{ and } h_1 = h_2 \text{ in } D_1.$$

---

1 **History of the Hahn–Banach theorem 14.1.** Theorem 14.1 was established independently by Hans HAHN in 1927 [51] and by Stefan BANACH in 1929 [8, p. 212].

Thus ordered, X is inductive (Definition 11.6 (a)). In effect, if A is a completely ordered subset of X, an upper bound (D_0, h_0) of A is defined by:

$$\begin{cases} D_0 \text{ is the union of } D \text{ such that } (D,h) \in A, \\ h_0 = h \text{ in any } D \text{ such that } (D,h) \in A. \end{cases}$$

Zorn's lemma (Theorem 11.7) therefore shows that X has a maximal element (V, L). In order that L has the expected properties, it suffices that $V = E$. We will prove it by contradiction, assuming that there exists

$$a \in E \setminus V. \tag{14.3}$$

Then, we define $(V_1, L_1) \succ (V, L)$ by:

$$V_1 = \{v + ta : v \in V,\ t \in \mathbb{R}\},$$
$$L_1(v + ta) = L(v) + t\alpha, \quad \forall v \in V,\ t \in \mathbb{R}$$

where

$$\alpha = \inf_{w \in V} p(w + a) - L(w). \tag{14.4}$$

This α is finite because $p(w+a) - L(w) \geq p(w) - p(-a) - L(w) \geq -p(-a)$ (since $p \geq L$ in V according to (14.2) since $(V, L) \in X$).

Let us show that

$$L_1 \leq p. \tag{14.5}$$

For $t > 0$, by taking $w = v/t$ in (14.4), it follows that

$$L_1(v + ta) = L(v) + t\alpha \leq L(v) + t(p(v/t + a) - L(v/t)) = p(v + ta).$$

Let us now bound $L_1(v - ta)$ for $t > 0$. For all w and w' in V, we have

$$L(w) + L(w') = L(w + w') \leq p(w + w') \leq p(w + a) + p(w' - a)$$

and therefore, for all $w' \in V$,

$$-\alpha = \sup_{w \in V} L(w) - p(w + a) \leq p(w' - a) - L(w').$$

By choosing $w' = v/t$, this allows the majoration

$$L_1(v - ta) = L(v) - t\alpha \leq L(v) + t(p(v/t - a) - L(v/t)) = p(v - ta).$$

Finally, for $t = 0$, we have $L_1(v) = L(v) \leq p(v)$ and thus (14.5) is verified.

In addition, $L_1 = L = \ell$ in F, thus $(V_1, L_1) \in X$. Since $(V_1, L_1) \succ (V, L)$, this contradicts the maximality of (V, L). As a result, the hypothesis (14.3) cannot be satisfied, whence $V = E$ and L has the required properties. □

14.2. Corollaries of the Hahn–Banach theorem

Let us complete the Hahn–Banach theorem with a continuity property.

THEOREM 14.2.– *Let E be a separated semi-normed space, p a continuous semi-norm on E and L a linear form on E such that:*

$$L(u) \leq p(u), \quad \forall u \in E.$$

Then:

$$L \in E',$$
$$|L(u)| \leq p(u), \quad \forall u \in E. \blacksquare$$

Proof. If $L(u) \geq 0$, its absolute value verifies $|L(u)| = L(u) \leq p(u)$ and, otherwise, it verifies $|L(u)| = L(-u) \leq p(-u) = p(u)$. This establishes the stated inequality.

Let $\{\| \ \|_{E;\nu} : \nu \in \mathcal{N}_E\}$ be the family of semi-norms of E. According to the characterization of the continuous semi-norms of Theorem 7.11 (a), there exist a finite set $N \subset \mathcal{N}_E$ and $c \in \mathbb{R}$ such that $p(u) \leq c \sup_{\nu \in N} \|u\|_{E;\nu}$ for all $u \in E$. Thus, $|L(u)| \leq c \sup_{\nu \in N} \|u\|_{E;\nu}$, and hence (Theorem 7.14 (a)) L is continuous, that is $L \in E'$. □

Let us give a corollary of the Hahn–Banach theorem when p is a semi-norm (that is when, in addition to (14.1), it verifies $p(-u) = p(u)$).

THEOREM 14.3.– *Let E be a separated semi-normed space, $e \in E$ and p a continuous semi-norm on E. Then, there exists $e' \in E'$ such that:*

$$\langle e', e \rangle = p(e);$$
$$|\langle e', u \rangle| \leq p(u), \quad \forall u \in E. \blacksquare$$

Proof. Let $F = \{te : t \in \mathbb{R}\}$. The linear form defined by $\ell(te) = tp(e)$ for all $t \in \mathbb{R}$ verifies $\ell \leq p$ in F therefore, according to the Hahn–Banach theorem 14.1, it has a linear extension e' such that $e'(e) = \ell(e) = p(e)$ and $e'(u) \leq p(u)$ for all $u \in E$. It thereof results (Theorem 14.2) that $|e'(u)| \leq p(u)$ and $e' \in E'$. □

Let us show that an element that any continuous linear form cancels out is zero.

THEOREM 14.4.– *Let E be a separated semi-normed space and $e \in E$. Then:*

$$[\, \langle e', e \rangle = 0, \quad \forall e' \in E' \,] \quad \Rightarrow \quad e = 0_E. \blacksquare$$

Proof. If $e \neq 0_E$, there exists a semi-norm of E such that $\|e\|_{E;\nu} \neq 0$. Such a semi-norm being continuous (Theorem 7.10), Theorem 14.3 shows that there exists $e' \in E'$ such that $\langle e', e\rangle = \|e\|_{E;\nu} \neq 0$. □

Let us show that any continuous linear form on a topological subspace can be extended to the whole space.

THEOREM 14.5.– *Let E and F be separated semi-normed spaces such that:*

$$E \text{ is a topological subspace of } F$$

and let $e' \in E'$. Then, there exists $f' \in F'$ such that:

$$\langle f', u\rangle = \langle e', u\rangle, \quad \forall u \in E. \blacksquare$$

Proof. Let $\{\| \ \|_{F;\nu} : \nu \in \mathcal{N}_F\}$ be the family of semi-norms of F. Following Definition 3.9 of a topological subspace, E can be endowed with the restrictions of these semi-norms. The characterization of E' in Theorem 13.2 (a) then shows that there exist a finite set $N \subset \mathcal{N}_F$ and $c > 0$ such that $|\langle e', u\rangle| \le c \sup_{\nu\in N} \|u\|_{F;\nu}$ for all $u \in E$. The Hahn–Banach theorem 14.1 then shows that e' has a linear extension f' such that, for all $u \in F$,

$$\langle f', u\rangle \le c \sup_{\nu\in N} \|u\|_{F;\nu}.$$

This inequality relative to $-u$ shows that $-\langle f', u\rangle$ also is bounded by the right-hand side, and thus, so is $|\langle f', u\rangle|$. As a result (Theorem 13.2 (a)), $f' \in F'$. □

14.3. Characterization of a dense subspace

Let us show that a vector subspace F is dense in a separated semi-normed space E if any continuous linear form on E canceling out on F cancels out on E[2].

THEOREM 14.6.– *Let E be a separated semi-normed space and*

$$F \text{ a vector subspace of } E.$$

Then:

$$\overline{F} = E \quad \Leftrightarrow \quad [\,\forall f' \in E',\ f'|_F = 0 \Rightarrow f' = 0\,]. \blacksquare$$

2 **History of Theorem 14.6.** This result is often deducted from the *geometric form* of the Hahn–Banach theorem or from the properties of bipolar sets. The more direct proof given here has been provided to me by Olivier BESSON.

Proof. *Direct part.* If $\overline{F} = E$, any $f' \in E'$ such that $f'|_F = 0$ is zero since then it is zero in the whole $\overline{F}$ according to Theorem 10.1 (a) (with $U = F$ and $X = E$).

Converse. If $\overline{F} \neq E$, let $e \in E \setminus \overline{F}$,

$$G = \{te + v : t \in \mathbb{R},\ v \in \overline{F}\}$$

and ℓ the linear form on G defined, for all $t \in \mathbb{R}$ and $v \in \overline{F}$, by

$$\ell(te + v) = t. \tag{14.6}$$

According to Theorem 3.15, we endow here E with a filtering family of semi-norms $\{\|\ \|_{E;\nu} : \nu \in \mathcal{N}_E\}$, which (Theorem 13.6 (a)) does not change E'.

Since $E \setminus \overline{F}$ is open, there exist (Theorem 3.17 (a)) $\nu \in \mathcal{N}_E$ and $\eta > 0$ such that $\|e - u\|_{E;\nu} \leq \eta$ implies $u \notin \overline{F}$. For all $v \in \overline{F}$ and $t \in \mathbb{R}$, $t \neq 0$, since $-v/t \in \overline{F}$, it thereof follows that $\|e + v/t\|_{E;\nu} > \eta$ from which

$$\ell(te + v) = t < \frac{t}{\eta}\left\|e + \frac{v}{t}\right\|_{E;\nu} = \frac{1}{\eta}\,\|te + v\|_{E;\nu}.$$

This is also true if $t = 0$ because $\ell(v) = 0$, following (14.6). According to the Hahn–Banach theorem 14.1, ℓ has therefore a linear extension f' such that, for all $u \in E$,

$$f'(u) \leq \frac{1}{\eta}\|u\|_{E;\nu}.$$

Since $\|\ \|_{E;\nu}$ is continuous (Theorem 7.10), there results thereof (Theorem 14.2) that $f' \in E'$. This f' is zero on F but does not cancel out on E. □

14.4. Dual of a subspace

Let us give properties of the dual of a subspace.

THEOREM 14.7.– *Let E and F be separated semi-normed spaces such that:*

$$E \underset{\rightarrow}{\subseteq} F.$$

Then, by denoting $f'|_E$ the restriction to E of $f' \in F'$:

(a) $\qquad f' \mapsto f'|_E$ *is continuous linear from F' into E'.*

(b) $\qquad$ *E is dense in F* $\quad\Leftrightarrow\quad$ *$f' \mapsto f'|_E$ is injective.* ▮

Proof. (a) *Membership of $f'|_E$ to E'.* Let $f' \in F'$. It is continuous from F into $\mathbb{R}$ therefore (Theorem 7.8 (a)) its restriction $f'|_E$ is continuous from E endowed with the topology of F into $\mathbb{R}$ and therefore (Theorem 7.7 (a)) from E endowed with its own topology into $\mathbb{R}$. Being in addition linear, $f'|_E \in E'$.

Linearity of $f' \mapsto f'|_E$. Indeed, $(f' + g')|_E = f'|_E + g'|_E$ and $(tf')|_E = t(f'|_E)$.

Continuity of $f' \mapsto f'|_E$. Any bounded set B of E is bounded in F (Theorem 3.8 (b)) and, following Definition 13.1 (b) of the semi-norms of a dual,

$$\|f'|_E\|_{E';B} = \sup_{e \in B} |\langle f'|_E, e\rangle_{E' \times E}| = \sup_{e \in B} |\langle f', e\rangle_{F' \times F}| = \|f'\|_{F';B}.$$

Hence, according to the characterization of continuous linear applications of Theorem 7.14 (a), the mapping $f' \mapsto f'|_E$ is continuous from F' into E'.

(b) This is an alternative writing of Theorem 14.6. □

Let us now give properties of the dual of a topological subspace (that is Definition 3.9 when E is endowed with the restrictions of the semi-norms of F).

THEOREM 14.8.– *Let E and F be separated semi-normed spaces such that:*

$$E \text{ is a topological vector subspace of } F.$$

Then, by denoting $f'|_E$ the restriction to E of $f' \in F'$:

(a) $\quad f' \mapsto f'|_E$ *is continuous linear and surjective from F' onto E'.*

(b) $$E \text{ is dense in } F \quad \Rightarrow \quad f' \mapsto f'|_E \text{ is bijective.}$$

(c) $$\left\{\begin{matrix} F \text{ is normed and} \\ E \text{ is dense in } F \end{matrix}\right\} \quad \Rightarrow \quad f' \mapsto f'|_E \text{ is an isomorphism.}$$

(d) $$\left\{\begin{matrix} F \text{ is normed,} \\ E \text{ is dense in } F \text{ and} \\ E \text{ is endowed with the norm of } F \end{matrix}\right\} \quad \Rightarrow \quad \left\{\begin{matrix} \text{for the dual norm } |\!|\!|\ |\!|\!| \text{ defined} \\ \text{in Theorem 13.9:} \\ |\!|\!| f'|_E |\!|\!|_{E'} = |\!|\!| f' |\!|\!|_{F'}. \end{matrix}\right.$$ ∎

Proof. (a) The mapping $f' \mapsto f'|_E$ is continuous linear according to Theorem 14.7 (a) and it is surjective according to Theorem 14.5.

(b) If E is dense in F, the mapping $f' \mapsto f'|_E$ is injective according to Theorem 14.7 (b). Being surjective according to (a), it is then bijective.

(d) If F is normed and E is endowed with the norm of F, the dual semi-norm defined in Theorem 13.9 (a) verifies

$$\|\|f'|_E\|\|_{E'} = \sup_{e\in E,\, e\neq 0} \frac{|\langle f'|_E, e\rangle_{E'\times E}|}{\|e\|_E} = \sup_{e\in E,\, e\neq 0} \frac{|\langle f', e\rangle_{F'\times F}|}{\|e\|_F} \leq$$

$$\leq \sup_{e\in F,\, e\neq 0} \frac{|\langle f', e\rangle_{F'\times F}|}{\|e\|_F} = \|\|f'\|\|_{F'}.$$

Conversely, if E is dense in F, it is sequentially dense in F (Theorem 4.6 (e)) and so, for any non-zero $e \in F$, there exists a sequence $(e_n)_{n\in\mathbb{N}} \subset E$ such that $e_n \to e$ in F. Then, there exists n_e such that, if $n \geq n_e$, then $e_n \neq 0$ and (Theorems 7.10 and 7.26 (b)) $1/\|e_n\|_F \to 1/\|e\|_F$, hence

$$\|\|f'\|\|_{F'} = \sup_{e\in F,\, e\neq 0} \frac{|\langle f', e\rangle_{F'\times F}|}{\|e\|_F} \leq \sup_{e\in F,\, e\neq 0,\, n\geq n_e} \frac{|\langle f', e_n\rangle_{F'\times F}|}{\|e_n\|_F} =$$

$$= \sup_{e\in F,\, e\neq 0,\, n\geq n_e} \frac{|\langle f'|_E, e_n\rangle_{E'\times E}|}{\|e_n\|_E} \leq \sup_{e\in E,\, e\neq 0,} \frac{|\langle f'|_E, e\rangle_{E'\times E}|}{\|e\|_E} = \|\|f'|_E\|\|_{E'}.$$

(c) The topologies of E' and F' are here generated by their dual norms defined in Theorem 13.9 (a), thus the equality (d) shows that the mapping $f \mapsto f'|_E$ is bicontinuous. Being linear, it is an isomorphism. □

14.5. Dual of an intersection

Let us show that the dual of a finite intersection of semi-normed spaces can be "represented" by the sum (of the restrictions at their intersection) of their dual spaces[3].

THEOREM 14.9.– *Let $E_1 \cap \cdots \cap E_d$ be an intersection of separated semi-normed spaces satisfying its definition 6.13. Then, by denoting E_i' the dual of E_i:*

(a) *For all $k \in (E_1 \cap \cdots \cap E_d)'$, there exists $h_1 \in E_1' \ldots, h_d \in E_d'$ such that:*

$$k = h_1|_{E_1\cap\cdots\cap E_d} + \cdots + h_d|_{E_1\cap\cdots\cap E_d}. \tag{14.7}$$

3 **History of Theorem 14.9.** The equality $(E_1 \cap E_2)' \mathrel{\overset{\rightarrow}{\leftarrow}} E_1'|_{E_1\cap E_2} + E_2'|_{E_1\cap E_2}$, topological as well as algebraic, is demonstrated by Jöran BERGH and Jörgen LÖFSTRÖM [12, th. 2.7.1, p. 32] for the normed spaces E_1 and E_2 that are dense in their intersection. It is implicit in an article by Jacques-Louis LIONS and Jaak PEETRE dating from 1964 [68, equality (3.2), p. 22].
Our generalization of the algebraic equality without the assumption of density and also to semi-normed spaces **seems to be new**.

(b) *For all $h_1 \in E_1', \dots, h_d \in E_d'$, the equality (14.7) defines $k \in (E_1 \cap \cdots \cap E_d)'$.* ▮

Another way of writing Theorem 14.9. The equality (14.7) can be written as

$$(E_1 \cap \cdots \cap E_d)' = E_1'|_{E_1 \cap \cdots \cap E_d} + \cdots + E_d'|_{E_1 \cap \cdots \cap E_d}. \tag{14.8}$$

(This sum of spaces is well defined because it satisfies Definition 6.20 with $F = (E_1 \cap \cdots \cap E_d)'$.) □

Proof. Let us denote $\mathcal{E} = E_1 \cap \cdots \cap E_d$.

(b) For every $i \in [\![1, d]\!]$, we have $\mathcal{E} \subsetneq E_i$ (Theorem 6.17), thus $h_i \in E_i'$ implies $h_i|_{\mathcal{E}} \in \mathcal{E}'$ (Theorem 14.7 (a)) and therefore $k \in \mathcal{E}'$.

(a) We define a vector subspace of $E_1 \times \cdots \times E_d$ by

$$\mathcal{F} = \{(v, \dots, v) : v \in \mathcal{E}\}$$

and we endow it with the semi-norms of $E_1 \times \cdots \times E_d$.

Given $k \in \mathcal{E}'$, we define $\ell \in \mathcal{F}'$ by

$$\langle \ell, (v, \dots, v) \rangle_{\mathcal{F}' \times \mathcal{F}} = \langle k, v \rangle_{\mathcal{E}' \times \mathcal{E}}.$$

Since $\mathcal{F}$ is a topological subspace of $E_1 \times \cdots \times E_d$, according to Theorem 14.5 (corollary of the Hahn–Banach theorem), ℓ has an extension $L \in (E_1 \times \cdots \times E_d)'$.

According to the representation of the dual of a product of spaces (Theorem 13.23), there exists $h \in E_1' \times \cdots \times E_d'$ such that, for all $(u_1, \dots, u_d) \in E_1 \times \cdots \times E_d$,

$$\langle L, (u_1, \dots, u_d) \rangle_{(E_1 \times \cdots \times E_d)' \times (E_1 \times \cdots \times E_d)} = \sum_{i=1}^{d} \langle h_i, u_i \rangle_{E_i' \times E_i}.$$

For all $v \in \mathcal{E}$, this gives

$$\langle k, v \rangle_{\mathcal{E}' \times \mathcal{E}} = \langle L, (v, \dots, v) \rangle_{\mathcal{F}' \times \mathcal{F}} = \sum_{i=1}^{d} \langle h_i, v \rangle_{E_1' \times E_i} = \sum_{i=1}^{d} \langle h_i|_{\mathcal{E}}, v \rangle_{\mathcal{E}' \times \mathcal{E}}$$

which proves (14.7). □

14.6. Dangerous identifications

The identification of an element to another consists in not differencing them. It is a voluntary confusion destined to facilitate understanding by simplifying the writing.

Permanent identifications. For example, the identifications of an integer number n to the rational number $n/1$ and of a rational number x to the real number $\dot{x}$ expressed by the inclusions (1.6), p. 4, make it possible to write $1 + 3/2 + \pi$, which is incorrect (since $\dot{1} \in \mathbb{N}$, $3/2$ is an equivalence class of $\mathbb{Z}/\mathbb{N}$, see (1.4), p. 4, and π is a semi-cut, see (1.5), p. 4), instead of $\dot{1/1} + \dot{3/2} + \pi$ which is, for its part, correct but less clear. These identifications are not problematic because they are permanent and because their compatibility with all of the operations defined on these spaces have been certified. Without them, a book like the present one would become unreadable.

The identification (13.6), p. 196 of a linear form $x' \in \mathbb{R}'$ to the real number $\langle x', 1 \rangle$ is no more problematic because it is also permanent and compatible with all the operations on $\mathbb{R}$ and $\mathbb{R}'$. As well as the identification of a continuous function in an open set Ω of $\mathbb{R}^d$ to a distribution, and therefore of $\mathcal{C}(\Omega)$ to a subset of $\mathcal{D}'(\Omega)$, that will be done in Volume II.

Occasional identifications. Some identifications cannot be systematic because they would be conflicting with each other. For instance, the "classical" functional framework $V \underset{\rightarrow}{\subsetneq} H \underset{\rightleftarrows}{=} H' \underset{\rightarrow}{\subsetneq} V'$ for solving parabolic equations makes use of the identification of the Hilbert space H to its dual (Theorem 14.10, below) and of the identification of linear forms on H to their restrictions to V (Theorem 14.11), which prohibit the identification of V to its dual, see (14.9) below.

More problematic, this functional framework may be incompatible with permanent identifications to distribution spaces. This is, for example, the case for the spaces V and H associated with Navier–Stokes equations. It is then necessary to choose between considering H as a subspace of V' or of $\mathcal{D}'$, otherwise we might obtain the — alas common — writing of the Navier–Stokes equations in the sense of distributions with a second member with values in V', which is impossible [SIMON, 106, proposition 3, p. 50].

As discussed in Volume V, the writing of the conventional proofs is just as simple without these identifications and is less ambiguous. That is to say, by utilizing $V \underset{\rightarrow}{\subsetneq} H$ and linear forms belonging to H' and V' for the variational abstract equation adapted to the Lions theorem, and by utilizing $V \underset{\rightarrow}{\subsetneq} H \underset{\rightarrow}{\subsetneq} \mathcal{D}'$ and right-hand side data in H and in a distribution space "corresponding" to V' for equations in the sense of distributions.

It is thus preferable to only make use of such occasional identifications, which we will not indulge in, when they contribute more to simplifying the writing rather than to introducing ambiguity. Furthermore, it is obvious that care should be taken to ensure their compatibility with the already made identifications, be they permanent or occasional.

Identification of a Hilbert space to its dual. Let us present the identification of a Hilbert space to its dual. **It may be impossible**, see (14.9) below, if other identifications incompatible with this one were carried out.

THEOREM 14.10.– *Let E be a Hilbert space and J the isomorphism of E on E' defined in the Riesz–Fréchet representation theorem 13.14.*

***If every** $u \in E$ **is identified to its image** $Ju \in E'$, it yields:*

(a)
$$E' \underset{\rightleftarrows}{=} E.$$

(b) *For all $u \in E$ and $v \in E$,*
$$\langle u, v \rangle_{E' \times E} = (u, v)_E.$$

(c) *For all $u \in E$ and for the dual norm $|||\ |||_{E'}$ defined in Theorem 13.9 (a),*
$$|||u|||_{E'} = \|u\|_E. \ ▮$$

Proof. The mapping J being an isomorphism from E onto E' (Theorem 13.14 (c)), these properties are obtained by replacing Ju by u in the said theorem. □

CAUTION. It is possible that, for a Hilbert space E,

$$E' \not\underset{\leftrightarrows}{=} E. \tag{14.9}$$

Proof. Let E and F be two distinct Hilbert spaces such that $E \underset{\to}{\subset} F$ and E is dense in F. If F is identified to its dual with Theorem 14.10 and if F' is identified to a subspace of E' with the dual identification given in Theorem 14.11 below, the following strict inclusion is obtained, which leads to (14.9),

$$E \underset{\to}{\subset} F \underset{\leftrightarrows}{=} F' \underset{\to}{\subset} E'. \tag{14.10}$$

□

Precise writing. The contradiction between (14.9) and Theorem 14.10 (a) disappears if the writing is not simplified by omitting the isomorphisms used for the identifications. By denoting J_E the isomorphism from E onto E', I the identity from E into F and R the restriction to E of the elements of F', it follows that

$$J_E E \underset{\leftrightarrows}{=} E', \quad J_F F \underset{\leftrightarrows}{=} F', \quad IE \underset{\to}{\subset} F, \quad RF' \underset{\to}{\subset} E' \tag{14.11}$$

therefore, Theorem 14.10 (a) and Property (14.10) are respectively written as

$$J_E E \underset{\leftrightarrows}{=} E', \quad (R \circ J_F \circ I)E \underset{\to}{\subset} (R \circ J_F)F \underset{\leftrightarrows}{=} RF' \underset{\to}{\subset} E'.$$

These properties are always true and are not contradictory because $R \circ J_F \circ I \neq J_E$. □

Identification of a dual to a subspace of another dual. Let us give the dual inclusion of a continuous inclusion with density. **It may be impossible**, see (14.13) below, if other identifications incompatible with this one were carried out.

THEOREM 14.11.– *Let E and F be separated semi-normed spaces such that:*

$$E \underset{\to}{\subset} F, \quad E \text{ is dense in } F. \tag{14.12}$$

***If every** $f' \in F'$ **is identified to its restriction** $f'|_E \in E'$, it yields:*

(a) $$F' \underset{\to}{\subset} E'.$$

(b) $$\langle f', e \rangle_{E' \times E} = \langle f', e \rangle_{F' \times F}, \quad \forall f' \in F', e \in E. \ □$$

Proof. The mapping $f' \mapsto f'|_E$ is injective here (Theorem 14.7 (b)) therefore the identification of f' to $f'|_E$ gives $F' \subset E'$ and (b). The identity from F' into E' is then (identified to) the mapping $f' \mapsto f'|_E$, which is continuous (Theorem 14.7 (a)), which gives (Theorem 7.16 (a)) $F' \underset{\to}{\subset} E'$. □

CAUTION. Even when (14.12) is verified, it is possible that

$$F' \not\underset{\to}{\subset} E'. \tag{14.13}$$

Proof. If E and F are two distinct Hilbert spaces verifying (14.12) and if each is identified to its dual with Theorem 14.10, the following strict inclusion is obtained, which leads to (14.13),

$$E' \underset{\leftrightarrows}{=} E \underset{\to}{\subset} F \underset{\leftrightarrows}{=} F'. \tag{14.14}$$

□

Precise writing. The contradiction between (14.13) and Theorem 14.11 (a) disappears if the isomorphisms used for the identifications are not omitted in the writing. With (14.11), Theorem 14.11 (a) and Property (14.14) are respectively written in the following way, which is always correct:

$$RF' \underset{\to}{\subset} E', \quad (J_F \circ I \circ J_E^{-1})E' \underset{\leftrightarrows}{=} (J_F \circ I)E \underset{\to}{\subset} J_F F \underset{\leftrightarrows}{=} F'. \ □$$

Let us show that, for topological subspaces, the dual inclusion, **when it is possible**, see (14.13), is an equality.

THEOREM 14.12.– *Let E and F be separated semi-normed spaces such that:*

$$E \text{ is a topological vector subspace of } F, \quad E \text{ is dense in } F.$$

***If every $f' \in F'$ is identified to its restriction** $f'|_E \in E'$, it yields:*

(a) $$F' = E', \quad F' \underset{\rightarrow}{\subsetneq} E'.$$

(b) $$F \text{ is normed} \quad \Rightarrow \quad F' \underset{\leftrightarrows}{=} E'.$$

(c) $$\begin{cases} F \text{ is normed and} \\ E \text{ is endowed with the restriction of the norm of } F \end{cases} \Rightarrow \begin{cases} \text{the dual norms of } E' \\ \text{and } F' \text{ are equal.} \end{cases} \square$$

Proof. (a) The mapping $f \mapsto f'|_E$ is here bijective and continuous linear (Theorem 14.8 (b) and (a)) from F' into E' and therefore the identification of f to $f'|_E$ leads to $F' = E'$ and $F' \underset{\rightarrow}{\subsetneq} E'$.

(b) If F is normed, the mapping $f \mapsto f'|_E$ is an isomorphism (Theorem 14.8 (c)) and therefore the present identification gives the topological equality.

(c) If E is endowed with the restriction of the norm of F, the mapping $f' \mapsto f'|_E$ is an isometry for the dual norms (Theorem 14.8 (d)) and therefore the present identification gives their equality. $\square$

Chapter 15

Weak Topology

15.1. Weak topology

Let us define the **weak topology**[1] of a separated semi-normed space.

DEFINITION 15.1.– *Let E be a separated semi-normed space. We denote by* E-weak *the vector space E endowed with the semi-norms, indexed by $e' \in E'$,*

$$\|e\|_{E\text{-weak};e'} \stackrel{\text{def}}{=} |\langle e', e\rangle_{E'\times E}|. \blacksquare$$

Recall that E' is the set of continuous linear forms on E, that is $\mathcal{L}(E;\mathbb{R})$. Let us compare the weak topology with the initial topology.

THEOREM 15.2.– *For every separated semi-normed space E:*

(a) E-weak *is a separated semi-normed space.*

(b) $$E \subsetneq E\text{-weak}. \blacksquare$$

Proof. Let us denote by $\{\| \ \|_{E;\nu} : \nu \in \mathcal{N}_E\}$ the family of semi-norms of E.

(a) The space E-weak is semi-normed (by Definition 15.1) and it is separated because, if $e \in E$-weak is such that $\|e\|_{E\text{-weak};e'} = 0$ for all $e' \in E'$, that is if $\langle e', e\rangle = 0$, then $e = 0_E$ according to Theorem 14.4 (corollary of the Hahn–Banach theorem).

1 **History of the notion of weak topology.** John VON NEUMANN defined the weak topology of a topological vector space in 1930 [77, p. 380].
The weak convergence of a sequence was used in $L^2(0,1)$ by David HILBERT and in $L^p(0,1)$ by Frigyes RIESZ but, as John VON NEUMANN has demonstrated [77], weakly convergent sequences are not sufficient to determine the weak topology of these spaces.

(b) Let $e' \in E'$. According to the characterization of E' of Theorem 13.2 (a), there exist a finite set $N \subset \mathcal{N}_E$ and $c \in \mathbb{R}$ such that $|\langle e', e\rangle| \le c \sup_{\nu \in N} \|e\|_{E;\nu}$ for all $e \in E$. That is,

$$\|e\|_{E\text{-weak};e'} \le c \sup_{\nu \in N} \|e\|_{E;\nu}. \tag{15.1}$$

According to Definition 3.6 (c) of a topological inclusion, this gives $E \subsetneq E$-weak. □

Observe that weak topology is preserved by topological equalities.

THEOREM 15.3.– *For every separated semi-normed space E:*

$$\begin{cases} E\text{-weak } \textit{and its semi-norms are preserved when the family} \\ \textit{of semi-norms of } E \textit{ is replaced by an equivalent family.} \end{cases}$$ ∎

Proof. When the family of semi-norms of E is replaced by an equivalent family, the semi-norms of E-weak are preserved because they are indexed (Definition 15.1) by E' which is preserved (Theorem 13.6 (a)). For its part, the vector space E-weak remains equal to E. □

Let us characterize **weakly convergent** sequences, that is, convergent for the weak topology.

THEOREM 15.4.– *Let E be a separated semi-normed space, $(u_n)_{n\in\mathbb{N}} \subset E$ and $u \in E$. Then:*

$$u_n \to u \textit{ in } E\text{-weak} \quad \Leftrightarrow \quad [\forall e' \in E',\ \langle e', u_n\rangle_{E'\times E} \to \langle e', u\rangle_{E'\times E}].$$ ∎

Proof. According to Definitions 2.6 (a) of the convergence of a sequence and 15.1 of the semi-norms of E-weak, $u_n \to u$ in E-weak if and only if, for all $e' \in E'$, it satisfies $\|u_n - u\|_{E\text{-weak};e'} = |\langle e', u_n - u\rangle| \to 0$, that is $\langle e', u_n - u\rangle \to 0$. □

Let us show that the weak topology of a Hilbert space can be expressed in terms of the scalar product.

THEOREM 15.5.– *Let E be a Hilbert space, $(\ ,\ )_E$ its scalar product and $(u_n)_{n\in\mathbb{N}}$ a sequence of E. Then:*

(a) *The topology of E-weak is generated by the semi-norms, indexed by $e \in E$,*

$$|||u|||_{E\text{-weak};e} = |(u, e)_E|.$$

More specifically, this family of semi-norms coincides with that of Definition 15.1 of E-weak.

(b) $$u_n \to u \textit{ in } E\text{-weak} \quad \Leftrightarrow \quad [\forall e \in E,\ (u_n, e)_E \to (u, e)_E].$$ ∎

Proof. (a) Denoting by J the isomorphism from E onto E' given by the Riesz–Fréchet representation theorem 13.14, there is, for all $u \in E$ and $e \in E$,

$$|||u|||_{E\text{-weak};e} = |(u,e)_E| = |\langle Je, u\rangle_{E'\times E}| = \|u\|_{E\text{-weak};Je}.$$

Since J is bijective (Theorem 13.14 (c)), these two families of semi-norms coincide and therefore define the same topology.

(b) According to Definition 2.6 (a) of the convergence of a sequence and since the semi-norms $|||\ |||_{E\text{-weak};e}$ coincide with those of E-weak by (a), $u_n \to u$ in E-weak if and only if, for all $e \in E$, it satisfies $|||u_n - u|||_{E\text{-weak};e} = |(u_n - u, e)_E| \to 0$, that is $(u_n - u, e)_E \to 0$. □

Default semi-norms of E-weak. We denote by $|||\ |||$ the semi-norms above to distinguish them from the semi-norms $\|\ \|$ of Definition 15.1 that do not require that E be Hilbertian.

Observe that the separation of E-weak by the family $\{|||\ |||_{E\text{-weak};e} : e \in E\}$ is immediate since $|||u|||_{E\text{-weak};u} = |(u,u)_E| = \|u\|_E$. In contrast, the separation by the family $\{\|\ \|_{E\text{-weak};e'} : e' \in E'\}$ refers to the difficult Hahn–Banach theorem (see the proof of Theorem 15.2 (a)). □

Let us show that the weak topology of $\mathbb{R}$ is generated by its norm.

Theorem 15.6.– *The following holds:*

$$\mathbb{R}\text{-weak} \underset{\leftrightarrow}{=} \mathbb{R}. \blacksquare$$

Proof. The identity I from $\mathbb{R}$ into itself is continuous linear, therefore it is an element of $\mathbb{R}'$ and, for all $x \in \mathbb{R}$,

$$\|x\|_{\mathbb{R}} = |x| = |\langle I, x\rangle| = \|x\|_{\mathbb{R}\text{-weak};I}. \tag{15.2}$$

Thus, according to Definition 3.6 (c) of a topological inclusion, $\mathbb{R}\text{-weak} \underset{\rightarrow}{\subseteq} \mathbb{R}$.

The converse inclusion is always true (Theorem 15.2 (b)). □

Equivalence of all topologies of separated semi-normed space on $\mathbb{R}$. Theorem 15.6 is a special case of the following property:

$$\left\{\begin{array}{l}\textit{Any family of semi-norms on } \mathbb{R} \textit{ containing a non-zero semi-norm}\\ \textit{(and thus all families making } \mathbb{R} \textit{ separated) are equivalent.}\end{array}\right. \tag{15.3}$$

Observe that this property gives back Theorem 15.6 because the family of semi-norms of $\mathbb{R}$-weak contains a non-zero semi-norm, namely $\|\ \|_{\mathbb{R}\text{-weak};I}$ given by (15.2), and so it is equivalent to the family of $\mathbb{R}$ (which is reduced to its norm).

Proof of (15.3). Following Definition 2.2 of a semi-norm, any semi-norm p on $\mathbb{R}$ verifies, by denoting $c = p(1)$, for all $x \in \mathbb{R}$,

$$p(x) = p(1x) = p(1)\,|x| = c\,|x|.$$

According to Definition 3.1 of equivalent families of semi-norms, the norm of $\mathbb{R}$ is therefore equivalent to any family containing a semi-norm $p \neq 0$ because this is equivalent to $c \neq 0$. □

15.2. Weak continuity and topological inclusions

A mapping from a separated semi-normed space E into another F is said to be **weakly continuous** if it is continuous from E-weak into F-weak.

Let us show that any continuous linear mapping is weakly continuous, which is a **Banach–Dieudonné theorem**[2].

THEOREM **15.7**.– *Let E and F be two separated semi-normed spaces. Then:*

$$\begin{cases} \textit{Any continuous linear mapping from } E \textit{ into } F \\ \textit{is continuous from } E\text{-weak } \textit{into } F\text{-weak.} \end{cases}$$ ▮

Proof. Let L be a continuous linear mapping from E into F and $f' \in F'$. The composite mapping $f' \circ L$, being linear and continuous (Theorem 12.12 (a)) from E into $\mathbb{R}$, is an element of E' and, for all $u \in E$,

$$\langle f', Lu \rangle_{F' \times F} = \langle f' \circ L, u \rangle_{E' \times E}.$$

By taking the absolute values, we get (Definition 15.1)

$$\| Lu \|_{F\text{-weak};f'} = \| u \|_{E\text{-weak};f' \circ L}.$$

This equality implies that L is continuous from E-weak into F-weak according to the characterization of continuous linear mappings of Theorem 7.14 (a). □

Equivalent formulation of the Banach–Dieudonné theorem 15.7. Let E and F be two separated semi-normed spaces. Then:

$$\mathcal{L}(E; F) \subset \mathcal{L}(E\text{-weak}; F\text{-weak}). \tag{15.4}$$

□

Converse of Theorem 15.7. Any weakly continuous linear mapping in Fréchet spaces is continuous according to a Dieudonné–Schwartz theorem[3]. Thus:

If E and F are Fréchet spaces: $\mathcal{L}(E; F) = \mathcal{L}(E\text{-weak}; F\text{-weak})$. □

The case of a multilinear mapping. Theorem 15.7 does not apply to multilinear mappings (even in normed spaces), see (16.13), p. 241. □

Let us show that E-weak is topologically included in F-weak if E is topologically included in F.

---

2 **History of Theorem 15.7.** Stefan BANACH proved in 1932 [10, p. 143] that any continuous linear mapping from a Banach space into another is sequentially continuous for weak topologies. Jean Alexandre DIEUDONNÉ established the general result in 1942 [34, th. 15, p. 122].

3 **History of a Dieudonné–Schwartz theorem.** Jean Alexandre DIEUDONNÉ and Laurent SCHWARTZ showed in 1949 that any weakly continuous linear mapping in Fréchet spaces is continuous [35, prop. 20 p. 91].

THEOREM 15.8.– *Let E and F be two separated semi-normed spaces. Then:*

(a) $$E \underset{\to}{\subsetneq} F \quad \Rightarrow \quad E\text{-weak} \underset{\to}{\subsetneq} F\text{-weak}.$$

(b) *E is a topological subspace of F*

$$\Rightarrow \quad E\text{-weak } \textit{is a topological subspace of } F\text{-weak.} \blacksquare$$

Proof. (a) This follows from the Banach–Dieudonné theorem 15.7 with respect to the identity mapping since the topological inclusion is equivalent (Theorem 7.16 (a)) to the continuity of the identity.

(b) The restriction to E of a semi-norm of F-weak, indexed by $f' \in F'$, is the semi-norm of E-weak indexed by the restriction $f'|_E \in E'$ because, for all $e \in E$,

$$\|e\|_{F\text{-weak};f'} = |\langle f', e\rangle_{F' \times F}| = |\langle f'|_E, e\rangle_{E' \times E}| = \|e\|_{E\text{-weak};f'|_E}.$$

Conversely, any semi-norm of E-weak is the restriction of a semi-norm of F-weak because any $e' \in E'$ is the restriction to E of a $f' \in F'$ (Theorem 14.5, corollary of the Hahn–Banach theorem). Thus, E-weak is a topological subspace (Definition 3.9) of F-weak. □

15.3. Weak topology of a product

Let us show that the weak topology of a product of semi-normed spaces coincides with the topology of the product of the spaces endowed with their weak topologies.

THEOREM 15.9.– *Let $E_1, E_2 \dots, E_d$ be separated semi-normed spaces. Then:*

$$(E_1 \times \cdots \times E_d)\text{-weak} \underset{\leftrightarrow}{=} E_1\text{-weak} \times \cdots \times E_d\text{-weak.} \blacksquare$$

Proof. Reminder. These two spaces are algebraically equal to $E_1 \times \cdots \times E_d$, and therefore to each other, because there is always (Definition 15.1) E-weak $= E$.

The semi-norms of the first are indexed by $g \in (E_1 \times \cdots \times E_d)'$ and are equal to

$$\|e\|_{(E_1 \times \cdots \times E_d)\text{-weak};g} = |\langle g, e\rangle_{(E_1 \times \cdots \times E_d)' \times (E_1 \times \cdots \times E_d)}|.$$

The topology of the second is generated (Theorem 6.9 (a)) by the semi-norms, indexed by $e'_1 \in E'_1 \dots, e'_d \in E'_d$,

$$|||e|||_{E_1\text{-weak} \times \cdots \times E_d\text{-weak};e'_1,\dots,e'_d} = \sum_{1 \le i \le d} \|e_i\|_{E_i\text{-weak};e'_i} = \sum_{1 \le i \le d} |\langle e'_i, e_i\rangle_{E'_i \times E_i}|.$$

Topological equality. For every $g \in (E_1 \times \cdots \times E_d)'$, there exist (Theorem 13.23 (b)) $e'_1 \in E'_1 \ldots, e'_d \in E'_d$ such that $\langle g, e\rangle = \sum_{1\le i\le d}\langle e'_i, e_i\rangle$. Then:

$$\|e\|_{(E_1\times\cdots\times E_d)\text{-weak};g} = |\langle g, e\rangle| \le$$
$$\le \sum_{1\le i\le d} |\langle e'_i, e_i\rangle| = |||e|||_{E_1\text{-weak}\times\cdots\times E_d\text{-weak};e'_1,\ldots,e'_d}.$$

This proves that E_1-weak $\times \cdots \times E_d$-weak $\underset{\rightarrow}{\subsetneq}(E_1 \times \cdots \times E_d)$-weak, by Definition 3.6 (c) of a topological inclusion.

Conversely, for every $i \in [\![, d]\!]$, given $e'_i \in E'_i$, an element $g_i \in (E_1 \times \cdots \times E_d)'$ is defined by $\langle g_i, e\rangle = \langle e'_i, e_i\rangle$. In fact, g_i, being the composite of continuous mappings $e \mapsto e_i$ (Theorem 7.28) and e', is continuous linear (Theorem 12.12 (a)). Then:

$$|||e|||_{E_1\text{-weak}\times\cdots\times E_d\text{-weak};e'_1,\ldots,e'_d} = \sum_{1\le i\le d} |\langle g_i, e\rangle| = \sum_{1\le i\le d} \|e\|_{(E_1\times\cdots\times E_d)\text{-weak};g_i}.$$

This proves that $(E_1 \times \cdots \times E_d)$-weak $\underset{\rightarrow}{\subsetneq} E_1$-weak $\times \cdots \times E_d$-weak. Thus, these two spaces are topologically equal. □

Let us show that the weak topology of $\mathbb{R}^d$ is generated by its norm.

THEOREM 15.10.– *The following holds:*

$$\mathbb{R}^d\text{-weak} \underset{\leftrightarrow}{=} \mathbb{R}^d. \blacksquare$$

Proof. Theorems 6.11 (b), 15.9, 15.6 and again 6.11 (b) successively give

$$\mathbb{R}^d\text{-weak} \underset{\leftrightarrow}{=} (\mathbb{R} \times \cdots \times \mathbb{R})\text{-weak} \underset{\leftrightarrow}{=} \mathbb{R}\text{-weak} \times \cdots \times \mathbb{R}\text{-weak} \underset{\leftrightarrow}{=} \mathbb{R} \times \cdots \times \mathbb{R} \underset{\leftrightarrow}{=} \mathbb{R}^d.$$
□

15.4. Weak topology of an intersection

Let us compare the weak topology of an intersection of semi-normed spaces with the topology of the intersection of spaces endowed with their weak topologies[4].

4 **History of Theorem 15.11.** This result appears to be **new**...at least to the author of these lines. And more especially as it is based on the characterization of the dual of an intersection of Theorem 14.9 which seems new in the general case, see note 3, p. 215.

THEOREM 15.11.– *Let $\bigcap_{i\in\mathcal{I}} E_i$ be an intersection of separated semi-normed spaces satisfying its Definition 6.13. Then:*

(a)
$$\Big(\bigcap_{i\in\mathcal{I}} E_i\Big)\text{-weak} = \bigcap_{i\in\mathcal{I}} E_i\text{-weak}.$$

(b)
$$\Big(\bigcap_{i\in\mathcal{I}} E_i\Big)\text{-weak} \subsetneqq \bigcap_{i\in\mathcal{I}} E_i\text{-weak}.$$

(c) *If $\mathcal{I} = [\![1, d]\!]$,*
$$\Big(\bigcap_{1\leq i\leq d} E_i\Big)\text{-weak} \underset{\leftrightarrow}{=} \bigcap_{1\leq i\leq d} E_i\text{-weak}. \blacksquare$$

Notations. The reader might ask here: what does the equality (c) bring more than (a)? It gives the topological equality, that is (Definition 3.6 (b)) the equivalence of the families of the semi-norms. The equality (a) means only algebraic equality, that is the equality of the sets. □

Proof of Theorem 15.11. Let us denote $\mathcal{E} = \bigcap_{i\in\mathcal{I}} E_i$.

(a) The spaces $\mathcal{E}$-weak and $\bigcap_{i\in\mathcal{I}} E_i$-weak are both algebraically equal to $\mathcal{E}$ and therefore to each other, because there is always (Definition 15.1) E-weak $= E$.

(b) *Reminder.* The semi-norms of $\mathcal{E}$-weak are indexed by $g \in \mathcal{E}'$ and equal to

$$\|e\|_{\mathcal{E}\text{-weak};g} = |\langle g, e\rangle_{\mathcal{E}'\times\mathcal{E}}|.$$

The topology of $\bigcap_{i\in\mathcal{I}} E_i$-weak is generated (Definitions 6.13 and 15.1) by the semi-norms, indexed by $j \in \mathcal{I}$ and $h_j \in E'_j$,

$$\|e\|_{\bigcap_{i\in\mathcal{I}} E_i\text{-weak};j,h_j} = \|e\|_{E_j\text{-weak};h_j} = |\langle h_j, e\rangle_{E'_j\times E_j}|.$$

Topological inclusion. For all $j \in \mathcal{I}$, we have $\mathcal{E} \underset{\to}{\subseteq} E_j$ (Theorem 6.17) therefore, for all $h_j \in E'_j$, it follows that (Theorem 14.7 (a)) $h_j|_{\mathcal{E}} \in \mathcal{E}'$ and

$$\|e\|_{\bigcap_{i\in\mathcal{I}} E_i\text{-weak};j,h_j} = |\langle h_j, e\rangle| = |\langle h_j|_{\mathcal{E}}, e\rangle| = \|e\|_{\mathcal{E}\text{-weak};h_j|_{\mathcal{E}}}.$$

This proves (b) following Definition 3.6 (c) of a topological inclusion.

(c) *The case when* $\mathcal{E} = E_1 \cap \cdots \cap E_d$. Given $g \in \mathcal{E}'$, there exist (Theorem 14.9 (a)) $h_1 \in E'_1 \ldots, h_d \in E'_d$ such that $g = h_1|_{\mathcal{E}} + \cdots + h_d|_{\mathcal{E}}$ and thus

$$\|e\|_{\mathcal{E}\text{-weak};g} = |\langle g, e\rangle| = \Big|\sum_{1\leq i\leq d} \langle h_i, e\rangle\Big| \leq \sum_{1\leq i\leq d} |\langle h_i, e\rangle| =$$
$$= \sum_{1\leq i\leq d} \|e\|_{E_i\text{-weak};h_i} = \sum_{1\leq i\leq d} \|e\|_{E_1\text{-weak}\cap\cdots\cap E_d\text{-weak};i,h_i}.$$

This proves that E_1-weak $\cap \cdots \cap E_d$-weak $\subsetneq \mathcal{E}$-weak. With (b), it follows therefore that these two spaces are topologically equal. □

15.5. Norm and semi-norms of a weak limit

Observe that the roles of E and E' can be exchanged in the equality defining the dual norm of a normed space.

THEOREM 15.12.– *Let E be a normed space, $\| \ \|_E$ its norm and $|\!|\!| \ |\!|\!|_{E'}$ the dual norm defined in Theorem 13.9 (a). Then, there is the* ***dual equality****, for all $e \in E$,*

$$\|e\|_E = \sup_{e' \in E',\, e' \neq 0} \frac{|\langle e', e\rangle|}{|\!|\!|e'|\!|\!|_{E'}}. \blacksquare$$

Proof. Let $e \in E$. By a corollary (Theorem 14.3) of the Hahn–Banach theorem, there exists $u' \in E'$ such that:

$$\langle u', e\rangle = \|e\|_E, \quad |\langle u', u\rangle| \leq \|u\|_E, \ \forall u \in E.$$

By definition, $|\!|\!|u'|\!|\!|_{E'} = \sup_{u \in E,\, u \neq 0} |\langle u', u\rangle| / \|u\|_E$ thus $|\!|\!|u'|\!|\!|_{E'} \leq 1$ and

$$\|e\|_E \leq \frac{\|e\|_E}{|\!|\!|u'|\!|\!|_{E'}} = \frac{\langle u', e\rangle}{|\!|\!|u'|\!|\!|_{E'}} \leq \sup_{e' \in E',\, e' \neq 0} \frac{|\langle e', e\rangle|}{|\!|\!|e'|\!|\!|_{E'}}.$$

The converse inequality results from the inequality $|\langle e', e\rangle| \leq |\!|\!|e'|\!|\!|_{E'} \|e\|_E$ of Theorem 13.9 (c). □

Let us bound the norm of a weak limit in a normed space.

THEOREM 15.13.– *Let E be a normed space, $(u_n)_{n \in \mathbb{N}} \subset E$, $u \in E$ and $c \in \mathbb{R}$ such that:*

$$\|u_n\|_E \leq c, \quad \forall n \in \mathbb{N},$$

$$u_n \to u \text{ in } E\text{-weak}.$$

Then:

$$\|u\|_E \leq c. \blacksquare$$

Proof. The dual norm verifies (Theorem 13.9 (c)) $|\langle e', e\rangle| \leq |\!|\!|e'|\!|\!|_{E'} \|e\|_E$ for all $e \in E$ and $e' \in E'$ thus, in particular,

$$|\langle e', u_n\rangle| \leq c\, |\!|\!|e'|\!|\!|_{E'}.$$

At the limit, $|\langle e', u\rangle| \le c\, |||e'|||_{E'}$ from which, with the dual equality (Theorem 15.12),

$$\|u\|_E = \sup_{e' \in E',\, e' \neq 0} \frac{|\langle e', u\rangle|}{|||e'|||_{E'}} \le c. \ \square$$

Another proof of Theorem 15.13 when E is a Hilbert space. Then:

$$0 \le (u_n - u, u_n - u)_E = \|u_n\|_E^2 + \|u\|_E^2 - 2(u_n, u)_E \le c^2 + \|u\|_E^2 - 2(u_n, u)_E$$

hence, at the limit,

$$0 \le c^2 + \|u\|_E^2 - 2(u, u)_E = c^2 - \|u\|_E^2. \ \square$$

To complement this, let us show that, in a Hilbert space, the weak convergence jointly with the convergence of the norm implies strong convergence.

THEOREM 15.14.– *Let E be a Hilbert space, $(u_n)_{n\in\mathbb{N}} \subset E$ and $u \in E$ such that:*

$$u_n \to u \text{ in } E\text{-weak},$$

$$\|u_n\|_E \to \|u\|_E.$$

Then:

$$u_n \to u \text{ in } E. \ \blacksquare$$

Proof. There is

$$\|u_n - u\|_E^2 = \|u_n\|_E^2 + \|u\|_E^2 - 2(u_n, u)_E = \|u_n\|_E^2 - \|u\|_E^2 + 2(u - u_n, u)_E$$

which tends to 0 since $(u_n, u)_E \to (u, u)_E$ (Theorem 15.5 (b)) and $\|u_n\|_E^2 \to \|u\|_E^2$ (Theorem 7.26 (a)). $\square$

Variant of Theorem 15.14. Let us show that, when the norm of the u_n is bounded by that of their limit, weak convergence leads again to strong convergence.

THEOREM 15.15.– *Let E be a Hilbert space, $(u_n)_{n\in\mathbb{N}} \subset E$ and $u \in E$ such that:*

$$u_n \to u \text{ in } E\text{-weak},$$

$$\|u_n\|_E \le \|u\|_E.$$

Then:

$$u_n \to u \text{ in } E. \ \square$$

Proof. Under the assumptions made,

$$0 \le \|u_n - u\|_E^2 = \|u_n\|_E^2 - \|u\|_E^2 + 2(u - u_n, u)_E \le 2(u - u_n, u)_E$$

which tends to 0. Therefore, $\|u_n - u\|_E^2 \to 0$. $\square$

Supplement to Theorem 15.15. In this theorem, $\|u_n\|_E \leq \|u\|_E$ can be replaced by

$$\lim_{m\to\infty} \sup_{n\geq m} \|u_n\|_E \leq \|u\|_E. \tag{15.5}$$

□

Let us bound any continuous semi-norm of a weak limit, in any semi-normed space.

THEOREM 15.16.– *Let E be a separated semi-normed space, $(u_n)_{n\in\mathbb{N}} \subset E$, $u \in E$, $c \in \mathbb{R}$ and p such that:*

$$p \textit{ is a continuous semi-norm on } E,$$

$$p(u_n) \leq c, \quad \forall n \in \mathbb{N},$$

$$u_n \to u \quad \textit{in } E\text{-weak.}$$

Then:

$$p(u) \leq c. \blacksquare$$

Proof. Let $\{\| \ \|_{E;\nu} : \nu \in \mathcal{N}_E\}$ be the family of semi-norms of E and (Theorem 6.36)

$$E/\mathrm{N}_p = \{\widetilde{v} : v \in E\} \quad \text{where} \quad \widetilde{v} = \{x \in E : p(x - v) = 0\}$$

the quotient space of E by the kernel N_p of p, which is normed by

$$\|\widetilde{v}\|_{E/\mathrm{N}_p} = p(v).$$

The semi-norm p being continuous, there exist (Theorem 7.11 (a)) a finite set $N \subset \mathcal{N}_E$ and $c \geq 0$ such that, for every $v \in E$, there is $p(v) \leq c\sup_{\nu\in N} \|v\|_{E;\nu}$, that is

$$\|\widetilde{v}\|_{E/\mathrm{N}_p} \leq c \sup_{\nu\in N} \|v\|_{E;\nu}.$$

The mapping $v \mapsto \widetilde{v}$ being linear (by Definition 6.35 of the operations in a quotient space), it results therefore (Theorem 7.14 (a)) that it is continuous from E into E/N_p. According to the Banach–Dieudonné theorem 15.7, it is continuous from E-weak into E/N_p-weak and *a fortiori* (Theorem 7.2 (a)) sequentially continuous in these spaces. Thus

$$\widetilde{u_n} \to \widetilde{u} \text{ in } E/\mathrm{N}_p\text{-weak.}$$

Since E/N_p is normed and $\|\widetilde{u_n}\|_{E/\mathrm{N}_p} = p(u_n) \leq c$, it therefore results that (Theorem 15.13) $\|\widetilde{u}\|_{E/\mathrm{N}_p} \leq c$. That is, $p(u) \leq c$. □

Chapter 16

Properties of Sets for the Weak Topology

16.1. Banach–Mackey theorem (weakly bounded sets)

Let us show that any weakly bounded set is bounded, which is the **Banach–Mackey theorem**[1].

THEOREM 16.1.– *For every separated semi-normed space E:*

$$\textit{Bounded in } E\text{-weak} \quad \Leftrightarrow \quad \textit{bounded in } E. \blacksquare$$

Proof. Bounded in $E \Rightarrow$ in E-weak. The inclusion $E \subsetneq E$-weak (Theorem 15.2 (b)) implies (Theorem 3.8 (b)) this property. (It results more directly from the majoration (15.1) p. 222).

Bounded in E-weak $\Rightarrow$ in E. Let U be a bounded subset of E-weak.

The case in which E is normed. The topology of E' is then defined (Theorem 13.9 (a)) by the dual norm

$$|||e'|||_{E'} = \sup_{e \in E,\, e \neq 0} \frac{|\langle e', e\rangle|}{\|e\|_E}.$$

For all $e \in E$, let us define a linear mapping T_e from E' into $\mathbb{R}$ by

$$T_e(e') = \langle e', e\rangle.$$

1 **History of the Banach–Mackey theorem 16.1.** George MACKEY proved a more general result, that is so called by Nicolas BOURBAKI [17, ch. IV, § 2, th. 3, p. 70] (Theorem 16.1 is recovered as a corollary [17, p. 70]). Stefan BANACH proved it in 1932 for any countable set (that is, for every sequence) of a normed space [10, p. 80]. Nelson DUNFORD extended it in 1938 to every subset of a normed space [37, th. 2, p. 308].

Since (Theorem 13.9 (c)) $|\langle e', e\rangle| \leq |||e'|||_{E'} \|e\|_E$, the mapping T_e is continuous (Theorem 7.14 (a)). On the other hand, since U is bounded in E-weak and since (Definition 15.1) $|\langle e', e\rangle| = \|e\|_{E\text{-weak};e'}$, it follows that, for all $e' \in E'$,

$$\sup_{e\in U} |T_e(e')| = \sup_{e\in U} \|e\|_{E\text{-weak};e'} < \infty.$$

The space E' being a Banach space (Theorem 13.9 (d) since E is normed here), the Banach–Steinhaus theorem 10.11 shows that $\{T_e : e \in U\}$ is equicontinuous and more specifically, that there exists $c \in \mathbb{R}$ such that:

$$\sup_{e\in U} |T_e(e')| \leq c\,|||e'|||_{E'}.$$

With the dual equality $\|e\|_E = \sup_{e'\in E',\, e'\neq 0} |\langle e', e\rangle|/|||e'|||_{E'}$ (Theorem 15.12), it therefore results that

$$\sup_{e\in U} \|e\|_E = \sup_{e\in U}\ \sup_{e'\in E',\, e'\neq 0} \frac{|T_e(e')|}{|||e'|||_{E'}} \leq c.$$

Therefore, U is bounded in E.

The general case. Let us reconsider the case where E is semi-normed. Let $\|\ \|_{E;\nu}$ be one of its semi-norms and

$$E/\mathrm{N} = \{\widetilde{v} : v \in E\} \quad \text{where} \quad \widetilde{v} = \{x \in E : \|x - v\|_{E;\nu} = 0\}$$

the quotient space (Definition 6.35) of E by the kernel N of $\|\ \|_{E;\nu}$. It is normed (Theorem 6.36) by

$$\|\widetilde{v}\|_{E/\mathrm{N}} = \|v\|_{E;\nu}.$$

Let $\lambda \in (E/\mathrm{N})'$. According to the characterization of duals of Theorem 13.2 (b), there exists $c \geq 0$ such that $|\langle \lambda, \cdot\rangle| \leq c\|\cdot\|_{E/\mathrm{N}}$. We define a mapping ℓ from E into $\mathbb{R}$ by: for all $e \in E$,

$$\langle \ell, e\rangle = \langle \lambda, \widetilde{e}\rangle.$$

It is linear because the mapping $e \mapsto \widetilde{e}$ is linear according to the property (6.13) of Theorem 6.35. In addition, $|\langle \ell, e\rangle| = |\langle \lambda, \widetilde{e}\rangle| \leq c\|\widetilde{e}\|_{E/\mathrm{N}} = c\|e\|_{E;\nu}$ thus (Theorem 13.2 (a)) $\ell \in E'$. In addition,

$$\|\widetilde{e}\|_{E/\mathrm{N}\text{-weak};\lambda} = |\langle \lambda, \widetilde{e}\rangle| = |\langle \ell, e\rangle| = \|e\|_{E\text{-weak};\ell}.$$

Since U is bounded in E-weak, it results that

$$\sup_{e\in U} \|\widetilde{e}\|_{E/\mathrm{N}\text{-weak};\lambda} = \sup_{e\in U} \|e\|_{E\text{-weak};\ell} < \infty.$$

The set $\widetilde{U} = \{\widetilde{e} : e \in U\}$ is therefore bounded in E/N-weak. Since E/N is normed, the case of such a space shows that $\widetilde{U}$ is bounded in E/N and therefore

$$\sup_{e \in U} \|e\|_{E;\nu} = \sup_{\widetilde{e} \in \widetilde{U}} \|\widetilde{e}\|_{E/\mathrm{N}} < \infty.$$

This is true for any semi-norm $\| \;\; \|_{E;\nu}$ of E, which proves that U is bounded in E. □

Let us show that E-weak has the same dual as E, which is a **Dieudonné's theorem**[2], and so that E-weak-weak coincides with E-weak.

THEOREM 16.2.– *For every separated semi-normed space E:*

(a) $$(E\text{-weak})' \underset{\leftrightarrows}{=} E'.$$

(b) $$E\text{-weak-weak} \underset{\leftrightarrows}{=} E\text{-weak}. \;\blacksquare$$

Proof. (a) Since $E \underset{\to}{\subsetneq} E$-weak (Theorem 15.2 (b)), it follows that (first equality of (13.2) and Theorem 7.7 (a))

$$(E\text{-weak})' = \mathcal{L}(E\text{-weak}; \mathbb{R}) \subset \mathcal{L}(E; \mathbb{R}) = E'.$$

Conversely, since continuity implies weak continuity (Banach–Dieudonné theorem 15.7) and $\mathbb{R}$-weak is topologically equal to $\mathbb{R}$ (Theorem 15.6), it follows that

$$E' = \mathcal{L}(E; \mathbb{R}) \subset \mathcal{L}(E\text{-weak}; \mathbb{R}\text{-weak}) = \mathcal{L}(E\text{-weak}; \mathbb{R}) = (E\text{-weak})'.$$

Thus $(E\text{-weak})' = E'$.

Furthermore, their families of semi-norms are identical because they are (Definition 13.1 (b)) respectively indexed by the bounded subsets of E-weak and E, which coincide according to Banach–Mackey theorem 16.1.

(b) By Definition 15.1, E-weak-weak is the vector space E-weak, that is E, endowed with the semi-norms $\|e\|_{E\text{-weak-weak};e'} = |\langle e', e \rangle|$ indexed by $e' \in (E\text{-weak})'$, that is by $e' \in E'$ according to (a). Its family of semi-norms is thus identical to that of E-weak. □

16.2. Gauge of a convex open set

Let us build the gauge of a convex open set containing the origin.

2 **History of Dieudonné's theorem 16.2 (a).** Jean Alexandre DIEUDONNÉ proved that $(E\text{-weak})' = E'$ in 1942 [34, th. 2 p. 114].

THEOREM 16.3.– *Let E be a separated semi-normed space and $U \subset E$ such that:*

$$U \textit{ is a convex open set containing } 0_E.$$

Then, there exists a unique continuous real mapping p on E, called **gauge** *of U, such that, for all $u \in E$, $v \in E$ and $t \geq 0$,*

$$p(u) \geq 0, \quad p(tu) = t\,p(u), \quad p(u+v) \leq p(u) + p(v) \tag{16.1}$$

and

$$U = \{u \in E : p(u) < 1\}. \tag{16.2}$$

■

Proof. *Preliminary*. Based on Theorem 3.15, let us endow E with a filtering family of semi-norms $\{\| \ \|_{E;\nu} : \nu \in \mathcal{N}_E\}$ generating its topology, which preserves the open sets (Theorem 3.4 (b)) and continuity (Theorem 7.5). Since U is open and contains 0_E, there exist (Theorem 3.17 (a)) $\nu \in \mathcal{N}_E$ and $\eta > 0$ such that:

$$v \in E, \ \|v\|_{E;\nu} \leq \eta \quad \Rightarrow \quad v \in U. \tag{16.3}$$

Construction of p. Given $u \in E$, let

$$p(u) = \inf \left\{ s \in \mathbb{R} : s > 0, \ \frac{u}{s} \in U \right\}. \tag{16.4}$$

This greatest lower bound exists in $\mathbb{R}$ (Theorem 1.2 (b)) because the set of these s is not empty since, according to (16.3), $u/s \in U$ as soon as $s \geq \|u\|_{E;\nu}/\eta$.

Verification of (16.1). Obviously, $p(u) \geq 0$ and $p(tu) = t\,p(u)$.

On the other hand, if $a > p(u)$ and $b > p(v)$, then $u/a \in U$ and $v/b \in U$, hence

$$\frac{u+v}{a+b} = \frac{a}{a+b}\frac{u}{a} + \frac{b}{a+b}\frac{v}{b} \in U$$

since U is convex. Thus, $p(u+v) \leq a + b$. By taking the greatest lower bounds with respect to a and then with respect to b, it follows that $p(u+v) \leq p(u) + p(v)$.

Verification of (16.2). If $u \in E$ and $p(u) < 1$, there exists $s < 1$ such that $u/s \in U$. Hence $u = s(u/s) + (1-s)0_E \in U$ since $0_E \in U$ and U is convex, which proves that

$$\{u \in E : p(u) < 1\} \subset U. \tag{16.5}$$

Conversely, let $u \in U$. Since U is open, there exist (Theorem 3.17 (a)) $\mu \in \mathcal{N}_E$ and $\epsilon > 0$ such that $\|v - u\|_{E;\mu} \leq \epsilon$ implies that $v \in U$. If $\|u\|_{E;\mu} \neq 0$, by denoting $r = \epsilon/\|u\|_{E;\mu}$, it results thereof that $u + ru = (1+r)u \in U$ and therefore

$p(u) \leq 1/(1+r) < 1$. This is still true if $\|u\|_{E;\mu} = 0$ because then $2u \in U$ and thus $p(u) \leq 1/2 < 1$. Hence the converse inclusion of (16.5), which is thus an equality.

Unicity of p. Let q be another gauge and $0 < r < 1$. Then, $p(u) \leq r$ implies that $q(u) \leq 1$ thus (Theorem 7.12) $q(u) \leq p(u)/r$ for all $u \in E$. This is for all $r < 1$, thus $q(u) \leq p(u)$. Since q and p have the same role, it also follows that $p(u) \leq q(u)$ and therefore $q = p$.

Continuity of p. According to (16.3), $\|v\|_{E;\nu} \leq \eta$ leads to $p(v) < 1$ and therefore (Theorem 7.12) $p(v) \leq \|v\|_{E;\nu}/\eta$ for all $v \in E$. Then, (16.1) gives

$$p(u+v) - p(u) \leq p(v) \leq \|v\|_{E;\nu}/\eta.$$

Hence $p(u) - p(u+v) = p((u+v)-v) - p(u+v) \leq \|-v\|_{E;\nu}/\eta = \|v\|_{E;\nu}/\eta$, and finally $|p(u+v) - p(u)| \leq \|v\|_{E;\nu}/\eta$ for all $v \in E$, which implies (Theorem 7.3 (b)) that p is continuous. □

Gauge versus semi-norm. In order for the gauge p to be a semi-norm, it is necessary (and it suffices) that, in addition to (16.1), it verifies $p(-u) = p(u)$ and thus that $-u \in U$ as soon as $u \in U$ (that is, U is *balanced*). □

Motivation for the study of gauges. We will use gauges to prove Mazur's theorem 16.4. □

16.3. Mazur's theorem (weakly closed convex sets)

Let us state **Mazur's theorem**[3] that says that any convex closed set is weakly closed (the converse is immediate).

THEOREM 16.4.– *For any subset of a separated semi-normed space* E*:*

$$\textit{Closed convex in } E\text{-weak} \quad \Leftrightarrow \quad \textit{closed convex in } E. \ \blacksquare$$

Proof. Let U be a convex subset of E. If it is closed in E-weak, it is closed in E according to Theorem 3.8 (c)) since $E \subsetneq E$-weak (Theorem 15.2 (b)).

Conversely, assume that U is not empty (if it is empty, it is closed in E and in E-weak, see Theorem 2.13) and closed in E. Let $x \in U$ and $z \notin U$. Then, $U - x$ is a

3 **History of Mazur's theorem 16.4.** Stanislaw MAZUR showed in 1933 [70, p. 80] that any convex closed subset of a normed space is weakly sequentially closed. Its proof gives the weak closure at the expense of minor modifications.
Stefan BANACH showed the property in 1932 for every vector subspace of a normed space [10, p. 58 and 134]. Jean Alexandre DIEUDONNÉ proved it in 1942 for any vector subspace of a semi-normed space [34, th. 3 p. 115].

closed convex of E containing 0_E and $z - x \notin U - x$ thus, as we will verify hereafter (Theorem 16.5), there exists $e' \in E'$ such that $e' \leq 1$ in $U - x$ and

$$\langle e', z - x \rangle = r > 1.$$

Then, following Definition 15.1 of the semi-norms of E-weak, if $v \in E$ verifies

$$\|v - z\|_{E\text{-weak};e'} = |\langle e', v - z \rangle| \leq \frac{r-1}{2},$$

it verifies

$$\langle e', v - x \rangle = \langle e', z - x \rangle + \langle e', v - z \rangle \geq r - \frac{r-1}{2} = 1 + \frac{r-1}{2} > 1$$

and thus $v \notin U$. This proves that the complement $E \setminus U$ of U is open (Definition 2.9 (b)) in E-weak. Therefore, U is closed in E-weak. □

The following consequence of the Hahn–Banach theorem remains to be verified.

THEOREM 16.5.– *Let E be a semi-normed space, $U \subset E$ and $z \in E$ such that:*

U is a closed convex set such that $0_E \in U$ and $z \notin U$.

There exists then $e' \in E'$ such that:

$$\langle e', u \rangle \leq 1, \quad \forall u \in U, \tag{16.6}$$

$$\langle e', z \rangle > 1. \blacksquare$$

Proof. Preliminary. Based on Theorem 3.15, let us endow E with a filtering family of semi-norms $\{\| \ \|_{E;\nu} : \nu \in \mathcal{N}_E\}$ generating its topology, which preserves its open sets (Theorem 3.4 (b)) and its dual E' (Theorem 13.6 (a)).

Since U is closed, its complement $E \setminus U$ is open, and therefore there exist (Theorem 3.17 (a)) $\nu \in \mathcal{N}_E$ and $\eta > 0$ such that:

$$v \in E, \ \|v - z\|_{E;\nu} \leq \eta \quad \Rightarrow \quad v \notin U. \tag{16.7}$$

Let us denote $B = \{v \in E : \|v\|_{E;\nu} < \eta/2\}$. Then,

$$U + B \text{ is a convex open set containing } 0_E,$$

$$(U + B) \cap (z + B) = \emptyset. \tag{16.8}$$

Construction of e'. Let p be the gauge of $U + B$ (given by Theorem 16.3) and let $F = \{tz : t \in \mathbb{R}\}$. A linear form ℓ on F is defined by: for all $t \in \mathbb{R}$,

$$\ell(tz) = tp(z).$$

It verifies $\ell \leq p$ on F (because $\ell(tz) = p(tz)$ if $t \geq 0$ and $\ell(tz) \leq 0 \leq p(tz)$ if $t \leq 0$). According to the Hahn–Banach theorem 14.1, ℓ has a linear extension e' that verifies $e' \leq p$ in the whole E, and thus (16.6) since $p < 1$ in $U + B$, that contains U.

The mapping q defined by $q(v) = p(v)+p(-v)$ is a semi-norm (it verifies $q(tv) = tq(v)$ for any $t \in \mathbb{R}$ whereas p only verifies it for $t \geq 0$) which bounds p and thus e'. It is continuous since the gauge p is (Theorem 16.3), and therefore (Theorem 14.2) $e' \in E'$.

On the other hand, $\|z\|_{E;\nu} > \eta > 0$ (following (16.7) with $v = 0_E$, since $0_E \in U$). Let us denote $r = \eta/(4\|z\|_{E;\nu})$. Since $\|-rz\|_{E;\nu} = \eta/4$, it follows that $-rz \in B$; thus, (16.8) gives $z-rz \notin U+B$ and hence (Theorem 16.3) $p(z-rz) = (1-r)p(z) \geq 1$ and finally,

$$\langle e', z\rangle = \ell(z) = p(z) \geq 1/(1-r) > 1. \ \square$$

Let us apply Mazur's theorem to the vector subspaces.

THEOREM 16.6.– *A vector subspace of a separated semi-normed space E is closed in E-weak if, and only if, it is closed in E.* ▮

Proof. This is a particular case of Mazur's theorem 16.4 since any vector subspace is convex. □

16.4. Šmulian's theorem (weakly compact sets)

Let us show that, for the weak topology of a metrizable space, compactness leads to sequential compactness, which is **Šmulian's theorem**[4].

THEOREM 16.7.– *Let*

E be a metrizable space.

Its sets have the following properties:

(a) *Compact in E-weak* $\Rightarrow$ *sequentially compact in E-weak.*

(b) $\left\{\begin{array}{l}\textit{Relatively compact}\\ \textit{in } E\text{-weak}\end{array}\right. \Rightarrow \left\{\begin{array}{l}\textit{relatively sequentially compact}\\ \textit{in } E\text{-weak.}\end{array}\right.$ ▮

4 **History of Šmulian's theorem.** Vitold Lvovitsch ŠMULIAN proved in 1940 [108] that compactness leads to sequential compactness in a Banach space endowed with the weak topology. Jean Alexandre DIEUDONNÉ and Laurent SCHWARTZ extended this result to Fréchet spaces in 1949 [35, prop. 17, p. 88].

Proof. Let $U \subset E$.

(a) Assume that U is compact in E-weak and let $(u_n)_{n\in\mathbb{N}}$ be a sequence of U.

Denote
$$X = \Big\{ \sum_{n\le m} c_n u_n : c_n \in \mathbb{Q},\ m \in \mathbb{N} \Big\}$$
the set of the linear rational combinations of the u_n. It is countable as an image (Theorem 1.1 (c)) of the set of all finite sequences of $\mathbb{Q}$, which is countable (Theorems 1.1 (g) and 1.6), by the mapping $(c_n)_{n\le m} \mapsto \sum_{n\le m} c_n u_n$.

Let F be the closure of X in E, endowed with the semi-norms of E. It is a topological subspace of E, separable (Definition 2.21 (a), since X is countable) and metrizable thus, as we will verify hereafter (Lemma 16.8):

$$\begin{cases} \text{There exists a countable set } W' \text{ sequentially} \\ \text{dense (Definition 2.19 (b)) in } F'\text{-}{*}\text{weak.} \end{cases} \qquad (16.9)$$

Let us denote $F_{W'}$ the space F endowed with the semi-norms $\|\cdot\|_{F_{W'};w'} = |\langle w', \cdot\rangle|$ indexed by $w' \in W'$. Let us show that

$$F_{W'} \text{ is a metrizable semi-normed space.} \qquad (16.10)$$

Its family of semi-norms being countable, it remains (Definition 4.1 (a)) to verify that $F_{W'}$ is separated (Definition 2.3 (c)). Let thus $v \in F_{W'}$ such that $\|v\|_{F_{W'};w'} = 0$ for all $w' \in W'$. For all $e' \in F'$, according to (16.9) there exists a sequence $(w'_n)_{n\in\mathbb{N}}$ of W' converging to e' in F'-$*$weak; since $\langle w'_n, v\rangle = 0$, at the limit $\langle e', v\rangle = 0$ which implies (Theorem 14.4, corollary of the Hahn–Banach theorem) $v = 0_E$. This ends the proof of (16.10).

In E-weak, U is compact and thus closed (Theorem 2.26 (a)) and F is closed (Theorem 16.6, corollary of Mazur's theorem); therefore, $U \cap F$ is a closed subset of the compact set U and therefore it is itself compact (Theorem 2.26 (a) and (c)). Since F-weak is a topological subspace of E-weak (Theorem 15.8 (b)), it results (Theorem 3.11 (b)) that $U \cap F$ is compact in F-weak. In addition,

$$F\text{-weak} \subsetneq F_{W'}$$

(Definition 3.6 (c)) since every semi-norm of the first is a semi-norm of the second. Therefore (Theorem 3.8 (b)), $U \cap F$ is compact in $F_{W'}$. The latter being metrizable, it follows that (Theorem 4.6 (b))

$$U \cap F \text{ is sequentially compact in } F_{W'}.$$

The identity of F-weak in $F_{W'}$ being continuous (Theorem 7.16 (a)), its restriction J to $U \cap F$ also is (Theorem 7.8 (a)). Since J is bijective and $U \cap F$ is compact in

F-weak, its inverse mapping J^{-1} is continuous (Theorem 8.11) and *a fortiori* sequentially continuous (Theorem 7.2 (a)) from the subset $U \cap F$ of $F_{W'}$ into F-weak. Since $U \cap F$ is sequentially compact in $F_{W'}$, it thereof results (Theorem 8.7 (b)) that its image $J^{-1}(U \cap F)$, that is $U \cap F$, is sequentially compact in F-weak.

Since F-weak is a topological subspace of E-weak, $U \cap F$ is thus sequentially compact in E-weak (Theorem 3.8 (b)). The sequence $(u_n)_{n\in\mathbb{N}}$ therefore has (Definition 2.24 (a)) a convergent subsequence in E-weak whose limit is an element of $U \cap F$, and thus of U, which proves that U is sequentially compact in E-weak.

(b) Let us assume that U is relatively compact in E-weak. Then (Definition 2.23 (b)), $\overline{U}$ is compact and therefore sequentially compact according to (a). Any sequence of U therefore has a subsequence that converges in (the subset $\overline{U}$ of) E-weak, which proves that U is relatively sequentially compact (Definition 2.24 (b)). □

It remains to verify (16.9), that being the following property.

LEMMA 16.8.– *For any separable and metrizable separated semi-normed space E, there exists a sequentially dense countable set in E'-∗weak.* ▮

Proof. The space E being metrizable, it can (Theorem 4.4) be endowed with an increasing sequence $(\| \ \|_{E;k})_{k\in\mathbb{N}}$ of semi-norms. Let $(v_i)_{i\in\mathbb{N}}$ be a sequence such that $\overline{(v_i)_{i\in\mathbb{N}}} = E$.

Let us fix $k \in \mathbb{N}$ and denote

$$A_k = \{e' \in E' : \forall e \in E, \ |\langle e', e\rangle| \le k \,\|e\|_{E;k}\}.$$

For a fixed $n \in \mathbb{N}$, let us show that there is a finite subset $B_{k,n}$ of A_k such that:

$$\forall e' \in A_k, \text{ there exists } e'_n \in B_{k,n} \text{ such that } \sup_{i\le n} |\langle e' - e'_n, v_i\rangle| \le \frac{1}{n}. \qquad (16.11)$$

To this end, we denote

$$X_{k,n} = \{(\langle e', v_0\rangle, \ldots, \langle e', v_n\rangle) \in \mathbb{R}^{n+1} : e' \in A_k\}.$$

It follows that $X_{k,n} \subset [-r, r]^{n+1}$ where $r = k \sup_{i\le n} \|v_i\|_{E;k}$. Let us choose an integer $s \ge r$ and divide the cube $[-s, s]^{n+1}$ into $(2sn)^{n+1}$ subcubes with sides of length $1/n$. In each of the subcubes intersecting $X_{k,n}$, choose a point x of $X_{k,n}$ and then choose $e'_x \in A_k$ such that $(\langle e'_x, v_0\rangle, \ldots, \langle e'_x, v_n\rangle) = x$. The set $B_{k,n}$ of these e'_x satisfies (16.11).

Let now $e' \in E'$. Following the characterization of E' of Theorem 13.2 (b), there exist $j \in \mathbb{N}$ and $c \ge 0$ such that $|\langle e', \cdot\rangle| \le c\,\| \cdot \|_{E;j}$, and thus $e' \in A_k$ for all

$k \geq \sup\{c, j\}$. Let us fix such a k and for each $n \in N$, let us choose $e'_n \in B_{k,n}$ verifying (16.11). Then:

$$e'_n \to e' \text{ in } E'\text{-}*\text{weak}.$$

As a matter of fact, given $u \in E$ and $\epsilon > 0$, the characterization of closure of Theorem 3.17 (c) gives the existence of $i \in \mathbb{N}$ such that $\|u - v_i\|_{E;k} \leq \epsilon/(3k)$, and thus $|\langle e', u - v_i \rangle| \leq \epsilon/3$ and $|\langle e'_n, u - v_i \rangle| \leq \epsilon/3$ (since e' and e'_n are elements of A_k), from which for $n \geq \sup\{i, 3/\epsilon\}$,

$$\|e'_n - e'\|_{E'\text{-}*\text{weak};u} = |\langle e'_n - e', u \rangle| =$$

$$= |\langle e'_n - e', v_i \rangle + \langle e'_n, u - v_i \rangle - \langle e', u - v_i \rangle| \leq \epsilon.$$

The set $B = \bigcup_{k \in \mathbb{N},\, n \in \mathbb{N}} B_{k,n}$ is therefore sequentially dense in E'-$*$weak. In addition, it is countable (Theorem 1.1 (e)), which proves Lemma 16.8. □

Šmulian's theorem versus metrizability. In a metrizable space, *relatively compact* is equivalent to *relatively sequentially compact* (Theorem 4.6 (c)) but this does not imply Šmulian's theorem 16.7 because the metrizability of E does not imply that of E-weak. However, the metrizability of E implies that of its relatively compact countable subsets endowed with the weak topology, which is the key point of the proof of Theorem 16.7. □

Usage of Šmulian's theorem. Šmulian's theorem 16.7 will help us to prove the extractability of any semi-reflexive metrizable space (Theorem 18.7). □

16.5. Semi-weak continuity of a bilinear mapping

Let us give a semi-weak sequential continuity property of continuous bilinear mappings.

THEOREM 16.9.– *Let E_1, E_2 and F be three separated semi-normed spaces, a mapping T and two sequences $(a_n)_{n \in \mathbb{N}} \subset E_1$ and $(b_n)_{n \in \mathbb{N}} \subset E_2$ such that:*

$$T \textit{ is continuous bilinear from } E_1 \times E_2 \textit{ into } F,$$

$$a_n \to a \textit{ in } E_1\text{-weak}, \quad b_n \to b \textit{ in } E_2.$$

Then:

$$T(a_n, b_n) \to T(a, b) \textit{ in } F\text{-weak}. \;\blacksquare$$

Proof. Let us decompose

$$T(a_n, b_n) - T(a, b) = T(a_n, b_n - b) + T(a_n - a, b). \tag{16.12}$$

The last term converges because the partial mapping $T(\,\cdot\,, b)$, being continuous linear from E_1 into F, is continuous from E_1-weak into F-weak (Banach–Dieudonné theorem 15.7) and thus it is sequentially continuous in these spaces (Theorem 7.2 (a)), from which

$$T(a_n - a, b) \to 0_F \quad \text{in } F\text{-weak}.$$

On the other hand, by denoting $\mathcal{N}_{E_1}$, $\mathcal{N}_{E_2}$ and $\mathcal{N}_F$ the sets respectively indexing the semi-norms of E_1, E_2 and F, for all $\mu \in \mathcal{N}_F$, there exist (Theorem 7.20 (a)) $c \geq 0$ and two finite sets $N_1 \subset \mathcal{N}_{E_1}$ and $N_2 \subset \mathcal{N}_{E_2}$ such that, for all $u_1 \in E_1$ and $u_2 \in E_2$,

$$\|T(u_1, u_2)\|_{F;\mu} \leq c \sup_{\nu_1 \in N_1} \|u_1\|_{E_1;\nu_1} \sup_{\nu_2 \in N_2} \|u_2\|_{E_2;\nu_2}.$$

The sequence $(a_n)_{n\in\mathbb{N}}$, being bounded in E_1-weak, is bounded in E_1 according to Banach–Mackey theorem 16.1, that is, for each $\nu_1 \in \mathcal{N}_{E_1}$,

$$\sup_{n\in\mathbb{N}} \|a_n\|_{E_1;\nu_1} = c_{\nu_1} < \infty.$$

Therefore, by denoting $m = \sup_{\nu_1 \in N_1} c_{\nu_1}$,

$$\|T(a_n, b_n - b)\|_{F;\mu} \leq cm \sup_{\nu_2 \in N_2} \|b_n - b\|_{E_2;\nu_2}.$$

The right-hand side, and therefore the left one, tends to 0. This is true for each $\mu \in \mathcal{N}_F$, thus

$$T(a_n, b_n - b) \to 0_F \quad \text{in } F.$$

Since $F \underset{\to}{\subsetneq} F$-weak (Theorem 15.2 (b)), this convergence also takes place in F-weak (Theorem 3.8 (a)). The right-hand side of (16.12) therefore converges to 0_F in F-weak which proves that $T(a_n, b_n) \to T(a, b)$ in F-weak. □

Equivalent formulation of Theorem 16.9. For a mapping from a product of separated semi-normed spaces into a separated semi-normed space:

Bilinear continuous from $E_1 \times E_2$ *into* F $\Rightarrow$ $\begin{cases} \textit{sequentially continuous from} \\ E_1\text{-weak} \times E_2 \textit{ in } F\text{-weak.} \end{cases}$ □

Optimality of Theorem 16.9. Under the assumptions of Theorem 16.9, even if E_1, E_2 and F are normed, T is **not necessarily continuous** from E_1-weak $\times$ E_2 into F-weak. In addition, T is also not necessarily sequentially continuous from E_1-weak $\times$ E_2-weak into F-weak: for example, the product is not sequentially continuous from $L^2((0,1))$-weak $\times$ $L^2((0,1))$-weak into $L^2((0,1))$-weak according to (16.14), hereafter. □

Non-continuity of multilinear mappings for weak topologies. Unlike linear mappings (see Banach–Dieudonné theorem 15.7), continuous multilinear mappings are not necessarily weakly continuous. More specifically, even for normed spaces E_1, E_2 and F:

Continuous multilinear from $E_1 \times E_2$ *into* F $\not\Rightarrow$ $\begin{cases} \textit{continuous from } (E_1 \times E_2)\text{-weak} \\ \textit{into } F\text{-weak.} \end{cases}$ (16.13)

Proof. The mapping $(f, g) \mapsto fg$ is continuous from $(L^2((0,1)))^2$ into $L^1((0,1))$, but it is not weakly continuous. In effect, the step function f_n in $(0,1)$ defined, for $n \in \mathbb{N}$, by

$$f_n = \begin{cases} 1 & \text{in the intervals }]i/n, (i+1/2)/n[,\ i = 0, 1\ldots, n-1, \\ -1 & \text{elsewhere,} \end{cases}$$

verifies, when $n \to \infty$, (the second property follows from $f_n^2 \equiv 1$)

$$f_n \to 0 \text{ in } L^2((0,1))\text{-weak}, \quad f_n^2 \not\to 0 \text{ in } L^1((0,1))\text{-weak}. \tag{16.14}$$

□

Chapter 17

Reflexivity

17.1. Reflexive spaces

Let us define reflexivity[1] and variants.

DEFINITION 17.1.– *Let E be a separated semi-normed space and $\{\| \ \|_{E;\nu} : \nu \in \mathcal{N}_E\}$ its semi-norms.*

(a) *E is **semi-reflexive** if, for all $e'' \in E''$, there exists $e \in E$ such that:*

$$\langle e'', e' \rangle_{E'' \times E'} = \langle e', e \rangle_{E' \times E}, \quad \forall e' \in E'.$$

(b) *E is **prereflexive** if, for every bounded set B' of E', there exist a finite set $N \subset \mathcal{N}_E$ and $c \in \mathbb{R}$ such that, for all $e \in E$,*

$$\sup_{e' \in B'} |\langle e', e \rangle_{E' \times E}| \leq c \sup_{\nu \in N} \|e\|_{E;\nu}.$$

(c) *E is **reflexive** if it is semi-reflexive and prereflexive.* ∎

Terminology. **Semi-reflexive** is a property conventionally defined by (a).

Prereflexive is a **new property**, so the reader will have to recall its definition if he utilizes it. Not being used in this form, it was not named.

---

1 **History of the notion of reflexivity.** The problem of reflexivity was raised by Hans HAHN in 1927 in [51], in which he examined the successive duals of a normed space.
Jean Alexandre DIEUDONNÉ and Laurent SCHWARTZ defined the reflexivity of a separated locally convex topological vector space in 1949, in the form given in Theorem 17.4 (d) [35, p. 78].

History of the notion of semi-reflexivity. This notion was introduced by Jean Alexandre DIEUDONNÉ and Laurent SCHWARTZ in 1949 [35, p. 78].

History of the notion of prereflexivity. This notion seems to be **new**.

Our definition of reflexive is new. It is equivalent (Theorem 17.4 (d)) to the conventional definition, which states that the canonical injection from E into E'' is an isomorphism. The introduction of prereflexivity makes it possible to break it down into two independent properties: one algebraic (semi-reflexivity) and the other topological (prereflexivity). □

Let us show that an inequality converse to that defining prereflexivity is always satisfied.

THEOREM 17.2.– *For every semi-norm $\| \ \|_{E;\nu}$ of a separated semi-normed space E, there exists a bounded set B' of E' such that, for all $e \in E$,*

$$\|e\|_{E;\nu} \leq \sup_{e' \in B'} |\langle e', e \rangle_{E' \times E}|. \ \blacksquare$$

Proof. Let $e \in E$. The semi-norm $\| \ \|_{E;\nu}$ being continuous (Theorem 7.10), there exists (Theorem 14.3 corollary of the Hahn–Banach theorem) $b' \in E'$ such that:

$$\langle b', e \rangle = \|e\|_{E;\nu}, \qquad |\langle b', u \rangle| \leq \|u\|_{E;\nu}, \ \forall u \in E. \tag{17.1}$$

Let B' be the set spanned by these b' when e spans E. It is bounded in E' since (according to Definition 13.1 (b) of the semi-norms of E'), for any bounded set B of E,

$$\sup_{b' \in B'} \|b'\|_{E';B} = \sup_{b' \in B'} \sup_{u \in B} |\langle b', u \rangle| \leq \sup_{u \in B} \|u\|_{E;\nu} < \infty.$$

In addition, the equality of (17.1) gives, since $b' \in B'$, the expected inequality:

$$\|e\|_{E;\nu} = |\langle b', e \rangle| \leq \sup_{e' \in B'} |\langle e', e \rangle|. \ \square$$

Let us define the **canonical injection** I from a space into its bidual.

DEFINITION 17.3.– *Let E be a separated semi-normed space. For every $e \in E$, we define $Ie \in E''$ by: for all $e' \in E'$,*

$$\langle Ie, e' \rangle_{E'' \times E'} \stackrel{\text{def}}{=} \langle e', e \rangle_{E' \times E}. \ \blacksquare$$

Justification.– The mapping Ie is linear and $|\langle Ie, e' \rangle| = |\langle e', e \rangle| = \|e'\|_{E';\{e\}}$, and therefore it is indeed an element of E'' according to the characterization of the bidual of Theorem 13.3 (b) with respect to $B = \{e\}$. □

Let us justify the designation of canonical *injection* and let us show that a reflexive space is isomorphic to its bidual.

THEOREM 17.4.– *Let E be a separated semi-normed space and I the canonical injection (Definition 17.3) from E into its bidual E''. Then:*

(a) *I is linear and injective.*

(b) *E is semi-reflexive $\Leftrightarrow$ I is bijective.*

(c) *E is prereflexive $\Leftrightarrow$ I is continuous.*

(d) *E is reflexive $\Leftrightarrow$ I is an isomorphism.* ▮

Proof. (a) The mapping I is linear by Definition 17.3 because every $e' \in E'$ is linear. It is injective because, if $Ie = 0$, then $\langle e', e \rangle = 0$ for all $e' \in E'$ and thus $e = 0_E$ according to a corollary (Theorem 14.4) of the Hahn–Banach theorem.

(b) Definition 17.1 (a) of the semi-reflexivity is equivalent to the surjectivity of I, and thus to its bijectivity since it is injective according to (a).

(c) The characterization of the semi-norms of E'' (Theorem 13.3 (a)) and Definition 17.3 of Ie yield

$$\|Ie\|_{E'';B'} = \sup_{e' \in B'} |\langle Ie, e' \rangle_{E'' \times E'}| = \sup_{e' \in B'} |\langle e', e \rangle_{E' \times E}|$$

thus the definition 17.1 (b) of prereflexivity is written as

$$\|Ie\|_{E'';B'} \leq c \sup_{\nu \in N} \|e\|_{E;\nu}$$

which is equivalent (Theorem 7.14 (a)) to the continuity of I.

(d) According to (b) and (c), reflexivity is equivalent to I being a continuous bijection and therefore bicontinuous, because its inverse mapping is continuous according to Theorem 17.2 (this point is detailed in the proof of Theorem 17.5 (a)). □

CAUTION. A space isomorphic to its bidual is not necessarily reflexive. It is reflexive if the canonical injection is an isomorphism. Even if E is a Banach space:

$$E \text{ is reflexive} \not\Leftarrow E \text{ is isomorphic to } E''. \qquad (17.2)$$

In effect, an isomorphism may exist, different from I, from E into E'', without E being reflexive [JAMES, 60]. □

Let us complement and clarify Theorem 17.4.

THEOREM 17.5.– *The canonical injection I of a separated semi-normed space E into its bidual E'' verifies:*

$$\text{(a)}\quad \begin{cases} I \text{ is bijective from } E \text{ into the vector space } I(E); \\ I^{-1} \text{ is continuous linear from } I(E) \text{ endowed with the topology of } E'' \text{ into } E. \end{cases}$$

$$\text{(b)}\quad E \text{ is semi-reflexive} \quad \Leftrightarrow \quad \begin{cases} I \text{ is bijective from } E \text{ into } E''; \\ I^{-1} \text{ is continuous linear from } E'' \text{ into } E. \end{cases}$$

$$\text{(c)}\quad E \text{ is prereflexive} \quad \Leftrightarrow \quad \begin{cases} I \text{ is an isomorphism from } E \text{ onto } I(E) \\ \text{endowed with the topology of } E''. \end{cases}$$

$$\text{(d)}\quad E \text{ is reflexive} \quad \Leftrightarrow \quad I \text{ is an isomorphism from } E \text{ onto } E''. \ \blacksquare$$

Proof. (a) Being injective (Theorem 17.4 (a)), the mapping I is bijective from E onto its image $I(E)$ which is a vector space (Theorem 7.17 (a)).

Since I is linear, so is its inverse mapping I^{-1} (Theorem 7.17 (b)). For every $e'' \in I(E'')$, the inequality of Theorem 17.2 with respect to $e = I^{-1}e''$ and Definitions 17.3 of the canonical injection and 13.1 (a) of the semi-norms of the bidual give

$$\|I^{-1}e''\|_{E;\nu} \leq \sup_{e' \in B'} |\langle e', I^{-1}e'' \rangle_{E' \times E}| = \sup_{e' \in B'} |\langle e'', e' \rangle_{E'' \times E'}| = \|e''\|_{E'';B'}.$$

According to the characterization of continuous linear mappings of Theorem 7.14 (a), this proves that I^{-1} is continuous from $I(E)$ endowed with the topology E'' into E.

(b) This follows from (a) since (Theorem 17.4 (b)) the semi-reflexivity of E is equivalent to $I(E) = E''$.

(c) This follows from (a) since (Theorem 17.4 (c)) the prereflexivity of E is equivalent to the continuity of I.

(d) This is Theorem 17.4 (d). □

Observe that reflexivity is preserved by topological equalities.

THEOREM 17.6.– *For every separated semi-normed space E:*

$$\begin{cases} \text{The reflexivity, semi-reflexivity and prereflexivity of } E \text{ are preserved} \\ \text{when its family of semi-norms is replaced by an equivalent family.} \ \blacksquare \end{cases}$$

Proof. Definition 17.1 (a) of semi-reflexivity is preserved because E' and its semi-norms are (Theorem 13.6 (a)), and therefore its dual E'' also is.

The inequality of Definition 17.1 (b) of prereflexivity is preserved, on the one hand (for the left-hand side) because E' and its semi-norms and thereby its bounded sets are, and on the other hand (for the right-hand side) according to Definition 3.6 (b) of the equivalent families of semi-norms.

Reflexivity (Definition 17.1 (c)) is thus also preserved. □

17.2. Sequential completion of a semi-reflexive space

Let us show that E and E-weak are sequentially complete spaces when E is semi-reflexive[2].

THEOREM 17.7.– *For every separated semi-normed space E:*

$$E \textit{ is semi-reflexive} \quad \Rightarrow \quad E \textit{ and } E\text{-weak} \textit{ are Neumann spaces.}$$ ∎

Proof. Let E be a semi-reflexive separated semi-normed space.

Sequential completion of E-weak. Let $(e_n)_{n\in\mathbb{N}}$ be a Cauchy sequence of E-weak. The set $B = \{e_n : n \in \mathbb{N}\}$ is bounded in E-weak (Theorem 2.7 (b)), and thus (Banach–Mackey theorem 16.1) it is bounded in E. Hence, for all $e' \in E'$,

$$|\langle e', e_n\rangle_{E'\times E}| \le \sup_{u\in B} |\langle e', u\rangle_{E'\times E}| = \|e'\|_{E';B}.$$

The sequence $(\langle e', e_n\rangle_{E'\times E})_{n\in\mathbb{N}}$ is a Cauchy sequence in $\mathbb{R}$ and therefore (Theorem 4.15) it has a limit, denoted by $\langle \ell, e'\rangle$. It verifies

$$|\langle \ell, e'\rangle| \le \|e'\|_{E';B}.$$

This mapping ℓ is linear, and thus, according to the characterization of E'' of Theorem 13.3 (b), it thereof results that $\ell \in E''$. Since E is semi-reflexive (Definition 17.1 (a)), there exists $e \in E$ such that $\langle \ell, e'\rangle_{E''\times E'} = \langle e', e\rangle_{E'\times E}$ for all $e' \in E'$. Then:

$$\langle e', e_n\rangle_{E'\times E} \to \langle \ell, e'\rangle_{E''\times E'} = \langle e', e\rangle_{E'\times E}.$$

That is (Theorem 15.4), $e_n \to e$ in E-weak. Thus, E-weak is sequentially complete.

Sequential completion of E. Let $(e_n)_{n\in\mathbb{N}}$ be now a Cauchy sequence of E. Since $E \subsetneqq E$-weak (Theorem 15.2 (b)), it is a Cauchy sequence of E-weak (Theorem 3.8 (a)) so, as it has just been seen, it has therein a limit e.

2 **History of Theorem 17.7.** Frigyes RIESZ showed that $L^p((0,1))$-weak is sequentially complete (if $1 < p < \infty$) in 1910 [91, p. 466].
Jean Alexandre DIEUDONNÉ showed that any semi-reflexive normed space is sequentially complete in 1942 [34, point 18, p. 125].

For every semi-norm $\| \ \|_{E;\nu}$ of E and every $\epsilon > 0$, there exists $n_0 \in \mathbb{N}$ such that, for all $m \geq n \geq n_0$,

$$\|e_n - e_m\|_{E;\nu} \leq \epsilon.$$

When $m \to \infty$, it follows that $e_n - e_m \to e_n - e$ in E-weak and therefore (Theorem 15.16)

$$\|e_n - e\|_{E;\nu} \leq \epsilon.$$

Hence, $e_n \to e$ in E which proves that E is sequentially complete. □

Some examples of weak spaces or duals that are Neumann spaces. Sequential completion being very fruitful (see the comment "*Motivation for the study of Neumann spaces*", p. 57) let us indicate a few cases where a separated semi-normed space E, or E-weak, its dual E', E'-weak, E'-$*$weak, or its bidual E''[3] is a Neumann space:

$$E \textit{ is normed or metrizable} \quad \Rightarrow \quad \begin{cases} E',\ E'' \textit{ and } E''\text{-}{*}\text{weak } \textit{are Neumann spaces;} \\ E'' \textit{ is a Fréchet space.} \end{cases} \tag{17.3}$$

$$E \textit{ is a Banach or a Fréchet space} \quad \Rightarrow \quad \begin{cases} E,\ E',\ E'\text{-}{*}\text{weak}, \\ E'' \textit{ and } E''\text{-}{*}\text{weak } \textit{are Neumann spaces;} \\ E'' \textit{ is a Fréchet space.} \end{cases} \tag{17.4}$$

$$E \textit{ is semi-reflexive} \quad \Rightarrow \quad E \textit{ and } E\text{-weak } \textit{are Neumann spaces.} \tag{17.5}$$

$$E \textit{ is a Hilbert or a reflexive space} \quad \Rightarrow \quad \begin{cases} E,\ E',\ E'',\ E''', \ldots, \\ E\text{-weak},\ E'\text{-weak},\ E''\text{-weak}\ldots, \\ E'\text{-}{*}\text{weak},\ E''\text{-}{*}\text{weak}\ldots \textit{ are Neumann spaces.} \end{cases} \tag{17.6}$$

Proof of (17.3). If E is metrizable, E' is a Neumann space (Theorem 13.8 (a)) and the other properties are given in [KÖTHE, 64, property (5), p. 301].

Proof of (17.4). If E is a Fréchet space, E'-$*$weak is a Neumann space (Theorem 13.8 (b)) and the other properties are given by (17.3) since E is metrizable.

Proof of (17.5). This is established in Theorem 17.7.

Proof of (17.6). If E is reflexive, it is a Neumann space along with E-weak (Theorem 17.7). In addition, E' is reflexive (Theorem 17.26) and therefore E' and E'-weak are also Neumann spaces. Furthermore, E'-$*$weak is topologically equal to E'-weak (Theorem 17.21), and therefore it is also a Neumann space. All the successive duals are then reflexive, and therefore Neumann spaces for their default, weak and $*$weak topologies. □

17.3. Prereflexivity of metrizable spaces

Let us show that any normed or metrizable space is prereflexive[4].

---

3 **History of a subset of (17.4) due to Grothendieck.** Alexander GROTHENDIECK showed in 1953 that E'' is a Fréchet space if E is, see also William F. DONOGHUE and Kennan T. SMITH [36].

4 **History of Theorem 17.8.** This result is **new** under this form since prereflexivity is a new concept that was only used implicitly.
Jean Alexandre DIEUDONNÉ showed in 1942 [34, § 18, p. 125] that the canonical injection I from a Banach space E into its bidual is bicontinuous from E onto $I(E)$ endowed with the topology of E", which is equivalent (Theorem 17.5 (c)) to prereflexivity. Jean Alexandre DIEUDONNÉ and Laurent SCHWARTZ showed it in 1949 [35, prop. 9 p. 78] for Fréchet spaces. Nicolas BOURBAKI showed it [17] for metrizable spaces, or more generally for bornological, barreled or infrabarreled ones.

THEOREM 17.8.– *For every semi-normed space:*

$$\text{Normed} \quad \Rightarrow \quad \text{metrizable} \quad \Rightarrow \quad \text{prereflexive.}$$

Proof. Let E be a separated semi-normed space. If it is normed, it is metrizable (Theorem 4.2).

If it is metrizable, it can be endowed (Theorem 4.4) with an increasing sequence of semi-norms $(|||\ |||_k)_{k\in\mathbb{N}}$ defining its topology, which preserves prereflexivity (Theorem 17.6). By Definition 17.1 (b) of this, the point is to show that, given a bounded set B' of E', there exist $n \in \mathbb{N}$ and $c \in \mathbb{R}$ such that, for all $u \in E$,

$$\sup_{e'\in B'} |\langle e', u\rangle| \leq c\,|||u|||_n. \tag{17.7}$$

If it was not true, for every $n \in \mathbb{N}$, there would exist an element v of E such that $n\,|||v|||_n \leq \sup_{e'\in B'} |\langle e', v\rangle|$. Then, $v_n = tv$, where $t = 1/\sup_{e'\in B'} |\langle e', v\rangle|$ would verify

$$n\,|||v_n|||_n \leq 1, \qquad \sup_{e'\in B'} |\langle e', v_n\rangle| = 1. \tag{17.8}$$

Let $k \in \mathbb{N}$. For all $n \geq k$ it would follow that $|||nv_n|||_k \leq |||nv_n|||_n \leq 1$ since $|||\ |||_k$ increases with k, hence

$$\sup_{n\in\mathbb{N}} |||nv_n|||_k \leq \sup\{|||v_1|||_k, |||2v_2|||_k, \ldots, |||(k-1)v_{k-1}|||_k, 1\} < \infty.$$

This would be true for all k, and therefore the set $X = \{nv_n : n \in \mathbb{N}\}$ would be bounded. Denote then, according to Definition 13.1 (b) of the semi-norms of E',

$$r = \sup_{e'\in B'} \|e'\|_{E';X} = \sup_{e'\in B'} \sup_{n\in\mathbb{N}} |\langle e', nv_n\rangle|.$$

Since B' is bounded, r would be finite. Then, $\sup_{e'\in B'} |\langle e', v_n\rangle| \leq r/n$ which would contradict (17.8) as soon as $n > r$; this proves (17.7) and therefore the prereflexivity of E. □

Other proof of Theorem 17.8 for a normed space. Let E be a normed space. For the dual norm of E', there is $|\langle e', e\rangle| \leq |||e'|||_{E'}\|e\|_E$ (Theorem 13.9 (c)) thus, for any bounded set B' of E',

$$\sup_{e'\in B'} |\langle e', e\rangle| \leq c\,\|e\|_E$$

where $c = \sup_{e'\in B'} |||e'|||_{E'}$ is finite. This proves that E is prereflexive (Definition 17.1 (b)). □

Let us observe that a semi-reflexive metrizable space is a Fréchet space.

THEOREM 17.9.– *For every semi-normed space:*

$$\text{Semi-reflexive metrizable} \quad \Leftrightarrow \quad \text{reflexive Fréchet space.}$$

Proof. A semi-reflexive space is sequentially complete (Theorem17.7). In addition, if it is metrizable, it is a Fréchet space (Definition 4.10 (b)) and it is prereflexive (Theorem 17.8) and thus reflexive (Definition 17.1). The converse is obvious. □

Let us observe that a semi-reflexive normed space is a Banach space.

THEOREM 17.10.– *For every vector space:*

Semi-reflexive normed space ⇔ *reflexive Banach space.* ∎

Proof. A semi-reflexive space is sequentially complete (Theorem 17.7). In addition, if it is normed, it is a Banach space (Definition 4.10 (c)) and it is prereflexive (Theorem 17.8) and thus reflexive (Definition 17.1). The converse is obvious. □

Generalization of Theorem 17.8. A separated semi-normed space is said to be **infrabarreled** if any bornivorous barrel (that is any set U convex, symmetric with respect to 0, closed and such that, for any bounded set B there exists $t > 0$ such that $U \supset tB$) is a neighborhood of 0, namely contains an open set containing 0.

Let us show that such a space is prereflexive[5].

THEOREM 17.11.– *Every bornological, or barreled, or infrabarreled separated semi-normed space is prereflexive.* □

Proof. Let B' be a bounded subset of E'. The set $X = \{e \in E : \sup_{e' \in B'} |\langle e', e\rangle| \leq 1\}$ is a bornivorous barrel. In effect, for any bounded set B of E,

$$r = \sup_{e' \in B'} \|e'\|_{E;B} = \sup_{e' \in B'} \sup_{u \in B} |\langle e', u\rangle|$$

is finite, thus $X \supset (1/r)B$.

If E is infrabarreled, X is therefore a neighborhood of 0_E, that is to say, $\{\| \ \|_{E;\nu} : \nu \in \mathcal{N}_E\}$ being the family of semi-norms of E, there exist $\eta > 0$ and a finite set $N \subset \mathcal{N}_E$ such that:

$$\{v \in E : \sup_{\nu \in N} \|v\|_{E;\nu} \leq \eta\} \subset X.$$

It follows thereof (Theorem 7.12) that $\sup_{e' \in B'} |\langle e', e\rangle| \leq (1/\eta) \sup_{\nu \in N} \|v\|_{E;\nu}$, which proves that E is prereflexive (Definition 17.1 (b)).

This is *a fortiori* verified if any barrel is a neighborhood of 0_E, that is if E is barreled, or if any convex, bornivorous symmetric set is a neighborhood of 0_E, namely if E is bornological. □

17.4. Reflexivity of Hilbert spaces

Let us show that Hilbert spaces are reflexive.

THEOREM 17.12.– *Every Hilbert space is reflexive.* ∎

---

5 **History of Theorem 17.11.** This result is **new** in this form; it is equivalent, see note 4, p. 248, to a result of Nicolas BOURBAKI [17].

Proof. Let E be a Hilbert space, J the isomorphism from E onto E' given by the Riesz–Fréchet theorem 13.14 and J' the isomorphism from E' onto E'' (it is well-defined since E' also is a Hilbert space according to Theorem 13.15 (c)). Then, for all $e'' \in E''$ and $e' \in E'$,

$$\langle e'', e' \rangle_{E'' \times E'} = (J'^{-1}e'', e')_{E'} = (J^{-1}J'^{-1}e'', J^{-1}e')_E = \\ = (J^{-1}e', J^{-1}J'^{-1}e'')_E = \langle e', J^{-1}J'^{-1}e'' \rangle_{E' \times E}$$

(these equalities follow, respectively, from Theorems 13.14 (a) with respect to J' and 13.15 (a), from Definition 5.1 (a) and from Theorems 13.14 (a) with respect to J). Thus, E verifies Definition 17.1 (a) of semi-reflexivity with $e = J^{-1}J'^{-1}e''$.

Being normed, it is in addition prereflexive (Theorem 17.8), and thus reflexive. □

Let us show that $\mathbb{R}$ and $\mathbb{R}^d$ are reflexive.

THEOREM 17.13.– *The spaces* $\mathbb{R}$ *and* $\mathbb{R}^d$ *are reflexive.* ▮

Proof. The spaces $\mathbb{R}$ and $\mathbb{R}^d$ are Hilbert spaces (Theorems 5.5 and 5.12), and as a result (Theorem 17.12), they are reflexive. □

Strong topology. Let us introduce the **strong topology** of a separated semi-normed space to be able to characterize hereafter prereflexivity.

We denote by E-strong the vector space E endowed with the semi-norms, indexed by the bounded sets B' of E',

$$\|e\|_{E\text{-strong};B'} \stackrel{\text{def}}{=} \sup_{e' \in B'} |\langle e', e \rangle|. \tag{17.9}$$

According to Theorems 17.2 and 15.2,

$$E\text{-strong} \subsetneq E \subsetneq E\text{-weak}. \tag{17.10}$$

□

Terminology. The name E-strong is justified by analogy with Definition 15.1 of E-weak and by (17.10). Its topology is also called **polar topology**.

Some authors call E-strong the space E itself when they want to distinguish it from E-weak ... which can be ambiguous if, for example, $E = F$-weak! More exactly, they call the topology of E *strong* whereas it would be better to call it the *initial topology*. □

Characterization of prereflexivity. Let us observe that:

E is prereflexive $\Leftrightarrow$ E-strong $\stackrel{=}{\rightleftarrows} E$

$\Leftrightarrow$ *any bounded family of E' is equicontinuous from E into $\mathbb{R}$.* □

Proof. First equivalence. Definition 17.1 (b) of prereflexivity is equivalent to $E \subsetneq E$-strong and therefore to the topological equality of these spaces according to (17.10).

Second equivalence. It results from the characterization of an equicontinuous family of linear mappings of Theorem 10.9 (a). □

17.5. Reflexivity of uniformly convex Banach spaces

Let us define uniform convexity[6]

DEFINITION 17.14.– *A normed space E is* **uniformly convex** *if, for all $\epsilon > 0$, there exists $\delta > 0$ such that, for all $u \in E$ and $v \in E$,*

$$\|u\|_E \leq 1,\ \|v\|_E \leq 1, \|u - v\|_E \geq \epsilon \quad \Rightarrow \quad \left\|\frac{u+v}{2}\right\|_E \leq 1 - \delta. \ \blacksquare$$

CAUTION. **Uniform convexity is not preserved** when the initial norm is replaced **by an equivalent norm.** For example, $\mathbb{R}^2$ is uniformly convex for its Euclidean norm $|x| = (x_1^2 + x_2^2)^{1/2}$, but no longer when it is endowed with the norm $|x|_1 = |x_1| + |x_2|$ or $|x|_\infty = \sup\{|x_1|, |x_2|\}$. □

We are going to show that uniformly convex Banach spaces are reflexive, which is **Milman's theorem**[7].

THEOREM 17.15.– *Any uniformly convex Banach space is reflexive.* ▮

The proof will use the following properties. Property (c) is **Goldstine's theorem**[8].

THEOREM 17.16.– *Let*

E be a normed space,

I the canonical injection (Definition 17.3) from E into its bidual E'', $|||\ |||_{E''}$ the bidual norm of E'' (defined in Theorem 13.10), $B_E = \{e \in E : \|e\|_E \leq 1\}$, and finally $B_{E''} = \{e'' \in E'' : |||e''|||_{E''} \leq 1\}$. Then, for all $e \in E$:

(a) $$|||Ie|||_{E''} = \|e\|_E.$$

(b) *$B_{E''}$ is closed in E''-$*$weak.*

(c) *$I(B_E)$ is dense in $B_{E''}$ for the topology of E''-$*$weak.*

(d) *E is a Banach space $\Rightarrow$ $I(B_E)$ is closed in E''.* ▮

Proof of Theorem 17.16. (a) The bidual norm of Ie is equal to (Theorem 13.10 (a) and Definition 17.3 of I)

$$|||Ie|||_{E''} = \sup_{e' \in E',\, e' \neq 0} \frac{|\langle Ie, e' \rangle_{E'' \times E'}|}{|||e'|||_{E'}} = \sup_{e' \in E',\, e' \neq 0} \frac{|\langle e', e \rangle_{E' \times E}|}{|||e'|||_{E'}}$$

6 **History of the notion of uniform convexity.** James Andrew CLARKSON defined the uniform convexity of a Banach space in 1936 [28].

7 **History of Milman's theorem 17.15.** David Pinhusovic MILMAN demonstrated the reflexivity of every uniformly convex Banach space in 1938 [73]. This result was also demonstrated, in 1939, by Bill PETTIS [86] and Shizuo KAKUTANI [62].

8 **History of Goldstine's theorem.** Herman GOLDSTINE proved Theorem 17.16 (c) in 1938 [47, p. 128].

which is equal to $\|e\|_E$ according to the dual equality of Theorem 15.12.

(b) The unit ball $B_{F'}$ for the dual norm of the dual of a normed space F is closed in F'-$*$weak (Theorem 13.19). The bidual norm being (Theorem 13.10 (a)) the dual norm of the dual norm, this result can be applied here to $F = E'$. This proves that $B_{E''}$ is closed in E''-$*$weak.

(c) According to Definition 2.19 (a) of density, the point is to show that, for the closure in E''-$*$weak, there is $I(B_E) \subset B_{E''} \subset \overline{I(B_E)}$.

The first inclusion is satisfied since, following (a), $|||Ie|||_{E''} = \|e\|_E$. Let us demonstrate the second. Let thus $e'' \in B_{E''}$, that is such that $|||e''|||_{E''} \leq 1$.

According to the characterization of closure of Theorem 2.17 (b) and Definition 13.4 (b) of the semi-norms of E''-$*$weak, the point now is to prove that, for every finite number of elements of E', say e'_1, $e'_2 \ldots$, e'_n, and for every $\eta > 0$, there exists $e \in B_E$ such that:

$$\sup_{i=1,\ldots,n} \|Ie - e''\|_{E''\text{-}*\text{weak};e'_i} = \sup_{i=1,\ldots,n} |\langle Ie - e'', e'_i\rangle_{E''\times E'}| \leq \eta.$$

By denoting $x_i = \langle e'', e'_i\rangle_{E''\times E'}$ and following Definition 17.3 of I, we have to find $e \in B_E$ such that:

$$\sup_{i=1,\ldots,n} |\langle e'_i, e\rangle_{E'\times E} - x_i| \leq \eta. \qquad (17.11)$$

Observe that, for any $s \in \mathbb{R}^n$,

$$|s \cdot x| = \Big|\Big\langle e'', \sum_{i=1,\ldots,n} s_i e'_i\Big\rangle_{E''\times E'}\Big| \leq \Big|\Big|\Big| \sum_{i=1,\ldots,n} s_i e'_i \Big|\Big|\Big|_{E'} \qquad (17.12)$$

because (Theorem 13.10 (c)) $|\langle e'', u'\rangle_{E''\times E'}| \leq |||e'''|||_{E''}|||u'|||_{E'}$ for all $u' \in E'$ and because here $|||e''|||_{E''} \leq 1$.

Let us show, by contradiction, that (17.11) is the result of (17.12). If (17.11) was not satisfied (by at least one $e \in B_E$ for each $\eta > 0$), x would not be an element of the closure $\overline{Z}$ in $\mathbb{R}^n$ of the set

$$Z = \{z \in \mathbb{R}^n : z = (\langle e'_i, e\rangle)_{i=1,\ldots,n},\ e \in B_E\}.$$

The set $\overline{Z}$ being convex (Theorem 2.35, because Z is) and closed in the Hilbert space $\mathbb{R}^n$, x would have (Theorem 5.8 (a)) a projection p over $\overline{Z}$. Let $s = x - p$. For all $z \in Z$, we would have

$$s \cdot x - s \cdot z = (x - p) \cdot (x - z) = |x - p|^2 + (x - p) \cdot (p - z) \geq |x - p|^2$$

since $(x - p) \cdot (p - z) \geq 0$ (Theorem 5.8 (b)). This inequality would also be verified by $-z$ that is also an element of Z. It would thus follow that $s \cdot x \geq |s \cdot z| + |x - p|^2$, and $p \neq x$ since $x \notin \overline{Z}$. Therefore, by the definition of Z and of the dual norm (Theorem 13.9 (b)),

$$s \cdot x > \sup_{z \in Z} |s \cdot z| = \sup_{e \in B_E} \left| \left\langle \sum_i s_i e_i', e \right\rangle_{E' \times E} \right| = \left|\left|\left| \sum_i s_i e_i' \right|\right|\right|_{E'}.$$

This would contradict (17.12) and as a result (17.11) is indeed satisfied.

(d) Let $(e_n)_{n \in \mathbb{N}}$ be a sequence of B_E such that $I(e_n) \to e''$ in E''.

Since it converges, $(I(e_n))_{n \in \mathbb{N}}$ is a Cauchy sequence (Theorem 2.7 (a)) in E'' and since $\|e_n - e_m\|_E = |||I(e_n) - I(e_m)|||_{E''}$ according to (a), the sequence $(e_n)_{n \in \mathbb{N}}$ is a Cauchy sequence of E.

If E is sequentially complete, $(e_n)_{n \in \mathbb{N}}$ has a limit e. Since e_n is an element of B_E that is closed (according to Justification of Definition 2.18 (a)) and thus (Theorem 2.10) sequentially closed, $e \in B_E$.

Since E is prereflexive (Theorem 17.8, because it is normed), I is continuous (Theorem 17.4 (c)) and therefore (Theorem 7.2 (a)) sequentially continuous. Hence $I(e_n) \to I(e)$ and therefore $I(e) = e''$. Thus, $e'' \in I(B_E)$. This shows that $I(B_E)$ is sequentially closed in E'', which coincides here with closed (Theorem 4.6 (a) since here E'' is normed). □

Proof of Milman's theorem 17.15. Let E be a uniformly convex Banach space. It is prereflexive (Theorem 17.8), and therefore (Definition 17.1) it remains to verify that it is semi-reflexive that is that, given $e'' \in E''$, there exists $e \in E$ such that $Ie = e''$ where I is the canonical injection (Definition 17.3) from E into E''. We will only consider the case where

$$|||e'''|||_{E''} = 1,$$

$||| \; |||_{E''}$ being the bidual norm defined in Theorem 13.10, because the general case results thereof by multiplying by a scalar.

It suffices to show that $e'' \in I(B_E)$ where $B_E = \{v \in E : \|v\|_E \leq 1\}$, namely that $e'' \in \overline{I(B_E)}$ since $I(B_E)$ is closed in E'' (Theorem 17.16 (d)). Based on the characterization of closure of Theorem 3.17 (c), for every $\epsilon > 0$, it suffices to find $u \in B_E$ such that:

$$|||Iu - e''|||_{E''} \leq \epsilon. \tag{17.13}$$

Let δ satisfy Definition 17.14 of uniform convexity with respect to this ϵ, and $e_0' \in E'$ be such that:

$$|||e_0'|||_{E'} \leq 1, \quad |\langle e'', e_0' \rangle| \geq 1 - \frac{\delta}{3}. \tag{17.14}$$

Such an e_0' does exist because $1 = |||e''|||_{E''} = \sup_{e' \in E',\, |||e'|||_{E'} \leq 1} |\langle e'', e' \rangle|$ according to Theorem 13.9 (b).

Based on Goldstine's theorem 17.16 (c), $I(B_E)$ is dense in $B_{E''}$ for the topology of E''-$*$weak, and therefore (following Definition 2.19 (a) of density and with the characterization of closure of Theorem 2.17 (b)) there exists $u \in B_E$ such that:

$$|||Iu - e''|||_{E''\text{-}*\text{weak};e'_0} \leq \frac{\delta}{3}. \tag{17.15}$$

Let us show by contradiction that it verifies (17.13). Otherwise, e'' would be an element of the set

$$\mathcal{O} = E'' \setminus B_{E''}(Iu, \epsilon).$$

The latter being open in E"-$*$weak since $B_{E''}$ and therefore $B_{E''}(Iu, \epsilon)$ are closed therein (Theorem 17.16 (b)), there would exist, according to Definition 2.9 (b) of an open set, a finite number of elements of E', say e'_1, $e'_2 \ldots$, e'_n, such that:

$$v'' \in E'', \quad \sup_{i=1,\ldots,n} |||v'' - e''|||_{E''\text{-}*\text{weak};e'_i} \leq \frac{\delta}{3} \quad \Rightarrow \quad v'' \in \mathcal{O}.$$

Using Goldstine's theorem as in (17.15), we would obtain $v \in B_E$ such that:

$$\sup_{i=0,1,\ldots,n} |||Iv - e''|||_{E''\text{-}*\text{weak};e'_i} \leq \frac{\delta}{3}.$$

Then, $Iv \in \mathcal{O}$. And

$$|||Iu + Iv - 2e''|||_{E''\text{-}*\text{weak};e'_0} = |\langle Iu + Iv - 2e'', e'_0\rangle| \leq \frac{2\delta}{3}$$

which, with (17.14), would imply that

$$1 - \delta < |\langle e'', e'_0\rangle| - \frac{\delta}{3} \leq \Big|\Big\langle I\Big(\frac{u+v}{2}\Big), e'_0\Big\rangle\Big| = \Big|\Big\langle e'_0, \frac{u+v}{2}\Big\rangle\Big| \leq \Big\|\frac{u+v}{2}\Big\|_E.$$

(the last inequality comes from $|\langle e'_0, e\rangle| \leq |||e'_0|||_{E'} \|e\|_E$, see Theorem 13.9 (c), and from $|||e'_0|||_{E'} \leq 1$, see (17.14)).

Since $\|u\|_E \leq 1$ and $\|v\|_E \leq 1$, the uniform convexity would yield $\|v - u\|_E < \epsilon$. And thus (Theorem 17.16 (a)) $|||Iv - Iu|||_{E''} < \epsilon$ which would contradict $Iv \in \mathcal{O}$. Consequently, (17.13) is indeed verified. □

A useless hypothesis for Goldstine's theorem. Some books state Theorem 17.16 (c) with the additional hypothesis that E is a Banach space ... and a proof that does not make use of it. □

Converse of Milman's theorem 17.15. A reflexive space is necessarily sequentially complete (Theorem 17.7) but it is not necessarily normed (for example $L^2_{\text{loc}}(\Omega)$ is reflexive and non-normable) and if it is, it is not necessarily uniformly convex. As a point of fact, there exist [DAY, 30] reflexive Banach spaces that are not uniformly convex (neither for their norm, nor for any equivalent norm). □

Uniform convexity of Hilbert spaces. Every Hilbert space is uniformly convex (this follows from the parallelogram law of Theorem 5.2 (b)), but we will not use it because we have already established that every Hilbert space is reflexive (Theorem 17.12). □

17.6. A property of the combinations of linear forms

An element u of a vector space E is a **linear combination** of elements $u_1, u_2, \dots,$ u_n of E if there exist real numbers c_i such that:

$$u = c_1 u_1 + c_2 u_2 + \cdots + c_n u_n.$$

A finite family $(u_1, u_2, \dots, u_n)$ of E is **free** if a linear combination $c_1 u_1 + \cdots + c_n u_n$ can only be equal to zero if all the c_i are equal to zero. An infinite family of E is said to be **free** if all of its finite subfamilies are free.

Vectors are said to be **linearly independent** if they form a free family.

Let us give a condition for a linear form to be a linear combination of linear forms which will be used to prove the characterization of semi-reflexives spaces of the Banach–Alaoglu–Bourbaki theorem 17.19.

THEOREM 17.17.– *Let $L_1, L_2 \dots, L_n$ and L be linear forms on a vector space E and $c \in \mathbb{R}$ such that, for all $e \in E$,*

$$|L(e)| \leq c \sup_{1 \leq i \leq n} |L_i(e)|.$$

Then, there exist real numbers c_i such that:

$$L = c_1 L_1 + c_2 L_2 + \cdots + c_n L_n. \blacksquare$$

Proof. This is a consequence of Theorem 17.18 below. □

We still have to establish the following result.

THEOREM 17.18.– *Let $L_1, L_2 \dots, L_n$ and L be linear forms on a vector space E such that:*

$$L(e) = 0 \textit{ at every point } e \textit{ where all } L_i(e) \textit{ are zero.}$$

Then, there exist real numbers c_i such that:

$$L = c_1 L_1 + c_2 L_2 + \cdots + c_n L_n. \blacksquare$$

Proof. Case $n = 1$. If $L_1 = 0$, then $L = 0 = L_1$. Otherwise, let u be such that $L_1(u) \neq 0$ and let $v = u / L_1(u)$. Then, $L_1(v) = 1$. Given $e \in E$, let us define $z \in E$ by

$$e = L_1(e)\, v + z.$$

Then, $L_1(z) = L_1(e) - L_1(e)\, L_1(v) = 0$. Hence, by assumption, $L(z) = 0$. Thus,

$$L(e) = L_1(e)\, L(v) + L(z) = c_1 L_1(e)$$

where $c_1 = L(v)$. This is true for all $e \in E$, which proves that $L = c_1 L_1$.

General case. If one of the L_i, say L_n, is a linear combination of the others, it is sufficient to show that $L = c_1 L_1 + \cdots + c_{n-1} L_{n-1}$. Through successive eliminations, we are brought back to the case in which the L_i are linearly independent.

In this case, we will reason by induction on n. Suppose therefore that the result is true for $n-1$. Let us show that, for $i = 1, 2, \ldots, n$, there exists $v_i \in E$ such that:

$$L_i(v_i) = 1, \quad L_j(v_i) = 0 \text{ if } j \neq i. \tag{17.16}$$

In effect, there exists $u_i \in E$ such that $L_i(u_i) \neq 1$ and $L_j(u_i) = 0$ if $j \neq i$; otherwise, according to the result with respect to $n-1$, L_i would be a linear combination of the L_j, $j \neq i$, which would contradict the hypothesis that the L_j are linearly independent. Then, $v_i = u_i / L_i(u_i)$ verifies (17.16).

Given $e \in E$, let us define $z \in E$ by

$$e = L_1(e)\, v_1 + \cdots + L_n(e)\, v_n + z.$$

Then:

$$L_i(e) = L_1(e)\, L_i(v_1) + \cdots + L_n(e)\, L_i(v_n) + L_i(z) = L_i(e) + L_i(z)$$

therefore $L_i(z) = 0$. This is true for all i from which, by assumption, $L(z) = 0$. Thus,

$$L(e) = L_1(e)\, L(v_1) + \cdots + L_n(e)\, L(v_n) = c_1 L_1(e) + \cdots + c_n L_n(e)$$

where $c_i = L(v_i)$. This proves the result for n. Being true for $n = 1$, it is true for all n. □

17.7. Characterizations of semi-reflexivity

Let us show that semi-reflexivity is equivalent to the weak compactness of bounded sets which is a **variant of the Banach–Alaoglu–Bourbaki theorem**[9].

THEOREM 17.19.– *For every separated semi-normed space E:*

$$E \text{ is semi-reflexive} \quad \Leftrightarrow \quad \begin{cases} \text{any bounded subset of } E \text{ is relatively} \\ \text{compact in } E\text{-weak.} \end{cases}$$ ▮

9 **History of Theorem 17.19.** The authorship and the designation of this result are not clear to the author of these lines. It is sometimes called (on *Wikipedia*, without accurate references) the **Banach–Alaoglu–Bourbaki theorem**, although this usually designates a result related to compactness in E'-$*$weak, see note 4, p. 199. The special case in which E is normed has been established by Jean Alexandre DIEUDONNÉ in 1942 [34, theorem 24, p. 130].

The proof will use the following properties.

LEMMA 17.20.– *Let E be a semi-normed space and let I be the canonical injection from E into E''. Then:*

(a) *I is continuous from E-*weak *into E''-*$*$weak.

(b) *E is semi-reflexive* $\Rightarrow$ *I is an isomorphism from E-*weak *onto E''-*$*$weak. ∎

Proof of Lemma 17.20. (a) Definitions 13.4 (b) of the semi-norms of a $*$weak dual (here E''-$*$weak $=$ $(E')'$-$*$weak), 17.3 of the canonical injection I and 15.1 of the semi-norms of E-weak give, for all $e \in E$ and $e' \in E'$,

$$\|Ie\|_{E''\text{-}*\text{weak};e'} = |\langle Ie, e'\rangle_{E''\times E'}| = |\langle e', e\rangle_{E'\times E}| = \|e\|_{E\text{-weak};e'}. \qquad (17.17)$$

Hence the continuity of I following the characterization of Theorem 7.14 (a).

(b) If E is semi-reflexive, I is a bijection from E onto E'' (Theorem17.4 (b)) and Equality (17.17) shows that its inverse mapping I^{-1} is itself also continuous since the semi-norms of E-weak are, such as those of E''-$*$weak, indexed by $e' \in E'$. □

Proof of Theorem 17.19. Direct part. Assume that E is semi-reflexive and let B be a bounded subset of E. The canonical injection I from E into E'' is then an isomorphism from E-weak onto E''-$*$weak (Lemma 17.20 (b)) so in order for B to be relatively compact in E-weak, it suffices (Theorem 8.8 (a)) that $I(B)$ is so in E''-$*$weak.

Now, following Definition 17.3 of I and 13.1 (b) of the semi-norms of E', for all $e' \in E'$,

$$\sup_{\ell\in I(B)} |\ell(e')| = \sup_{u\in B} |\langle Iu, e'\rangle_{E''\times E'}| = \sup_{u\in B} |\langle e', u\rangle_{E'\times E}| = \|e'\|_{E';B}.$$

This implies (Theorem 13.20) that $I(B)$ is relatively compact in E"-$*$weak.

Converse. Suppose that any bounded subset of E is relatively compact in E-weak, and let $e'' \in E''$. To establish the semi-reflexivity of E, it suffices (Theorem 17.4 (b)) to show that

$$e'' \in I(E)$$

where I is the canonical injection from E into E''. According to the characterization of E'' of Theorem 13.3 (b), there exist a bounded set B of E and $c \in \mathbb{R}$ such that, for all $e' \in E'$,

$$|\langle e'', e'\rangle| \leq c \sup_{x\in B} |\langle e', x\rangle|. \qquad (17.18)$$

Let us denote by $\{\| \ \|_{E;\nu} : \nu \in \mathcal{N}_E\}$ the family of semi-norms of E and let

$$U = \{u \in E : \forall \nu \in \mathcal{N}_E,\ \|u\|_{E;\nu} \leq c \sup_{x\in B} \|x\|_{E;\nu}\}.$$

It suffices to show that e'' is an element of the set $U'' = I(U)$.

Since $cB \subset U$, the inequality (17.18) leads to, with Definition 17.3 of the canonical injection, for all $e' \in E'$,

$$|\langle e'', e'\rangle| \leq \sup_{cx\in cB} |\langle e', cx\rangle| \leq \sup_{u\in U} |\langle e', u\rangle| = \sup_{u\in U} |\langle Iu, e'\rangle| = \sup_{u''\in U''} |\langle u'', e'\rangle|. \tag{17.19}$$

Let

$$A' = \{e' \in E' : \forall u'' \in U'',\ |\langle u'', e'\rangle| \leq 1\}$$

and

$$A'' = \{v'' \in E'' : \forall e' \in A',\ |\langle v'', e'\rangle| \leq 1\}.$$

The inequality (17.19) showing that $e'' \in A''$, it remains to verify that

$$A'' \subset U''. \tag{17.20}$$

If this were not the case, there would exist $z'' \in A'' \setminus U''$, that is $z'' \notin U''$, such that:

$$e' \in E',\ \sup_{u''\in U''} |\langle u'', e'\rangle| \leq 1 \quad \Rightarrow \quad |\langle z'', e'\rangle| \leq 1. \tag{17.21}$$

Furthermore, U is bounded in E and therefore, by assumption, it is relatively compact in E-weak. In addition, it is convex and closed in E and so in E-weak according to Mazur's theorem 16.4. Being relatively compact and closed in E-weak, U is compact therein (Theorem 2.26 (a)).

Since I is continuous from E-weak into E''-$*$weak (Lemma 17.20 (a)), its image $U'' = I(U)$ is compact (Theorem 8.7 (a)) and thus closed (Theorem 2.26 (a)) in E''-$*$weak. In addition, U'' is convex, symmetric and contains $0_{E''}$; therefore, according to a corollary (Theorem 16.5) of the Hahn–Banach theorem, and since $z'' \notin U''$, there would exist a continuous linear form ℓ on E''-$*$weak such that:

$$\ell \leq 1 \text{ on } U'', \quad |\ell(z'')| > 1. \tag{17.22}$$

The characterization of the continuous linear mappings of Theorem 7.14 (a) and Definition 13.4 (b) of the semi-norms of a dual $*$weak would result in the existence of a finite number of $e'_i \in E'$ and of $a \in \mathbb{R}$ such that:

$$|\langle \ell, e''\rangle| \leq a \sup_i \|e''\|_{E''\text{-}*\mathrm{weak};e'_i} = a \sup_i |\langle e'', e'_i\rangle|.$$

There would therefore exist (Theorem 17.17) real numbers c_i such that:

$$\langle \ell, e''\rangle = \sum_i c_i \langle e'', e'_i\rangle = \langle e'', e'\rangle$$

where $e' = \sum_i c_i e'_i$. As a result, (17.22) would be written as

$$\sup_{u'' \in U''} |\langle u'', e' \rangle| \leq 1, \quad |\langle z'', e' \rangle| > 1.$$

This would contradict (17.21), which proves (17.20). Thus, E is indeed semi-reflexive since

$$e'' \in A'' \subset U'' = I(U) \subset I(E). \ \square$$

Usage of the variant of the Banach–Alaoglu–Bourbaki theorem of Theorem 17.19. This characterization of the semi-reflexivity by the weak compactness of bounded sets will be helpful in demonstrating the extractability of any semi-reflexive metrizable space (Theorem 18.7). □

Let us show that semi-reflexivity is equivalent to the equality of weak and ∗weak duals.

THEOREM 17.21.– *For every separated semi-normed space E:*

$$E \text{ is semi-reflexive} \quad \Leftrightarrow \quad E'\text{-weak} \underset{\Leftrightarrow}{\equiv} E'\text{-}{*}\text{weak}. \ \blacksquare$$

Proof. Direct part. Let $e' \in E'$ and $e \in E$. Definitions 13.4 (b) of the semi-norms of E'-∗weak, 17.3 of the canonical injection into the bidual and 13.4 (a) of the semi-norms of E'-weak give

$$\|e'\|_{E'\text{-}*\text{weak};e} = |\langle e', e \rangle_{E' \times E}| = |\langle Ie, e' \rangle_{E'' \times E'}| = \|e'\|_{E'\text{-weak};Ie}.$$

If E is semi-reflexive, I is a bijection from E onto E'' (Theorem 17.4 (b)), and therefore the families of the semi-norms of E'-∗weak and of E'-weak coincide since they are respectively indexed by $e \in E$ and $e'' \in E''$.

Converse. If $E'\text{-}*\text{weak} \underset{\to}{\subseteq} E'\text{-weak}$, thus in particular if $E'\text{-weak} \underset{\Leftrightarrow}{\equiv} E'\text{-}*\text{weak}$, then following Definition 3.6 (c) of a topological inclusion, for every $e'' \in E''$, there exist elements $e_1, \ldots, e_n$ of E and $c \in \mathbb{R}$ such that, for all $e' \in E'$,

$$\|e'\|_{E'\text{-weak};e''} \leq c \sup_{1 \leq i \leq n} \|e'\|_{E'\text{-}*\text{weak};e_i} \leq c \sum_{1 \leq i \leq n} \|e'\|_{E'\text{-}*\text{weak};e_i},$$

that is to say, $|\langle e'', e' \rangle| \leq c \sum_{1 \leq i \leq n} |\langle e', e_i \rangle|$. Then (Theorem 17.17), there exist real numbers $c_1, \ldots, c_n$ such that, for all $e' \in E'$,

$$\langle e'', e' \rangle = \sum_{1 \leq i \leq n} c_i \langle e', e_i \rangle = \langle e', e \rangle$$

where $e = \sum_{1 \leq i \leq n} c_i e_i \in E$. Therefore, E is semi-reflexive (Definition 17.1 (a)). □

17.8. Reflexivity of a subspace

Let us show that any closed subspace of a semi-reflexive space is semi-reflexive which is a **Dieudonné–Schwartz theorem**[10].

THEOREM 17.22.– *Let F be a separated semi-normed space and*

$$E \text{ a closed topological subspace of } F.$$

Then:

$$F \text{ is semi-reflexive} \quad \Rightarrow \quad E \text{ is semi-reflexive.} \ ▮$$

Proof. Let B be a bounded subset of E. Since E is a topological subspace of F, B is bounded in F (Theorem 3.11 (b)). If F is semi-reflexive, B is then relatively compact in F-weak according to the variant of the Banach–Alaoglu–Bourbaki theorem given in Theorem 17.19.

Besides, E-weak is a topological subspace of F-weak (Theorem 15.8 (b)) and it is closed in F-weak since it is in F (Theorem 16.6 corollary of Mazur's theorem). Therefore (Theorem 3.13 (a)), B is relatively compact in E-weak. The converse part of Theorem 17.19 shows then that E is semi-reflexive. □

17.9. Reflexivity of the image of a space

Let us define the transpose of a continuous linear mapping.

DEFINITION 17.23.– *Let E and F be two separated semi-normed spaces and $L \in \mathcal{L}(E; F)$. The **transpose** of L is the linear mapping $L^* \in \mathcal{L}(F'; E')$ defined, for all $f' \in F'$, by*

$$L^* f' \stackrel{\text{def}}{=} f' \circ L.$$

That is, for every $e \in E$,

$$\langle L^* f', e \rangle_{E' \times E} \stackrel{\text{def}}{=} \langle f', Le \rangle_{F' \times F}. \ ▮$$

Justification.– Theorem 12.12 (b) shows that the mapping $f' \mapsto f' \circ L$, that is L^*, is continuous linear from $\mathcal{L}(F; \mathbb{R})$ into $\mathcal{L}(E; \mathbb{R})$, that is from F' into E'. □

Let us remark that the transpose of an isomorphism is an isomorphism.

10 **History of the Dieudonné–Schwartz theorem 17.22.** Jean Alexandre DIEUDONNÉ and Laurent SCHWARTZ proved that every closed subspace of a semi-reflexive space is semi-reflexive in 1949 [35, prop. 11, p. 81].
Jean Alexandre DIEUDONNÉ established it for normed spaces in 1942 [34, corollary, p. 131].

THEOREM 17.24.– *Let E and F be separated semi-normed spaces. If L is an isomorphism from E onto F, then L^* is an isomorphism from F' onto E', and*

$$(L^*)^{-1} = (L^{-1})^*. \blacksquare$$

Proof. If L is an isomorphism from E onto F, then $L^{-1} \in \mathcal{L}(F;E)$, and Definition 17.23 of its transpose yields $(L^{-1})^* \in \mathcal{L}(E';F')$ and $(L^{-1})^*e' = e' \circ L^{-1}$ for all $e' \in E'$. Hence, again with Definition 17.23,

$$L^*((L^{-1})^*e') = L^*(e' \circ L^{-1}) = (e' \circ L^{-1}) \circ L = e'$$

which proves that L^* is invertible and $(L^*)^{-1} = (L^{-1})^*$. This mapping being continuous, as L^* itself is, the latter is an isomorphism from F' onto E'. □

Let us show that isomorphisms preserve reflexivity.

THEOREM 17.25.– *Let E and F be two isomorphic separated semi-normed spaces. Then:*

(a) E *is semi-reflexive* $\Leftrightarrow$ F *is semi-reflexive.*

(b) E *is prereflexive* $\Leftrightarrow$ F *is prereflexive.*

(c) E *is reflexive* $\Leftrightarrow$ F *is reflexive.* $\blacksquare$

Proof. Preliminary. Let J be an isomorphism from E onto F. Let us first show that the canonical injection I_F from F into its bidual is related to the canonical injection I_E from E into its bidual by

$$I_F = J^{**} \circ I_E \circ J^{-1}. \tag{17.23}$$

Definition 17.23 of the transpose with respect to $L = J^*$ gives, for all $e'' \in E''$ and $f' \in F'$,

$$\langle J^{**}e'', f' \rangle_{F'' \times F'} = \langle e'', J^*f' \rangle_{E'' \times E'}.$$

This equality, successively added to Definitions 17.3 of I_E, 17.23 of J^* and 17.3 of I_F gives, for all $f \in F$ and $f' \in F'$,

$$\begin{aligned}\langle J^{**} \circ I_E \circ J^{-1}(f), f' \rangle_{F'' \times F'} &= \langle I_E \circ J^{-1}(f), J^*f' \rangle_{E'' \times E'} \\ &= \langle J^*f', J^{-1}(f) \rangle_{E' \times E} \\ &= \langle f', J \circ J^{-1}(f) \rangle_{F' \times F} \\ &= \langle f', f \rangle_{F' \times F} = \langle I_F f, f' \rangle_{F'' \times F'}\end{aligned}$$

which proves (17.23).

Let us in addition remark, according to Theorem 17.24, that J^* is an isomorphism from F' onto E': hence, again according to this theorem, J^{**} is an isomorphism from E'' onto F''.

(a) If E is semi-reflexive, I_E is bijective (Theorem 17.4 (b)) and therefore $J^{**} \circ I_E \circ J^{-1}$, that is to say, I_F is a composite of bijections and is therefore itself bijective; hence, again following Theorem 17.4 (b), F is semi-reflexive. And conversely since J^{-1} is an isomorphism from F onto E.

(b) If E is prereflexive, I_E is continuous (Theorem 17.4 (c)) and therefore $J^{**} \circ I_E \circ J^{-1}$, that is to say, I_F is a composite of continuous mappings and is therefore continuous (Theorem 7.9 (a)); hence, again by Theorem 17.4 (c), F is prereflexive. And conversely since J^{-1} is an isomorphism from F onto E.

(c) If E is reflexive, that is (Definition 17.1 (c)) semi-reflexive and prereflexive, then F also is according to (a) and (b). And conversely. □

17.10. Reflexivity of the dual

Let us show that the dual of a reflexive space is reflexive.

THEOREM 17.26.– *For every separated semi-normed space E:*

$$E \textit{ is reflexive} \quad \Rightarrow \quad E' \textit{ is reflexive.} \blacksquare$$

Proof. Let us assume that E is reflexive. The point is (Definition 17.1) to show that its dual E' is prereflexive and semi-reflexive.

Prereflexivity of E'. Let B'' be a bounded subset of E'', I the canonical injection from E into E'' and $B = I^{-1}(B'')$. Here, I^{-1} is continuous linear (Theorem 17.4 (d)) thus (Theorem 7.15) B is bounded in E. For all $e' \in E'$, Definitions 17.3 of I and 13.1 (b) of the semi-norms of E' give

$$\sup_{e'' \in B''} |\langle e'', e' \rangle_{E'' \times E'}| = \sup_{e \in B} |\langle Ie, e' \rangle_{E'' \times E'}| = \sup_{e \in B} |\langle e', e \rangle_{E' \times E}| = \|e'\|_{E';B}.$$

This proves that E' is prereflexive (Definition 17.1 (b)).

Semi-reflexivity of E'. Let $e''' \in E'''$. According to the characterization of the bidual (since $E''' = (E')''$) of Theorem 13.3 (b), there exist a bounded set B' of E' and $c \in \mathbb{R}$ such that, for all $e'' \in E''$,

$$|\langle e''', e'' \rangle_{E''' \times E''}| \leq c \sup_{e' \in B'} |\langle e'', e' \rangle_{E'' \times E'}|. \tag{17.24}$$

Since E is prereflexive (Definition 17.1 (b)), there exist a finite set $N \subset \mathcal{N}_E$ and $c' \in \mathbb{R}$ such that $\sup_{e' \in B'} |\langle e', e \rangle_{E' \times E}| \leq c' \sup_{\nu \in N} \|e\|_{E;\nu}$ for all $e \in E$. Inequality (17.24) with respect to Ie and Definition 17.3 of Ie therefore gives, for all $e \in E$,

$$|\langle e''', Ie \rangle_{E''' \times E''}| \leq c \sup_{e' \in B'} |\langle Ie, e' \rangle_{E'' \times E'}| =$$

$$= c \sup_{e' \in B'} |\langle e', e \rangle_{E' \times E}| \leq cc' \sup_{\nu \in N} \|e\|_{E;\nu}.$$

The linear form $e \mapsto \langle e''', Ie \rangle$ is thus (Theorem 13.2 (a)) an element u' of E'. That is, for every $e \in E$,

$$\langle e''', Ie \rangle_{E''' \times E''} = \langle u', e \rangle_{E' \times E} = \langle Ie, u' \rangle_{E'' \times E'}.$$

Moreover, E being semi-reflexive, I is a bijection from E onto E'' (Theorem 17.4 (b)) and therefore this equality gives, for all $e'' \in E''$,

$$\langle e''', e'' \rangle_{E''' \times E''} = \langle e'', u' \rangle_{E'' \times E'}.$$

This proves that E' is semi-reflexive (Definition 17.1 (a)). So, E' is reflexive. □

Converse of Theorem 17.26. The converse of Theorem 17.26 is false if E is not sequentially complete (this is a necessary condition for reflexivity, see Theorem 17.7), what can happen when E' is reflexive, even if E and E' are normed. As a matter of fact:

$$E \textit{ is normed and } E' \textit{ is a reflexive Banach space} \quad \not\Rightarrow \quad E \text{ is reflexive.} \tag{17.25}$$

Proof of (17.25). Let E be a dense topological vector subspace, distinct from the whole space, of a reflexive Banach space F. Then, E is normed, thus E' endowed with the dual norm is a Banach space (Theorem 13.9 (d)) and is reflexive (Theorem 17.25 (c)) because it is isomorphic to F' (Theorem 14.8 (c)) which is reflexive (Theorem 17.26).

Yet E, being not closed in F, is not sequentially closed (Theorem 4.6 (a)) and thus is not sequentially complete (Theorem 4.13), which implies (Theorem 17.10) that it is not semi-reflexive and therefore not reflexive. □

Converse of Theorem 17.26 for a Fréchet space. Let us establish this converse property when E is a Fréchet space.

THEOREM 17.27.– *Let*

E be a Fréchet space.

Then:

$$E \textit{ is reflexive} \quad \Leftrightarrow \quad E' \textit{ is reflexive.} \ \square$$

Proof. The direct part being given by Theorem 17.26, it remains to establish the converse. Since E is metrizable, it is prereflexive (Theorem 17.8) thus (Theorem 17.5 (c)) the canonical injection I from E into its bidual is an isomorphism from E onto the topological subspace $I(E)$ of E''.

Since E is sequentially complete, so is $I(E)$ (Theorem 7.18). Thus (Theorem 4.13) $I(E)$ is sequentially closed in E''. In addition, E'' is metrizable, see, for example, [KÖTHE, 64, property (5), p. 301]. Therefore (Theorem 4.6 (a)), $I(E)$ is closed in E''.

If E' is reflexive, its dual E'' is reflexive (Theorem 17.26) as well as its closed subspace $I(E)$ (Theorem 17.22). Its image E by the isomorphism I^{-1} is thus reflexive (Theorem 17.25 (c)). □

Let us observe that the dual of a reflexive space is sequentially complete.

THEOREM 17.28.– *For every separated semi-normed space:*

$$E \textit{ is reflexive} \quad \Rightarrow \quad E', E'\text{-weak} \textit{ and } E'\text{-}{*}\text{weak} \textit{ are Neumann spaces.} \ \blacksquare$$

Proof. If E is reflexive, so is E' (Theorem 17.26), thus (Theorem 17.7) E' and E'-weak are Neumann spaces. Furthermore, E'-$*$weak is topologically equal to E'-weak (Theorem 17.21), and therefore it is also a Neumann space. □

Chapter 18

Extractable Spaces

18.1. Extractable spaces

Let us define extractability.

DEFINITION 18.1.– *A separated semi-normed space E is* ***extractable*** *if any bounded sequence of E has a subsequence that converges in E-*weak. ▮

Terminology. **Extractable** is a **new term** and as such the reader will have to recall its definition if he employs it. Although their property had been extensively used in the literature, extractable spaces were not named and were not being studied as such. □

Let us characterize extractability.

THEOREM 18.2.– *Let E be a separated semi-normed space. Then:*

$$E \text{ is extractable} \quad \Leftrightarrow \quad \begin{cases} \text{any bounded subset of } E \text{ is relatively} \\ \text{sequentially compact in } E\text{-weak} \end{cases}$$

$$\Leftrightarrow \quad \begin{cases} \text{any bounded sequence of } E\text{-weak has a} \\ \text{subsequence that converges in } E\text{-weak.} \end{cases}$$ ▮

Proof. The first equivalence is the rewriting of Definition 18.1 with Definition 2.24 (b) of a relatively sequentially compact set.

The second equivalence follows from Definition 18.1 since bounded in E is equivalent to bounded in E-weak according to the Banach–Mackey theorem 16.1. □

Let us observe that extractability is preserved by topological equivalences.

THEOREM 18.3.– *Any extractable separated semi-normed space remains so when its family of semi-norms is replaced by an equivalent family.* ▮

Proof. This results from the fact that topological equivalences preserve the bounded sequences of E (Theorem 3.4 (a)) and the space E-weak (Theorem 15.3) and therefore its convergent sequences (Theorem 3.4 (a), again). □

Let us observe that the extractability of E is equivalent to that of E-weak.

THEOREM 18.4.– *Let E be a separated semi-normed space. Then:*

$$E \text{ is extractable} \quad \Leftrightarrow \quad E\text{-weak is extractable.}$$ ▮

Proof. This follows from the second equivalence of Theorem 18.2 according to the topological equality $(E\text{-weak})\text{-weak} \underset{\leftrightarrows}{=} E\text{-weak}$ (Theorem 16.2 (b)). □

18.2. Extractability of Hilbert spaces

Let us observe that $\mathbb{R}$ and $\mathbb{R}^d$ are extractable.

THEOREM 18.5.– *The spaces $\mathbb{R}$ and $\mathbb{R}^d$ are extractable.* ▮

Proof. According to the Bolzano–Weierstrass theorems 4.17 and 5.14, every bounded sequence of $\mathbb{R}$ or $\mathbb{R}^d$ has a convergent subsequence. This is also the case for the weak topology since $\mathbb{R}\text{-weak} \underset{\leftrightarrows}{=} \mathbb{R}$ and $\mathbb{R}^d\text{-weak} \underset{\leftrightarrows}{=} \mathbb{R}^d$ (Theorems 15.6 and 15.10). □

Let us show more generally that every Hilbert space is extractable.

THEOREM 18.6.– *Let*

E be a Hilbert space.

Then:

$$\begin{cases} \textit{Any bounded sequence of } E \textit{ or of } E\text{-weak has} \\ \textit{a subsequence that converges in } E\text{-weak.} \end{cases}$$ ▮

Proof. Preliminaries. It suffices to consider a sequence $(u_n)_{n\in\mathbb{N}}$ which is bounded in E-weak because, since $E \underset{\rightarrow}{\subsetneq} E\text{-weak}$ (Theorem 15.2 (b)), this is the case of any bounded sequence in E (Theorem 3.8 (a)).

On the other hand, let J be the canonical isomorphism from E onto its dual E'. According to Definitions 13.4 (b) and 15.1 of the semi-norms of E'-$*$weak and of E-weak and to the definition of J in Theorem 13.14 (a), for all $u \in E$ and $e \in E$,

$$\|Ju\|_{E'\text{-}*\text{weak};e} = |\langle Ju, e\rangle_{E'\times E}| = |(u,e)_E| = \\ = |\langle Je, u\rangle_{E'\times E}| = \|u\|_{E\text{-weak};Je}. \quad (18.1)$$

The case in which E is separable. The sequence $(Ju_n)_{n\in\mathbb{N}}$ is bounded in E'-$*$weak from (18.1); therefore, according to the Banach–Bourbaki theorem 13.16 (it can be applied because E here is a separable Fréchet space, see Theorem 5.4), it has a subsequence $(Ju_{\sigma(n)})_{n\in\mathbb{N}}$ that converges in E'-$*$weak. Let v' its limit and $u = J^{-1}v'$. Then, $Ju_{\sigma(n)} \to Ju$ in E'-$*$weak.

For all $e' \in E'$, the equality (18.1) with $e = J^{-1}e'$ gives

$$\|u_{\sigma(n)} - u\|_{E\text{-weak};e'} = \|Ju_{\sigma(n)} - Ju\|_{E'\text{-}*\text{weak};J^{-1}e'}$$

which tends to 0. This proves that $u_{\sigma(n)}$ converges to u in E-weak.

The general case. Let $(u_n)_{n\in\mathbb{N}}$ be a bounded sequence of E-weak. Denote

$$X = \Big\{\sum_{n\le m} c_n u_n : c_n \in \mathbb{Q},\ m \in \mathbb{N}\Big\}$$

the set of the finite rational combinations of the u_n and F the closure of X in E. Then, F is a closed topological subspace of E and thus (Theorem 5.6) it is Hilbertian. It is separable (Definition 2.21 (a)) because X, being the image of the set of the finite sequences of $\mathbb{Q}$, which is countable (Theorems 1.1 (g) and 1.6), by the mapping $(c_n)_{n\le m} \mapsto \sum_{n\le m} c_n u_n$, is countable (Theorem 1.1 (c)).

The previous case then shows that $(u_n)_{n\in\mathbb{N}}$ has a convergent subsequence in the space F-weak. The identity from F into E, being continuous (Theorem 7.16 (a)), is weakly continuous (Banach–Dieudonné theorem 15.7), so this convergence takes place in E-weak. □

18.3. Extractability of semi-reflexive spaces

Let us show that every semi-reflexive metrizable space is extractable.

THEOREM 18.7.– *Let E be a separated semi-normed space such that:*

E is metrizable and semi-reflexive.

Then:

$$\begin{cases}\textit{Any bounded sequence of } E \textit{ or } E\text{-weak has}\\ \textit{a subsequence that converges in } E\text{-weak.}\end{cases}$$ ▮

Proof. If E is semi-reflexive, any bounded set is relatively compact in E-weak according to the variant of the Banach–Alaoglu–Bourbaki theorem given in Theorem 17.19. In addition, if E is metrizable, relatively compact implies relatively sequentially compact in E-weak following Šmulian's theorem 16.7 (b). Under these conditions, every bounded sequence of E-weak therefore has a subsequence that converges in E-weak.

Since $E \underset{\rightarrow}{\subseteq} E$-weak (Theorem 15.2 (b)), it is also the case for any bounded sequence of E because it is bounded in E-weak (Theorem 3.8 (a)). □

Let us show that a semi-reflexive space whose dual is a separable Fréchet space is extractable.

THEOREM 18.8.– *Let E be a separated semi-normed space such that:*

$$E \text{ is semi-reflexive and } E' \text{ is a separable Fréchet space.}$$

Then:

$$\begin{cases} \textit{Any bounded sequence of } E \textit{ or of } E\text{-weak has} \\ \textit{a subsequence that converges in } E\text{-weak.} \end{cases}$$ ∎

Proof. It suffices to consider a sequence $(u_n)_{n\in\mathbb{N}}$ which is bounded in E-weak because, since $E \underset{\rightarrow}{\subseteq} E$-weak (Theorem 15.2 (b)), this is the case of any bounded sequence in E (Theorem 3.8 (a)).

Let I be the canonical injection from E in E''. For all $e' \in E'$, Definitions 13.4 (b) of the semi-norms $*$weak of a dual, 17.3 of I and 15.1 of the semi-norms of E-weak successively give

$$\|Iu_n\|_{(E')'\text{-}*\text{weak};e'} = |\langle Iu_n, e'\rangle_{E''\times E'}| = |\langle e', u_n\rangle_{E'\times E}| = \|u_n\|_{E\text{-weak};e'}.$$

Therefore, $(Iu_n)_{n\in\mathbb{N}}$ is bounded in $(E')'$-$*$weak.

The space E' being assumed to be a separable Fréchet space, the Banach–Bourbaki theorem 13.16 shows that there exists a subsequence $(Iu_{\sigma(n)})_{n\in\mathbb{N}}$ that converges in $(E')'$-$*$weak to a limit $\ell'' \in (E')'$. Then, for all $e' \in E'$,

$$\langle e', u_{\sigma(n)}\rangle_{E'\times E} = \langle Iu_{\sigma(n)}, e'\rangle_{E''\times E'} \to \langle \ell'', e'\rangle_{E''\times E'}.$$

Since E is semi-reflexive, there exists (Definition 17.1 (a)) $u \in E$ such that the last term is equal to $\langle e', u\rangle_{E'\times E}$ for all e', and thus (Theorem 15.4) $u_{\sigma(n)} \to u$ in E-weak. □

Comparison of Theorems 18.7 and 18.8. Theorem 18.8 does not result from Theorem 18.7 because E' may be a separable Fréchet space without E being metrizable: for example, the space $L^2_c(\mathbb{R})$ of square integrable functions with compact support is not metrizable and yet its dual is a separable Fréchet space (it is isomorphic to $L^2_{\text{loc}}(\mathbb{R})$).

Conversely, Theorem 18.7 does not result from Theorem 18.8 because, when E is metrizable but not normable, E' is not metrizable. □

Comparison of the proofs of Theorems 18.7 and 18.8. Contrary to the proof of Theorem 18.8 based on Ascoli's theorem, via the Banach–Bourbaki theorem 13.16, which is rather basic, that of Theorem 18.7 makes use, via the variant of the Banach–Alaoglu–Bourbaki theorem given in Theorem 17.19, of delicate results. These include Tychonoff's compactness theorem 11.9, itself based on Zorn's theorem 11.7. □

Extractable versus semi-reflexive. *Extractable* is a sequential counterpart of *semi-reflexive* since this property is equivalent to the relative compactness of bounded sets for the weak topology (Theorem 17.19, variant of the Banach–Alaoglu–Bourbaki theorem), whereas extractable is equivalent to their relative sequential compactness (Theorem 18.2). □

Let us show that any uniformly convex Banach space is extractable[1].

THEOREM 18.9.– *Let*

$$E \textit{ be a uniformly convex Banach space.}$$

Then:

$$\begin{cases} \textit{Any bounded sequence of } E \textit{ or of } E\text{-weak } \textit{has} \\ \textit{a subsequence that converges in } E\text{-weak}. \end{cases}$$ ▮

Proof. Every uniformly convex Banach space, being reflexive according to Milman's theorem 17.15, is *a fortiori* (Definition 17.1 (c)) semi-reflexive. Being in addition metrizable (Theorem 4.2), it is extractable following Theorem 18.7. □

18.4. Extractability of a subspace or of the image of a space

Let us show that closed subspaces of an extractable space are extractable.

THEOREM 18.10.– *Every closed topological subspace of an extractable separated semi-normed space is extractable.* ▮

Proof. Let E be a closed topological subspace of a separated semi-normed space F. Any bounded sequence of E is bounded in F (Theorem 3.11 (a)); therefore, if F is extractable, it has a subsequence that has a limit e in F-weak. Besides, E-weak is a topological subspace of F-weak (Theorem 15.8 (b), since E is a topological subspace of F) which is closed in F-weak (Theorem 16.6, corollary of Mazur's theorem, since E is closed in F). The subsequence thus converges (Theorem 3.11 (a)) to e in E-weak, which proves that E is extractable. □

Supplement to Theorem 18.10. For a topological subspace of a separated semi-normed space to be extractable, it is necessary that it be sequentially closed. In effect, it must be sequentially complete (Theorem 18.16) which is stronger (Theorem 4.13). □

Let us show that isomorphisms preserve extractability.

1 **History of Theorem 18.9.** Frigyes RIESZ showed in 1910 [91, p. 466] that, if $1 < p < \infty$, any bounded sequence of $L^p((0, 1))$ has a weakly convergent subsequence.

THEOREM 18.11.– *Every separated semi-normed space isomorphic to an extractable space is extractable.* ▮

Proof. Let E be an extractable separated semi-normed space, F another separated semi-normed space and T an isomorphism from E onto F.

Let $(v_n)_{n\in\mathbb{N}}$ be a bounded sequence of F. The mapping T^{-1}, being continuous linear from F into E, transforms bounded sets into bounded sets (Theorem 7.15) and therefore $(T^{-1}(v_n))_{n\in\mathbb{N}}$ is bounded in E. Since E is extractable, it has a subsequence $(T^{-1}(v_{\sigma(n)}))_{n\in\mathbb{N}}$ that converges in E-weak.

Being continuous linear from E into F, T is continuous from E-weak into F-weak (Banach–Dieudonné theorem 15.7). It is then sequentially continuous in these spaces (Theorem 7.2 (a)), thus $(T(T^{-1}(v_{\sigma(n)}))_{n\in\mathbb{N}} = (v_{\sigma(n)})_{n\in\mathbb{N}}$ converges in F-weak, which proves that E is extractable. □

18.5. Extractability of a product or of a sum of spaces

Let us show that a product of extractable spaces is extractable.

THEOREM 18.12.– *Every product $E_1 \times E_2 \times \cdots \times E_d$ of extractable separated semi-normed spaces is extractable.* ▮

Proof. Let $((u_{1,n}, \ldots, u_{d,n}))_{n\in\mathbb{N}}$ be a bounded sequence of $E_1 \times \cdots \times E_d$. The sequence $(u_{i,n})_{n\in\mathbb{N}}$ is bounded in E_i; thus, if E_i is extractable, it has a subsequence $(u_{i,\sigma_i(n)})_{n\in\mathbb{N}}$ that converges in E_i-weak.

If this is true for all i, by successively extracting, for $i = 1, 2, \ldots, d-1$, the subsequence $(\sigma_{i+1}(n))_{n\in\mathbb{N}}$ from $(\sigma_i(n))_{n\in\mathbb{N}}$, we obtain a strictly increasing sequence $(\sigma_d(n))_{n\in\mathbb{N}}$ of $\mathbb{N}$. Since $(u_{i,\sigma_d(n)})_{n\in\mathbb{N}}$ is a subsequence of $(u_{i,\sigma_i(n)})_{n\in\mathbb{N}}$, it converges (Theorem 2.8 (a)) in E_i-weak. Then, $(u_{1,\sigma_d(n)}, \ldots, u_{d,\sigma_d(n)})$ converges (Theorem 6.4) in E_1-weak $\times \cdots \times E_d$-weak, that is in $(E_1 \times \cdots \times E_d)$-weak since these two spaces are topologically equal (Theorem 15.9). This proves that $E_1 \times \cdots \times E_d$ is extractable. □

Let us show that a direct sum of extractable spaces is extractable.

THEOREM 18.13.– *Every direct sum $E_1 \oplus E_2 \oplus \cdots \oplus E_d$ of extractable separated semi-normed spaces is extractable.* ▮

Proof. The direct sum $E_1 \oplus \cdots \oplus E_d$, being isomorphic (Theorem 7.29) to the product $E_1 \times \cdots \times E_d$ which is extractable (Theorem 18.12), is also extractable (Theorem 18.11). □

18.6. Extractability of an intersection of spaces

Let us show that a finite intersection of extractable spaces included in a same space is extractable.

THEOREM 18.14.– *Let E_1, E_2..., E_d and F be separated semi-normed spaces such that for all $i \in [\![1, d]\!]$,*

$$E_i \subsetneq F,$$

$$E_i \textit{ is extractable.}$$

Then:

$$\bigcap_{1 \leq i \leq d} E_i \textit{ is extractable.} \blacksquare$$

Proof. Let $(u_n)_{n \in \mathbb{N}}$ be a bounded sequence of $\bigcap_{1 \leq i \leq d} E_i$ and $j \in [\![1, d]\!]$. Since $\bigcap_{1 \leq i \leq d} E_i \subsetneq E_j$ (Theorem 6.17), this sequence is bounded in E_j (Theorem 3.8 (a)); thus, if the latter is extractable, it has a subsequence $(u_{\sigma_j(n)})_{n \in \mathbb{N}}$ which converges to a limit ℓ_j in E_j-weak. Since E_j-weak $\subsetneq F$-weak (Theorem 15.8 (a)), then (Theorem 3.8 (a)) the convergence also takes place in F-weak.

If this is true for all j, let us successively extract, for $j = 1, 2..., d - 1$, the subsequence $(u_{\sigma_{j+1}(n)})_{n \in \mathbb{N}}$ from $(u_{\sigma_j(n)})_{n \in \mathbb{N}}$. Since these two sequences converge in F-weak, they have the same limit (Theorem 2.8 (a)). Thus, $\ell_{j+1} = \ell_j = \cdots = \ell_1$. The subsequence $(u_{\sigma_d(n)})_{n \in \mathbb{N}}$ of $(u_n)_{n \in \mathbb{N}}$ then converges (to ℓ_1) in each E_i-weak and therefore (Theorem 6.15) in $\bigcap_{1 \leq i \leq d} E_i$-weak, that is in $(\bigcap_{1 \leq i \leq d} E_i)$-weak since these two spaces are topologically equal (Theorem 15.11 (c)). This proves that the space $\bigcap_{1 \leq i \leq d} E_i$ is extractable. □

18.7. Sequential completion of extractable spaces

Let us show that every extractable space is "weakly" sequentially complete.

THEOREM 18.15.– *If a separated semi-normed space E is extractable, E-weak is a Neumann space.* ▮

Proof. Let $(e_n)_{n \in \mathbb{N}}$ be a Cauchy sequence of E-weak. Any Cauchy sequence being bounded (Theorem 2.7 (b)), it is bounded in E-weak and thus in E according to the Banach–Mackey theorem 16.1. If E is extractable, a subsequence thus converges in E-weak, which implies (Theorem 2.8 (b)) that the whole sequence $(e_n)_{n \in \mathbb{N}}$ converges. This proves that E is sequentially complete. □

Let us show that every extractable space is sequentially complete.

THEOREM 18.16.– *Every extractable separated semi-normed space is a Neumann space.* ∎

Proof. Let E be an extractable separated semi-normed space and $(e_n)_{n\in\mathbb{N}}$ a Cauchy sequence of E. Since $E \underset{\rightarrow}{\subsetneq} E$-weak (Theorem 15.2 (b)), this sequence is a Cauchy sequence in E-weak (Theorem 3.8 (a)), so according to Theorem 18.15, it has a limit e. Following Definition 2.6 (b) of a Cauchy sequence, for any semi-norm $\|\ \|_{E;\nu}$ of E and every $\epsilon > 0$, there exists $n_0 \in \mathbb{N}$ such that, for all $m \geq n \geq n_0$,

$$\|e_n - e_m\|_{E;\nu} \leq \epsilon.$$

Since $e_n - e_m \rightarrow e_n - e$ in E-weak when $m \rightarrow \infty$ it thereof results that (Theorem 15.16)

$$\|e_n - e\|_{E;\nu} \leq \epsilon.$$

Hence, $e_n \rightarrow e$ in E, which proves that E is sequentially complete. □

PART 4

DIFFERENTIAL CALCULUS

Chapter 19

Differentiable Mappings

19.1. Differentiable mappings

Let us define differentiability[1].

DEFINITION 19.1.– *Let E and F be two separated semi-normed spaces, $\{\| \ \|_{E;\nu} : \nu \in \mathcal{N}_E\}$ and $\{\| \ \|_{F;\mu} : \mu \in \mathcal{N}_F\}$ be their families of semi-norms, X an open subset of E and T a mapping from X into F.*

(a) *T is **differentiable at a point** $u \in X$ if there exists a mapping*

$$\mathrm{d}T(u) \in \mathcal{L}(E; F)$$

and if, for all $\mu \in \mathcal{N}_F$, there exists a finite set $N \subset \mathcal{N}_E$ such that, for all $\epsilon > 0$, there exists $\eta > 0$ such that $z \in E$ and $\sup_{\nu \in N} \|z\|_{E;\nu} \leq \eta$ implies that $u + z \in X$ and

$$\|T(u+z) - T(u) - \mathrm{d}T(u)z\|_{F;\mu} \leq \epsilon \sup_{\nu \in N} \|z\|_{E;\nu}.$$

1 **History of the notion of derivative of a real function.** EUCLID, in his *Elements* [40, book III, p. 16], was already looking for the tangent to a curve. Pierre DE FERMAT, in 1636, looked for it for a curve of equation $y = x^m$ by means of a calculation foreshadowing that of the derivative. Isaac NEWTON introduced in 1671 the *fluxion* of a function $y = f(x)$ that he denoted $\dot{y}$ [79, p. 76]. In 1675, Gottfried VON LEIBNIZ, denoted it by $\mathrm{d}y/\mathrm{d}x$ [67] in the *calculus* that he was developing. The notion of derivative was made rigorous in 1821, under the form of our Theorem 22.3, p. 314, by Augustin CAUCHY [25, p. 22].

History of the notion of differential. Maurice FRÉCHET defined the differentiable mappings from a normed space into another in 1906 [43].

The generalization for a semi-normed space E seems to be new. It is based on the use of semi-norms (in $\|T(u+z) - T(u) - \mathrm{d}T(u)z\|_{F;\mu} \leq \epsilon \sup_{\nu \in N} \|z\|_{E;\nu}$) which, contrary to an open set, are numerically evaluable! A metric d is not adequate either, because $d(tu, 0)$ is not proportional to t. This is the reason why, when only resorting to topologies, metrics and norms, differentiability is defined only in normed spaces, or in special cases such as *scalar differentiability* that corresponds to differentiability for the weak topology.

(b) $\mathrm{d}T(u)$ *is the* ***differential*** *of* T *at* u.

(c) T *is* ***differentiable*** *if it is differentiable at any point of* X.

(d) T *is* ***continuously differentiable*** *if it is differentiable and* $\mathrm{d}T$ *is continuous (from* X *into* $\mathcal{L}(E; F)$*).* ▮

When it is desirable to specify the variable with respect to the one we differentiate, the differential is denoted by $\mathrm{d}_u T(u)$.

Terminology. Although the differential is frequently called the **derivative**, it appears as a possible source of confusion, since they live in different spaces as explained hereafter. The ambiguity of this terminology is related to history: Maurice FRÉCHET generalized the derivative by introducing the differential, which was for a long time called **derivative in the Fréchet sense**. □

Derivative versus differential. The derivative $f'(u)$ of a mapping f from $\mathbb{R}$ into F at a point u is (Definition 22.1) an element of F, which is more pleasant than the differential $\mathrm{d}f(u)$ which is an element of $\mathcal{L}(\mathbb{R}; F)$. They are related by $f'(u) = \mathrm{d}f(u)1$.

For a mapping defined on $\mathbb{R}^d$, the gradient plays a role similar to that of the derivative. On the other hand, on an infinite dimension space, the author does not know any analogue, except if E is a Hilbert space and $F = \mathbb{R}$ in which case there is a unique element $\nabla T(u)$ of E such that $\mathrm{d}T(u).z = (\nabla T(u), z)_E$ according to the Riesz–Fréchet theorem 13.13. □

Usefulness of assuming the domain of a differentiable mapping as open. Differentiability is not defined in any subset X of a separated semi-normed space because then the differential would not be necessarily unique. For example, for X reduced to a point, any mapping would be differentiable and would admit any linear mapping as a differential in the sense of Definition 19.1 (since then $z = 0$).

However, the notion of differentiability can be extended (through continuity) to the closure of an open set, see Volume II. □

Let us show that the differential is unique.

THEOREM 19.2.– *The differential of a mapping is unique at any point where it exists.* ▮

Proof. Let us make use of the notations of Definition 19.1 and assume that T is differentiable at u. Let $L = \mathrm{d}T_2(u) - \mathrm{d}T_1(u)$ be the difference of two possible differentials, $\mu \in \mathcal{N}_F$ and $\epsilon > 0$. Let N_1, η_1, N_2, η_2 be such that they satisfy Definition 19.1 (a) respectively for $\mathrm{d}T_1$ and $\mathrm{d}T_2$. Then, $\sup_{\nu \in N_1 \cup N_2} \|z\|_{E;\nu} \leq \inf(\eta_1, \eta_2)$ implies

$$\|Lz\|_{F;\mu} \leq \sum_{i=1,\,2} \|T(u+z) - T(u) - \mathrm{d}T_i(u)z\|_{F;\mu} \leq 2\epsilon \sup_{\nu \in N_1 \cup N_2} \|z\|_{E;\nu}.$$

Since T is linear, it results thereof that $\|Lz\|_{F;\mu} \leq 2\epsilon \sup_{\nu \in N_1 \cup N_2} \|z\|_{E;\nu}$ for all $z \in E$. This holds for all $\epsilon > 0$ and thus for $\epsilon = 0$. Therefore, $\|Lz\|_{F;\mu} = 0$ for all $\mu \in \mathcal{N}_F$, hence $Lz = 0_F$ since F is separated. This holds for all $z \in E$, thus $L = 0_{\mathcal{L}(E;F)}$. □

Let us observe that the definition of differentiability is simplified with a filtering family of semi-norms: the upper envelope of a finite number of semi-norms is replaced by a single semi-norm.

THEOREM 19.3.– *With the notations of Definition 19.1 (a),*

$$\textit{if the family } \{\| \ \|_{E;\nu} : \nu \in \mathcal{N}_E\} \textit{ of semi-norms of } E \textit{ is filtering,}$$

then T is differentiable at the point u if and only if: there exists $\mathrm{d}T(u) \in \mathcal{L}(E;F)$ and for all $\mu \in \mathcal{N}_F$, there exists $\nu \in \mathcal{N}_E$ such that, for all $\epsilon > 0$, there exists $\eta > 0$ such that $z \in E$, $\|z\|_{E;\nu} \leq \eta$ implies $u + z \in X$ and

$$\|T(u+z) - T(u) - \mathrm{d}T(u)z\|_{F;\mu} \leq \epsilon \, \|z\|_{E;\nu}. \ \blacksquare$$

Proof. Definition 19.1 (a) of differentiability leads to the stated property because the set $N \subset \mathcal{N}_E$ that is involved therein being finite, if the family of semi-norms of E is filtering (Definition 3.14), there exists $\nu \in \mathcal{N}_E$ such that $\sup_{\sigma \in N} \| \ \|_{E;\sigma} \leq \| \ \|_{E;\nu}$.

Conversely, the stated property is a particular case of Definition 19.1 (a) with $N = \{\nu\}$. $\square$

Characterization of differentiability. In terms of limit (see Definition 22.2), the characterization of the differentiability of Theorem 19.3 with respect to a filtering family of semi-norms is expressed as follows:

$$\frac{\|T(u+z) - T(u) - \mathrm{d}T(u)z\|_{F;\mu}}{\|z\|_{E;\nu}} \to 0 \text{ when } \|z\|_{E;\nu} \neq 0, \ \|z\|_{E;\nu} \to 0. \ \square$$

Differentiability in the Gâteaux sense. A mapping T of an open set X from a separated semi-normed space E into a separated semi-normed space F, is said to be **differentiable in the Gâteaux sense** at a point $u \in X$ if, for all $v \in E$, the mapping $t \mapsto T(u + tv)$ is differentiable at the point 0 (of a subset) of $\mathbb{R}$ and if the application that to v associates the derivative $\mathrm{d}T(u + tv)/\mathrm{d}t(0)$ is continuous linear from E into F.

We do not use this notion, which is weaker than differentiability, because the composite of differentiable Gâteaux mappings is not necessarily Gâteaux differentiable, namely the important chain rule theorem 19.17 no longer applies. $\square$

19.2. Differentiality, continuity and linearity

Let us show that a differentiable mapping is locally Lipschitzian.

THEOREM 19.4.– *Let E and F be two separated semi-normed spaces, $\{\| \ \|_{E;\nu} : \nu \in \mathcal{N}_E\}$ and $\{\| \ \|_{F;\mu} : \mu \in \mathcal{N}_F\}$ be their families of semi-norms, X is an open subset of E, $u \in X$ and*

$$T \textit{ a mapping from } X \textit{ into } F \textit{ differentiable at } u.$$

Then, for all $\mu \in \mathcal{N}_F$, there exist a finite set $N \subset \mathcal{N}_E$, $\eta > 0$ and $c \geq 0$ such that, for all $z \in E$ such that $\sup_{\nu \in N} \|z\|_{E;\nu} \leq \eta$, it follows that $u + z \in X$ and

$$\|T(u+z) - T(u)\|_{F;\mu} \leq c \sup_{\nu \in N} \|z\|_{E;\nu}. \ \blacksquare$$

Proof. Definition 19.1 with respect to $\epsilon = 1$ provides a finite set $N_1 \subset \mathcal{N}_E$ and $\eta_1 > 0$ such that $\sup_{\nu \in N_1} \|z\|_{E;\nu} \leq \eta_1$ implies

$$\|T(u+z) - T(u) - \mathrm{d}T(u)z\|_{F;\mu} \leq \sup_{\nu \in N_1} \|z\|_{E;\nu}.$$

Since $\mathrm{d}T(u)$ is continuous linear, there exist (Theorem 7.14 (a)) a finite set $N_2 \subset \mathcal{N}_E$ and $c_2 \geq 0$ such that:

$$\|\mathrm{d}T(u)z\|_{F;\mu} \leq c_2 \sup_{\nu \in N_2} \|z\|_{E;\nu}.$$

Hence, the property stated for $N = N_1 \cup N_2$, $\eta = \eta_1$ and $c = 1 + c_2$. □

Let us observe that this property is simplified with a filtering family of semi-norms: the upper envelope of a finite number of semi-norms is replaced by a single semi-norm.

THEOREM 19.5.– *Under the assumptions of Theorem 19.4,*

if the family $\{\| \ \|_{E;\nu} : \nu \in \mathcal{N}_E\}$ *of semi-norms of E is filtering,*

then the differentiability of T *at the point* u *results in, for all* $\mu \in \mathcal{N}_F$*, the existence of* $\nu \in \mathcal{N}_E$, $\eta > 0$ *and* $c \geq 0$ *such that,* $z \in E$, $\|z\|_{E;\nu} \leq \eta$, *implies* $u + z \in X$ *and*

$$\|T(u+z) - T(u)\|_{F;\mu} \leq c\|z\|_{E;\nu}. \ \blacksquare$$

Proof. The set $N \subset \mathcal{N}_E$ that is involved in Theorem 19.4 being finite, if the family of semi-norms of E is filtering (Definition 3.14), there exists $\nu \in \mathcal{N}_E$ such that $\sup_{\sigma \in N} \| \ \|_{E;\sigma} \leq \| \ \|_{E;\nu}$. Hence, the stated property. □

Let us observe that differentiability implies continuity.

THEOREM 19.6.– *For every mapping from an open subset of a separated semi-normed space into a separated semi-normed space:*

(a) *Differentiable at a point* $\Rightarrow$ *continuous at this point.*

(b) *Differentiable* $\Rightarrow$ *continuous.* ▮

Proof. If a mapping T is differentiable at a point u, the inequality of Theorem 19.4 implies its continuity (Definition 7.1 (a)) at u. □

Uniform continuity. The mean value theorem 19.14 (a) and (b) implies that:

$$\left\{\begin{array}{l}\textit{Differentiable with bounded differential in a}\\ \textit{convex open subset of a normed or Fréchet space}\end{array}\right. \Rightarrow \textit{uniformly continuous.} \ \square$$

Let us observe that any constant mapping is differentiable.

THEOREM 19.7.– *Let E and F be two separated semi-normed spaces, X an open subset of E, T a mapping from X into F and $a \in F$ such that, for all $u \in X$,*

$$T(u) = a.$$

Then, T is continuously differentiable and, for all $u \in X$,

$$\mathrm{d}T(u) = 0_{\mathcal{L}(E;F)}. \blacksquare$$

Proof. Definition 19.1 is satisfied because $T(u+z) - T(u) - 0_{\mathcal{L}} z = 0_F$. □

Let us show that any continuous linear mapping is differentiable.

THEOREM 19.8.– *Let E and F be two separated semi-normed spaces and*

$$L \in \mathcal{L}(E;F).$$

Then, for all $u \in E$:

(a) *L is continuously differentiable.*

(b) $$\mathrm{d}L(u) = L. \blacksquare$$

Proof. Definition 19.1 is satisfied because $L(u+z) - L(u) - Lz = 0_F$. □

19.3. Differentiation and change of topology or restriction

Let us observe that differentiability is preserved by topological equalities.

THEOREM 19.9.– *Differentiability and the differential are preserved when, in Definition 19.1, the family of semi-norms of E or F is replaced by an equivalent family.* ▮

Proof. These properties follow from Theorems 19.10 and 19.11 below by Definitions 3.1 of equivalent families and 3.6 (c) of topological inclusion. □

Let us give the effect of changing the space of values on a differentiable mapping.

THEOREM 19.10.– *Let E, F_1 and F_2 be separated semi-normed spaces such that:*

$$F_1 \subsetneqq F_2,$$

X be an open subset of E and $u \in X$. Then, for any mapping from X into F_1:

(a) *Differentiable at u from X into F_1 $\Rightarrow$ differentiable at u from X into F_2.*

(b) *Differentiable from X into F_1* $\Rightarrow$ *differentiable from X into F_2.*

(c) *Continuously differentiable from X into F_1*
$\Rightarrow$ *continuously differentiable from X into F_2.* ▮

Proof. These properties follow from Definitions 3.6 (c) of topological inclusion and 19.1 of the notions of differentiability. □

Let us now give the effect of changing the space of definition.

THEOREM 19.11.– *Let E_1, E_2 and F be separated semi-normed spaces such that:*

$$E_1 \subsetneq E_2,$$

$X \subset E_1$ be an open subset of E_2 (and thus of E_1) and $u \in X$. Then, by denoting X_{E_1} the subset X of the semi-normed space E_1 and X_{E_2} that of E_2, every mapping from X into F verifies:

(a) *Differentiable at u from X_{E_2} into F* $\Rightarrow$ *differentiable at u from X_{E_1} into F.*

(b) *Differentiable from X_{E_2} into F* $\Rightarrow$ *differentiable from X_{E_1} into F.*

(c) *Continuously differentiable from X_{E_2} into F*
$\Rightarrow$ *continuously differentiable from X_{E_1} into F.* ▮

Other formulation. In terms of topology, X_{E_1} designates the set X endowed with the topology of E_1. □

Proof of Theorem 19.11. These properties follow from Definitions 3.6 (c) of topological inclusion and 19.1 of the notions of differentiability. The set X is open in E_1 according to Theorem 7.16 (a). □

Another proof of Theorems 19.10 and 19.11. These properties result from the fact that, if $E_1 \subsetneq E_2$, the identity of E_1 in E_2 is continuous linear (Theorem 7.16 (a)) and thus the composite by the latter preserves the differentiability (Theorem 19.19). □

Let us observe that the restrictions of differentiable mappings are differentiable.

THEOREM 19.12.– *Let E and F be separated semi-normed spaces, X and U two open subsets of E such that $U \subset X$, T a mapping from X into F and $u \in U$. Then, denoting by $T|_U$ the restriction of T to U:*

(a) *T is differentiable at u* $\Rightarrow$ $\begin{cases} T|_U \text{ is differentiable at } u \text{ and} \\ \qquad \mathrm{d}(T|_U)(u) = (\mathrm{d}T)|_U(u). \end{cases}$

(b) *T is differentiable* $\Rightarrow$ *$T|_U$ is differentiable.*

(c) *T is continuously differentiable.* $\Rightarrow$ *$T|_U$ is continuously differentiable.* ▮

Proof. This follows from Definition 19.1 of the notions of differentiability. □

It is said that T **is differentiable on** U, where U is an open subset of its domain of definition, if the restriction of T to U is differentiable. Continuously differentiable on U is defined in the same manner.

19.4. Mean value theorem

Let us give a **mean value theorem** that relates the variation of a mapping between two points to the intermediate values of the differential on the segment between these points.

THEOREM 19.13.– *Let E and F be two separated semi-normed spaces, X an open subset of E, T a mapping from X into F, $\| \ \|_{F;\mu}$ a semi-norm of F, $u \in X$ and $z \in E$ such that:*

$$T \text{ is differentiable}, \quad [u, u+z] \subset X.$$

Then:

(a) $$\|T(u+z) - T(u)\|_{F;\mu} \leq \sup_{0 \leq t \leq 1} \|\mathrm{d}T(u+tz)z\|_{F;\mu}.$$

(b) $$\|T(u+z) - T(u) - \mathrm{d}T(u)z\|_{F;\mu} \leq \sup_{0 \leq t \leq 1} \|\mathrm{d}T(u+tz)z - \mathrm{d}T(u)z\|_{F;\mu}. \ \blacksquare$$

Reminder. We denote $[u, u+z] = \{u + tz : 0 \leq t \leq 1\}$ the segment connecting u to $u+z$. □

Remark. The upper bounds on the right-hand side of the inequalities (a) and (b) are not necessarily finite. □

Proof of Theorem 19.13. (a) Let us proceed in two steps.

1) *Decomposition of* $[0,1]$ *into elementary intervals.* Denote $\{\| \ \|_{E;\nu} : \nu \in \mathcal{N}_E\}$ the semi-norms of E and let $\sigma > 0$.

The differentiability of T at the point $u + tz$ provides (Definition 19.1) a finite set $N_t \subset \mathcal{N}_E$ and, by choosing $\epsilon = \sigma / \sup_{\nu \in N_t} \|z\|_{E;\nu}$, a number $\eta > 0$ such that:

$$\sup_{\nu \in N_t} \|sz\|_{E;\nu} \leq \eta \quad \Rightarrow \quad \|T(u+tz+sz) - T(u+tz) - \mathrm{d}T(u+tz)sz\|_F \leq \sigma|s|.$$

Denoting

$$g(t) = T(u+tz), \quad \delta(t) = \mathrm{d}T(u+tz)z \tag{19.1}$$

and $\theta_t = \eta / \sup_{\nu \in N_t} \|z\|_{E;\nu}$, it follows that

$$|s| \leq \theta_t \quad \Rightarrow \quad \|g(t+s) - g(t) - s\delta(t)\|_{F;\mu} \leq \sigma\,|s|. \tag{19.2}$$

If N_t is such that $\sup_{\nu \in N_t} \|z\|_{E;\nu} = 0$, we cannot choose ϵ and η_t as above but Definition 19.1 gives (19.2) with $\theta_t = \infty$ (and 0 in the right-hand side).

The open sets $\mathcal{O}_t = (t - \theta_t, t + \theta_t)$ cover the compact set $[0, 1]$ when t spans it and therefore there is a finite subcover $\{\mathcal{O}_t\}_{t \in \mathcal{T}}$. There thus exist t_i and a_i such that:

$$\left\{ \begin{array}{c} 0 \leq t_1 \leq t_2 \leq \cdots \leq t_n \leq 1, \quad 0 = a_1 \leq a_2 \leq \cdots \leq a_n \leq a_{n+1} = 1, \\ a_i \leq t_i \leq a_{i+1}, \quad |a_i - t_i| \leq \theta_{t_i}, \quad |a_{i+1} - t_i| \leq \theta_{t_i}. \end{array} \right\} \tag{19.3}$$

In effect, this is achieved by choosing t_1 as the element (or one of the elements) of $\{t \in \mathcal{T} : t - \theta_t < 0\}$ that maximizes $t + \theta_t$ then, for each $i \geq 1$, by choosing t_{i+1} as the element (or one of the elements) of $\{t \in \mathcal{T} : t - \theta_t < t_i + \theta_{t_i}\}$ that maximizes $t + \theta_t$ and by stopping when $t_n + \theta_{t_n} \geq 1$, which necessarily happens. Then, $t_{i+1} \geq t_i$ and $\mathcal{O}_{t_{i+1}} \cap \mathcal{O}_{t_i}$ is not empty and as such it contains a point a_{i+1} such that $t_i \leq a_{i+1} \leq t_{i+1}$, and the last thing to do is to choose $a_1 = 0$ and $a_{n+1} = 1$.

2) *Computation of the derivative.* By denoting $M = \sup_{0 \leq t \leq 1} \|\delta(t)\|_{F;\mu}$, (19.2) implies

$$|s| \leq \theta_t \quad \Rightarrow \quad \|g(t + s) - g(t)\|_{F;\mu} \leq (M + \sigma)|s|.$$

For $t = t_i$ and, respectively $s = a_i - t_i$ and $s = a_{i+1} - t_i$, it follows that

$$\|g(a_i) - g(t_i)\|_{F;\mu} \leq (M + \sigma)(t_i - a_i),$$
$$\|g(a_{i+1}) - g(t_i)\|_{F;\mu} \leq (M + \sigma)(a_{i+1} - t_i),$$

and thus

$$\|g(a_{i+1}) - g(a_i)\|_{F;\mu} \leq (M + \sigma)(a_{i+1} - a_i).$$

By summing from $i = 1$ to n, it yields that

$$\|g(1) - g(0)\|_{F;\mu} \leq \sum_{1 \leq i \leq n} \|g(a_{i+1}) - g(a_i)\|_{F;\mu} \leq M + \sigma.$$

Being true for every $\sigma > 0$, this is also true for $\sigma = 0$, which gives the inequality stated.

(b) The inequality (b) is the inequality (a) with respect to the mapping T_1 defined by $T_1(v) = T(v) - \mathrm{d}T(u)v$ because $\mathrm{d}T_1(v)z = \mathrm{d}T(v)z - \mathrm{d}T(u)z$. (In effect, $\mathrm{d}T(u)$ being linear, its differential $\mathrm{d}(\mathrm{d}T(u))(v)$ is $\mathrm{d}T(u)$ itself at every point v according to Theorem 19.8 (b)). □

Let us estimate the variation of $T(u)$ in terms of the variation z of u.

THEOREM 19.14.– *Let E and F be two separated semi-normed spaces, $\{\| \ \|_{E;\nu} : \nu \in \mathcal{N}_E\}$ and $\{\| \ \|_{F;\mu} : \mu \in \mathcal{N}_F\}$ the families of semi-norms of E and F, X an open subset of E, T a differentiable mapping from X into F and $C \subset X$ such that:*

$$C \text{ is convex}, \quad \{\mathrm{d}T(v) : v \in C\} \text{ is bounded in } \mathcal{L}(E; F).$$

Then:

(a) *If E is a Fréchet space, for every $\mu \in \mathcal{N}_F$, there exist a finite set $N \subset \mathcal{N}_E$ and $c \in \mathbb{R}$ such that:*

$$u \in C,\ u + z \in C \quad \Rightarrow \quad \|T(u+z) - T(u)\|_{F;\mu} \leq c \sup_{\nu \in N} \|z\|_{E;\nu}.$$

(b) *If E is normed, for every $\mu \in \mathcal{N}_F$ and for the semi-norm $|||\ |||_{\mathcal{L}(E;F);\mu}$ defined in Theorem 12.9:*

$$u \in C,\ u+z \in C \quad \Rightarrow \quad \|T(u+z)-T(u)\|_{F;\mu} \leq \sup_{v \in C} |||\mathrm{d}T(v)|||_{\mathcal{L}(E;F);\mu} \|z\|_E.$$

(c) *If E and F are normed, for the norm $|||\ |||_{\mathcal{L}(E;F)}$ defined in Theorem 12.10:*

$$u \in C,\ u + z \in C \quad \Rightarrow \quad \|T(u+z) - T(u)\|_F \leq \sup_{v \in C} |||\mathrm{d}T(v)|||_{\mathcal{L}(E;F)} \|z\|_E. \ \blacksquare$$

Proof. Preliminaries. If $u \in C$ and $u + z \in C$, then $u + tz \in C$ for all $t \in [0, 1]$ since C is convex and therefore the mean value theorem 19.13 (a) gives

$$\|T(u+z) - T(u)\|_{F;\mu} \leq \sup_{v \in C} \|\mathrm{d}T(v)z\|_{F;\mu}. \tag{19.4}$$

(a) The set $\{\mathrm{d}T(v) : v \in C\}$ being bounded, by Definition 12.1 (a) of the semi-norms of $\mathcal{L}(E; F)$, for each $z \in E$,

$$\sup_{v \in C} \|\mathrm{d}T(v)z\|_{F;\mu} = \sup_{v \in C} \|\mathrm{d}T(v)\|_{\mathcal{L}(E;F);\{z\},\mu} < \infty. \tag{19.5}$$

Each $\mathrm{d}T(v)$ being continuous linear, the Banach–Steinhaus theorem 10.11 gives, if E is a Fréchet space, the existence of a finite set $N \subset \mathcal{N}_E$ and $c \in \mathbb{R}$, such that $\|\mathrm{d}T(v)z\|_{F;\mu} \leq c \sup_{\nu \in N} \|z\|_{E;\nu}$ for all $v \in C$ and $z \in E$. The inequality (19.4) then implies the inequality stated.

(b) If E is normed, Theorem 12.9 (c) gives $\|Lz\|_{F;\mu} \leq |||L|||_{\mathcal{L}(E;F);\mu} \|z\|_E$ for any mapping $L \in \mathcal{L}(E; F)$, so (19.4) leads to the inequality stated.

(c) If E and F are normed, Theorem 12.10 (c) gives $\|Lz\|_F \leq |||L|||_{\mathcal{L}(E;F)}\|z\|_E$ for all $L \in \mathcal{L}(E;F)$, and therefore (19.4) again leads to the inequality stated. □

Another proof of the mean value theorem. It is faster to prove the results of Theorem 19.13 by using the equality $T(u+z) - T(u) = \int_0^1 \mathrm{d}T(u+tz)z\,\mathrm{d}t$, but it requires more restrictive assumptions, for example that E and F be Banach spaces and that T be continuously differentiable [LANG, 66, corr. 1 and 2, p. 103]... and that the integral be defined on these spaces, which we will only do in Volume II. □

Supplements to Theorem 19.14. The results of Theorem 19.14 are still true if, instead of assuming that C is convex, it is assumed in (a), (b) and (c) that the segment $[u, u+z]$ is included in C.

On the other hand, for any semi-normed space E, we do not know how to estimate $T(u+z) - T(u)$ in terms of semi-norms of z as in (a), (b) or (c). Under the general assumptions of Theorem 19.14, we only know how to obtain (19.4) that, with (19.5), can also be written as

$$u \in C,\ u+z \in C \quad \Rightarrow \quad \|T(u+z) - T(u)\|_{F;\mu} \leq \sup_{v \in C} \|\mathrm{d}T(v)\|_{\mathcal{L}(E;F);\{z\},\mu}. \ \square$$

19.5. Bounds on a real differentiable mapping

Let us give the **mean value theorem** for a real mapping.

THEOREM 19.15.– *Let E be a separated semi-normed space, X an open subset of E, T a mapping from X into $\mathbb{R}$, $u \in X$ and $z \in E$ such that:*

$$T \textit{ is differentiable}, \quad [u, u+z] \subset X.$$

Then:

$$\inf_{0 \leq t \leq 1} \mathrm{d}T(u+tz)z \leq T(u+z) - T(u) \leq \sup_{0 \leq t \leq 1} \mathrm{d}T(u+tz)z. \ \blacksquare$$

Proof. Let us return to the proof of Theorem 19.13 (a). Here $\| \ \|_{F;\nu}$ is the absolute value $| \ |$ and therefore (19.2) implies: given $\sigma > 0$ and $t \in [0,1]$, there exists $\theta_t > 0$ such that:

$$|s| \leq \theta_t \quad \Rightarrow \quad g(t+s) - g(t) \leq s\delta(t) + |s|\sigma$$

where $g(t) = T(u+tz)$ and $\delta(t) = \mathrm{d}T(u+tz)z$ according to (19.1). For the t_i and the a_i defined by (19.3), this gives

$$g(t_i) - g(a_i) \leq (t_i - a_i)(\delta(t_i) + \sigma),$$
$$g(a_{i+1}) - g(t_i) \leq (a_{i+1} - t_i)(\delta(t_i) + \sigma)$$

from which by adding these two inequalities then by summing from $i = 1$ to n,

$$g(1) - g(0) \leq \sup_{0 \leq t \leq 1} \delta(t) + \sigma.$$

This holds for all $\sigma > 0$ and thus for $\sigma = 0$. That is to say,

$$T(u+z) - T(u) \leq \sup_{0 \leq t \leq 1} dT(u+tz)z.$$

This inequality with respect to the mapping $-T$ gives the other (the first) inequality stated. □

Let us show that the differential of a real mapping vanishes at extrema.

THEOREM 19.16.– *Let E be a separated semi-normed space, X an open subset of E, T a mapping from X into $\mathbb{R}$ and $u \in X$ such that:*

T has a maximum or a minimum at u and is differentiable there.

Then:

$$dT(u) = 0. \blacksquare$$

Proof. Suppose that there exists $x \in E$ such that:

$$dT(u)x \neq 0.$$

Definition 19.1 of differentiability (with $z = tx$) would then provide a finite set $N \subset \mathcal{N}_E$ and by choosing $\epsilon = |dT(u)x|/(2\sup_{\nu \in N} \|x\|_{E;\nu})$, a number $\eta > 0$ such that $|t| \sup_{\nu \in N} \|x\|_{E;\nu} \leq \eta$ would imply

$$|T(u+tx) - T(u) - tdT(u)x| \leq \frac{1}{2}|tdT(u)x|. \tag{19.6}$$

Then, choosing t of the same sign as $dT(u)x$, it would yield

$$T(u+tx) \geq T(u) + \frac{1}{2}tdT(u)x > T(u).$$

Therefore, T would have no maximum at u.

If N was such that $\sup_{\nu \in N} \|x\|_{E;\nu} = 0$, it would be impossible to choose ϵ as above but Definition 19.1 would give (19.6) with 0 in the right-hand side, hence the same conclusion.

Similarly, we would obtain $T(u+tx) < T(u)$ by choosing, in (19.6), t with the opposite sign to $dT(u)x$. Therefore, T would not also have a minimum in u. As a result, $dT(u)x = 0$ for all $x \in E$. □

The case of local extrema. More generally, the differential dT vanishes at local extrema, that is at any point u included in an open subset U of X such that $T|_U$ has a maximum or a minimum at u. This is the result of Theorem 19.16 as $T|_U$ is differentiable at u with $dT(u)$ as a differential (Theorem 19.12 (a)). □

19.6. Differentiation of a composite mapping

Let us give the **chain rule theorem**[2] on differentiation of composite mappings.

THEOREM 19.17.– *Let E, F and G be three separated semi-normed spaces, two open subsets X of E and Y of F, two mappings T from X into Y and S from Y into G, and $u \in X$ such that:*

$$T \text{ is differentiable at } u,$$

$$S \text{ is differentiable at } T(u).$$

Then:

(a) $$S \circ T \text{ is differentiable at } u.$$

(b) $$\mathrm{d}(S \circ T)(u) = \mathrm{d}S(T(u)) \circ \mathrm{d}T(u). \ ∎$$

Complement. In (b), $\mathrm{d}T(u) \in \mathcal{L}(E; F)$, $\mathrm{d}S(T(u)) \in \mathcal{L}(F; G)$ and $\mathrm{d}(S \circ T)(u) \in \mathcal{L}(E; G)$. □

Proof. Let $z \in E$ such that $u + z \in X$. Let us decompose

$$(S \circ T)(u + z) - (S \circ T)(u) - \mathrm{d}S(T(u))(\mathrm{d}T(u)z) = A + B \qquad (19.7)$$

where

$$A = S(T(u + z)) - S(T(u)) - \mathrm{d}S(T(u))(T(u + z) - T(u)),$$
$$B = \mathrm{d}S(T(u))(T(u + z) - T(u) - \mathrm{d}T(u)z).$$

Let us endow, according to Theorem 3.15, E and F with filtering families of semi-norms $\{\| \ \|_{E;\nu} : \nu \in \mathcal{N}_E\}$ and $\{\| \ \|_{F;\mu} : \mu \in \mathcal{N}_F\}$ defining their topologies, which preserves (Theorem 19.9) the differentiability. Let $\| \ \|_{G;\lambda}$ be a semi-norm of G and $\epsilon > 0$.

Since S is differentiable at $w = T(u)$, there exists (Theorem 19.3) $\mu_1 \in N_F$ such that, for all $\alpha > 0$, there exists $\beta > 0$ such that $\|y\|_{F;\mu_1} \leq \beta$ implies

$$\|S(w + y) - S(w) - \mathrm{d}S(w)y\|_{G;\lambda} \leq \alpha \|y\|_{F;\mu_1}.$$

Since T is differentiable at u, it is locally Lipschitzian, and more specifically (Theorem 19.5), there exist $\nu_1 \in N_E$, $\gamma > 0$ and $c_1 > 0$ such that $\|z\|_{E;\nu_1} \leq \gamma$ implies

$$\|T(u + z) - T(u)\|_{F;\mu_1} \leq c_1 \|z\|_{E;\nu_1}.$$

2 **Originality of the chain rule theorem 19.17.** The generalization of the differentiation of composite mappings to non-normed spaces is new, as is all the extension of calculus to semi-normed spaces, see note 1, p. 275.

Let us choose $\alpha = \epsilon/(2c_1)$. Then, $\|z\|_{E;\nu_1} \leq \inf\{\gamma, \beta/c_1\}$ leads to

$$\|A\|_{G;\lambda} \leq \frac{\epsilon}{2c_1}\|T(u+z) - T(u)\|_{F;\mu_1} \leq \frac{\epsilon}{2}\|z\|_{E;\nu_1}. \tag{19.8}$$

On the other hand, since $\mathrm{d}S(w)$ is continuous linear, there exist (Theorem 7.14 (b)) $\mu_2 \in \mathcal{N}_F$ and $c_2 \geq 0$ such that, for all $y \in F$,

$$\|\mathrm{d}S(w)y\|_{G;\lambda} \leq c_2\|y\|_{F;\mu_2}.$$

Let $\nu_2 \in \mathcal{N}_E$ and $\eta_2 > 0$ such that (Theorem 19.3) $\|z\|_{E;\nu_2} \leq \eta_2$ implies

$$\|T(u+z) - T(u) - \mathrm{d}T(u)z\|_{F;\mu_2} \leq (\epsilon/2c_2)\|z\|_{E;\nu_2}$$

and thus

$$\|B\|_{G;\lambda} \leq \frac{\epsilon}{2}\|z\|_{E;\nu_2}. \tag{19.9}$$

The family of semi-norms which E has been endowed with being filtering, there is $\nu \in \mathcal{N}_E$ such that $\| \ \|_{E;\nu} \geq \sup\{\| \ \|_{E;\nu_1}, \| \ \|_{E;\nu_2}\}$. For $\|z\|_{E;\nu} \leq \inf\{\gamma, \beta/c_1, \eta_2\}$, (19.8) and (19.9) give

$$\|A + B\|_{G;\lambda} \leq \epsilon\|z\|_{E;\nu}.$$

According to (19.7) and Definition 19.1 of differentiability, this proves that $S \circ T$ is differentiable at u and

$$\mathrm{d}(S \circ T)(u)z = \mathrm{d}S(T(u))(\mathrm{d}T(u)z). \ \square$$

Let us show that the composite of continuously differentiable mappings is continuously differentiable if the initial space E is metrizable or if the intermediate space F is normed.

THEOREM 19.18.– *Let E, F and G be three separated semi-normed spaces, two open sets X of E and Y of F, and two mappings T from X into Y and S from Y into G such that:*

$$E \text{ is metrizable or } F \text{ is normed,} \tag{19.10}$$

$$S \text{ and } T \text{ are continuously differentiable.}$$

Then:

$$S \circ T \text{ is continuously differentiable.} \ \blacksquare$$

Proof. The point is to verify that the differential given by the chain rule theorem 19.17, $\mathrm{d}(S \circ T)(u) = \mathrm{d}S(T(u)) \circ \mathrm{d}T(u)$, continuously depends on u. Yet, the mapping

$A_1 : u \mapsto (\mathrm{d}T(u), \mathrm{d}S(T(u))$ is continuous from X into $\mathcal{L}(E;F) \times \mathcal{L}(F;G)$.

In effect, $\mathrm{d}T$ is continuous from E into $\mathcal{L}(E;F)$ by assumption and the composite mapping $\mathrm{d}S \circ T$ is continuous from X into $\mathcal{L}(F;G)$ (Theorem 7.9 (a) because T is continuous from X into F and $\mathrm{d}S$ is continuous from F into $\mathcal{L}(F;G)$).

On the other hand, the mapping $A_2 : (L, M) \mapsto M \circ L$ is sequentially continuous from $\mathcal{L}(E;F) \times \mathcal{L}(F;G)$ into $\mathcal{L}(E;G)$ (Theorem 12.12 (d)). The composite mapping $A_2 \circ A_1 : u \mapsto \mathrm{d}S(T(u)) \circ \mathrm{d}T(u)$ is thus sequentially continuous from X into $\mathcal{L}(E;G)$ (Theorems 7.2 (a) and 7.9 (c)). If E is metrizable, then it is continuous (Theorem 9.1 (a)).

If F is normed, this is still true because the mapping A_2 is then continuous (Theorem 12.12 (f)), and as a result the composite mapping $A_2 \circ A_1$ also is (Theorem 7.9 (a)). □

QUESTION: The case in which E is not metrizable and F is not normed. Is the assumption (19.10) necessary in Theorem 19.18, that is in order for the differential of a composite of continuously differentiable mappings to be continuous? The author of these lines does not know. He only knows how to avoid it when S or T is linear (Theorems 19.19 (c) and 19.20 (c), below). □

Let us observe that composition by a linear continuous mapping preserves differentiability and continuous differentiability, even if the initial space E is not metrizable and if the intermediate space F is not normed.

THEOREM 19.19.– *Let E, F and G be three separated semi-normed spaces, X an open subset of E, T a mapping from X into F, $u \in X$ and L such that:*

$$L \in \mathcal{L}(F;G),$$

$$T \text{ is differentiable at } u.$$

Then:

(a) $$L \circ T \text{ is differentiable at } u.$$

(b) $$\mathrm{d}(L \circ T)(u) = L \circ \mathrm{d}T(u).$$

(c) *T is continuously differentiable $\Rightarrow$ $L \circ T$ is continuously differentiable.* ▮

Proof. (a) *and* (b). This is a particular case of the chain rule theorem 19.17 since (Theorem 19.8) L is differentiable and $\mathrm{d}L(v) = L$.

(c) The differential $\mathrm{d}(L \circ T)$ is the mapping $u \mapsto L \circ \mathrm{d}T(u)$, that is to say the composite of the mapping $u \mapsto \mathrm{d}T(u)$ from X into $\mathcal{L}(E;F)$ with the mapping $M \mapsto L \circ M$ from $\mathcal{L}(E;F)$ into $\mathcal{L}(E;G)$ which is continuous (Theorem 12.12 (c)).

If $\mathrm{d}T$ is continuous, namely if T is continuously differentiable, the composite mapping $u \mapsto L \circ \mathrm{d}T(u)$ also is (Theorem 7.9 (a)), that is $L \circ T$ is continuously differentiable. □

Other proof of Theorem 19.19. This result can be deducted, more directly, from the characterization of continuous linear mappings of Theorem 7.14 (a) and from Definition 19.1 of differentiability. □

Let us give similar properties for the composition on the right with a continuous linear mapping.

THEOREM 19.20.– *Let E, F and G be three separated semi-normed spaces, two open sets X of E and Y of F, S a mapping from Y into G, $u \in X$ and L such that:*

$$L \in \mathcal{L}(E; F), \quad L(X) \subset Y,$$

$$S \text{ is differentiable at } L(u).$$

Then:

(a) $$S \circ L \text{ is differentiable at } u.$$

(b) $$\mathrm{d}(S \circ L)(u) = \mathrm{d}S(Lu) \circ L.$$

(c) *S is continuously differentiable* $\Rightarrow$ *$S \circ L$ is continuously differentiable.* ▮

Proof. (a) and (b). This is a special case of the chain rule theorem 19.17 since (Theorem 19.8) L is differentiable and $\mathrm{d}L(u) = L$.

(c) The differential $\mathrm{d}(S \circ L)$ is the mapping $u \mapsto \mathrm{d}S(Lu) \circ L$, that is the composite of mappings $L|_X$ from X into Y which is continuous (Theorem 7.8 (a)), $\mathrm{d}S$ from Y into $\mathcal{L}(F; G)$, and $M \mapsto M \circ L$ from $\mathcal{L}(F; G)$ into $\mathcal{L}(E; G)$ which is continuous (Theorem 12.12 (b)).

If $\mathrm{d}S$ is continuous, namely if S is continuously differentiable, the composite mapping $u \mapsto \mathrm{d}S(Lu) \circ L$ also is (Theorem 7.9 (a)), that is $S \circ L$ is continuously differentiable. □

19.7. Differential of an inverse mapping

Let us show that, if a bijection is differentiable as well as its inverse mapping, the differential of the inverse is the inverse of the differential.

THEOREM 19.21.– *Let E and F be two separated semi-normed spaces, two open sets X of E and Y of F, T a mapping from X into F and $u \in X$ such that:*

$$T \text{ is a bijection from } X \text{ onto } Y,$$

$$T \text{ is differentiable at } u \text{ and } T^{-1} \text{ is differentiable at } T(u).$$

Then:

(a) $$\mathrm{d}T(u) \textit{ is an isomorphism from } E \textit{ onto } F.$$

(b) $$(\mathrm{d}T(u))^{-1} = \mathrm{d}T^{-1}(T(u)). \blacksquare$$

Proof. The chain rule theorem 19.17 (b) with respect to $T^{-1} \circ T = I_E$ gives, since a linear mapping is its own differential (Theorem 19.8 (b)),

$$\mathrm{d}T^{-1}(T(u)) \circ \mathrm{d}T(u) = \mathrm{d}I_E(u) = I_E.$$

It thereof results that $\mathrm{d}T(u)$ is a bijection from E onto F and that its inverse mapping is $\mathrm{d}T^{-1}(T(u))$ which is linear and continuous (following Definition 19.1 (a) of the differential). Hence (a) and (b). □

Comparison of Theorem 19.21 to the inverse mapping theorem 19.22. These theorems are, roughly, converse: the first gives the invertibility of $\mathrm{d}T(u)$ when T^{-1} exists and is differentiable at u, whereas the second gives the existence and the differentiability of T^{-1} at u when $\mathrm{d}T(u)$ is invertible. □

19.8. Inverse mapping theorem

Let us show that a mapping continuously differentiable in Banach spaces whose differential at a point is invertible is itself invertible in the neighborhood of the said point, which is the **inverse mapping theorem**.

THEOREM 19.22.– *Let*

$$E \textit{ and } F \textit{ be Banach spaces,}$$

X an open subset of E, $u \in X$ and T a mapping from X into F such that:

$$T \textit{ is continuously differentiable from } X \textit{ into } F,$$

$$\mathrm{d}T(u) \textit{ is a bijection from } E \textit{ onto } F.$$

Then, there exist two open sets $\mathcal{X}$ of E and $\mathcal{Y}$ of F such that $u \in \mathcal{X}$ and:

(a) $$T \textit{ is a bijection from } \mathcal{X} \textit{ onto } \mathcal{Y}.$$

(b) $$T^{-1} \textit{ is continuously differentiable from } \mathcal{Y} \textit{ into } E.$$

(c) $$\mathrm{d}T(v) \textit{ is an isomorphism from } E \textit{ onto } F, \quad \forall v \in \mathcal{X}.$$

(d) $$\mathrm{d}T^{-1}(y) = (\mathrm{d}T(T^{-1}(y)))^{-1}, \quad \forall y \in \mathcal{Y}. \blacksquare$$

Proof. Remember that $B_E(v,r)$ denotes a closed ball of E, $\mathring{B}_E(v,r)$ an open ball and that (Theorem 12.10 (a)) the topology of $\mathcal{L}(E;F)$ is generated by the norm $|||L|||_{\mathcal{L}(E;F)} = \sup_{e\in E,\, e\neq 0} \|Lu\|_F/\|u\|_E$.

(c) *Invertibility of* $\mathrm{d}T(v)$. Since $\mathrm{d}T(u)$ is linear, continuous and bijective from E into F which are Banach spaces, the Banach–Schauder theorem 9.10 then shows that $\mathrm{d}T(u)$ is an isomorphism from E onto F.

The set $\mathcal{I}(E;F)$ of the isomorphisms from E onto F being open in $\mathcal{L}(E;F)$ (Theorem 12.15 (a)), there exists $\epsilon > 0$ such that, for any $L \in \mathcal{L}(E;F)$,

$$|||L - \mathrm{d}T(u)|||_{\mathcal{L}(E;F)} \leq \epsilon \quad \Rightarrow \quad L \in \mathcal{I}(E;F).$$

Since $\mathrm{d}T$ is continuous from X into $\mathcal{L}(E;F)$ (by assumption), there exists $\eta > 0$ such that:

$$\|v - u\|_E \leq \eta \quad \Rightarrow \quad |||\mathrm{d}T(v) - \mathrm{d}T(u)|||_{\mathcal{L}(E;F)} \leq \epsilon.$$

Thus, $\mathrm{d}T(v)$ is an isomorphism from E onto F, for all $v \in B_E(u,\eta)$.

(a) *Invertibility of* T. Let us return, by stages, to the invertibility of a mapping S from E into itself, that we will establish by the Banach fixed-point theorem.

1) *Reduction to a mapping* S. Denote $\mathcal{O} = \mathring{B}_E(0_E,\eta)$ and, for all $v \in \mathcal{O}$,

$$\mathcal{T}(v) = T(v+u) - T(u). \tag{19.11}$$

Let

$$S = (\mathrm{d}T(u))^{-1} \circ \mathcal{T}. \tag{19.12}$$

The mapping $\mathcal{T}$ is continuously differentiable from $\mathcal{O}$ into F and $(\mathrm{d}T(u))^{-1}$ is continuous linear from F into E. Therefore, the composite mapping S is (Theorem 19.19 (c)) continuously differentiable from $\mathcal{O}$ into E. In addition, since $\mathrm{d}\mathcal{T}(v) = \mathrm{d}T(v+u)$, there follows (Theorem 19.19 (b)) that

$$\mathrm{d}S(v) = (\mathrm{d}T(u))^{-1} \circ \mathrm{d}T(v+u). \tag{19.13}$$

Observe that, by denoting I_E the identity of E,

$$S(0_E) = (\mathrm{d}T(u))^{-1}(0_F) = 0_E, \quad \mathrm{d}S(0_E) = (\mathrm{d}T(u))^{-1} \circ \mathrm{d}T(u) = I_E. \tag{19.14}$$

2) *Injectivity of* S. Let us denote

$$R = I_E - S. \tag{19.15}$$

The mapping R is continuously differentiable from $\mathcal{O}$ into E and $\mathrm{d}R(v) = I_E - \mathrm{d}S(v)$, hence $\mathrm{d}R(0_E) = 0_{\mathcal{L}(E;E)}$. There thus exists $r > 0$, which can be chosen smaller than η, such that $\|v\|_E \leq r$ implies $|||\mathrm{d}R(v)|||_{\mathcal{L}(E;E)} \leq 1/2$.

The mean value theorem 19.14 (c) therefore yields, for all v and v' in $B_E(0_E, r)$,

$$\|R(v) - R(v')\|_E \leq \frac{1}{2}\|v - v'\|_E \tag{19.16}$$

hence, according to (19.15),

$$\|S(v) - S(v')\|_E = \|(v - v)' - (R(v) - R(v'))\|_E \geq$$
$$\geq \|v - v'\|_E - \|R(v) - R(v')\|_E \geq \frac{1}{2}\|v - v'\|_E. \tag{19.17}$$

Therefore, S is injective.

3) *Surjectivity of* S. Let $x \in E$ (fixed) such that $\|x\|_E < r/2$ and $v \in E$ (variable) such that $\|v\|_E \leq r$. Denote

$$P(v) = x + R(v).$$

Since $R(0_E) = -S(0_E) = 0_E$, the inequality (19.16) gives $\|R(v)\|_E \leq \|v\|_E/2 \leq r/2$, from which $\|P(v)\|_E < r$.

Therefore, P is a mapping from the open ball $\mathring{B}_E(0_E, r)$ into itself. It is a strict contraction because (19.16) implies that $\|P(v) - P(v')\|_E \leq \|v - v'\|_E/2$. According to the Banach theorem 9.5, it has a fixed point z such that $z = P(z) = x + R(z)$ and thus

$$S(z) = z - R(z) = x.$$

This shows that $S(\mathring{B}_E(0_E, r))$ contains $\mathring{B}_E(0_E, r/2)$. With the injectivity established above, it thereof follows that

$$S \text{ is a bijection from } \mathcal{O}_1 = S^{-1}(\mathring{B}_E(0_E, r/2)) \text{ onto } \mathcal{Q} = \mathring{B}_E(0_E, r/2).$$

4) *Invertibility of* T. Since $\mathrm{d}T(u)$ is bijective, it follows thereof that $\mathrm{d}T(u) \circ S$, that is T according to (19.12), is bijective from $\mathcal{O}_1$ onto $\mathrm{d}T(u)(\mathcal{Q})$. Therefore, following (19.11),

$$T \text{ is bijective from } \mathcal{X} = \mathcal{O}_1 + u \text{ onto } \mathcal{Y} = \mathrm{d}T(u)(\mathcal{Q}) + T(u).$$

5) *Properties of* $\mathcal{X}$ *and* $\mathcal{Y}$. Since S is continuous and its domain of definition $\mathcal{O}$ is open, the preimage $\mathcal{O}_1$ of the open set $\mathcal{Q}$ is open (Theorem 8.2 (a)). Thus, $\mathcal{X} = \mathcal{O}_1 + u$ is open in E.

Since $(\mathrm{d}T(u))^{-1}$ is continuous linear, the image $\mathrm{d}T(u)(\mathcal{Q})$ of the open set $\mathcal{Q}$ by its inverse mapping is open (Theorem 8.8 (d)). Therefore, $\mathcal{Y} = \mathrm{d}T(u)(\mathcal{Q}) + T(u)$ is open in F.

(b) *and* (d). *Differentiability of* T^{-1}. Consider again S.

1) *Differentiability of* S^{-1}. Let us show that S^{-1} is differentiable at 0_E and

$$dS^{-1}(0_E) = I_E. \tag{19.18}$$

Let $x \in \mathcal{Q}$ and $z = S^{-1}(x)$. Since, see (19.14), $S^{-1}(0_E) = 0_E$,

$$\|S^{-1}(x) - S^{-1}(0_E) - I_E x\|_E = \|z - x\|_E. \tag{19.19}$$

Moreover, $S(z) - S(0_E) - dS(0_E)z = x - z$ because $S(z) = x$ and, according to (19.14), $S(0_E) = 0_E$ and $dS(0_E) = I_E$. And S is differentiable at 0_E (it was established before (19.13)) thus, given $\epsilon > 0$, there exists $\eta > 0$ such that $\|z\|_E \leq \eta$ implies

$$\|S(z) - S(0_E) - dS(0_E)z\|_E = \|x - z\|_E \leq \frac{\epsilon}{2}\|z\|_E.$$

In addition, (19.17) with $v' = 0_E$ gives $\|x\|_E = \|S(z)\|_E \geq \|z\|_E/2$.

The right-hand side of (19.19) is therefore bounded by $\epsilon\|x\|_E$ when $\|z\|_E \leq \eta$. This proves the differentiability of S^{-1} at 0_E and (19.18).

2) *Differentiability* of T^{-1} at $T(u)$. As it was established in (c), $dT(u)^{-1}$ is continuous linear from F into E. The differentiability of S^{-1} at 0_E thus implies (Theorem 19.20 (a)) that of $S^{-1} \circ dT(u)^{-1}$, that is $\mathcal{T}^{-1}$ according to (19.12), at 0_F and (Theorem 19.20 (b))

$$\mathrm{d}\mathcal{T}^{-1}(0_F) = \mathrm{d}S^{-1}(dT(u)^{-1}(0_F)) \circ \mathrm{d}T(u)^{-1} = \mathrm{d}T(u)^{-1} \tag{19.20}$$

since $dT(u)^{-1}(0_F) = 0_E$ and $\mathrm{d}S^{-1}(0_E) = I_E$ according to (19.18).

In addition, according to (19.11), $\mathcal{T}(T^{-1}(y) - u) = y - T(u)$ hence

$$T^{-1}(y) = u + \mathcal{T}^{-1}(y - T(u)).$$

The differentiability of $\mathcal{T}^{-1}$ at 0_F and the equality (19.20) thus implies the differentiability of T^{-1} at $T(u)$ and

$$\mathrm{d}T^{-1}(T(u)) = \mathrm{d}\mathcal{T}^{-1}(0_F) = \mathrm{d}T(u)^{-1}. \tag{19.21}$$

3) *Conclusion.* The mapping $dT(v)$ is, as it was established in (c), an isomorphism from E onto F at any point of $B_E(u, \eta)$, a ball that contains $\mathcal{X}$ according to its construction above. Thus, the assumptions of the theorem are verified at any point v of $\mathcal{X}$, and therefore (19.21) gives the differentiability of T^{-1} at any point y of $\mathcal{Y}$ and

$$\mathrm{d}T^{-1}(y) = \mathrm{d}T(T^{-1}(y))^{-1}. \tag{19.22}$$

It remains to show that the mapping $y \mapsto \mathrm{d}(T^{-1})(y)$ is continuous at any point of $\mathcal{Y}$. The right-hand side of (19.22) being a composite of mappings, it suffices

(Theorem 7.9 (a)) that each of them be continuous. This is the case because:
— T^{-1} is continuous (Theorem 19.6 (b)) since it is differentiable.
— $\mathrm{d}T$ is continuous by assumption.
— $L \mapsto L^{-1}$ is continuous (Theorem 12.15 (b)) from $\mathcal{I}(E;F)$ into $\mathcal{L}(F;E)$, and therefore at any point $\mathrm{d}T(v)$ where $v \in \mathcal{X}$ since this is then an isomorphism as it was established in (c). □

Inverse mapping theorem for a real function. When $E = F = \mathbb{R}$, that is for a real function of a real variable, an inverse mapping is much more simple to obtain than by using Theorem 19.22, see Theorem 22.9. □

Chapter 20

Differentiation of Multivariable Mappings

20.1. Partial differentiation

Let us define the partial differentials of a multivariable mapping.

DEFINITION 20.1.– *Let E_1, $E_2 \ldots$, E_d and F be separated semi-normed spaces, X an open subset of $E_1 \times \cdots \times E_d$, T a mapping from X into F and $i \in [\![1, d]\!]$.*

(a) *T has **a partial differential** $\mathrm{d}_i T(u_1, \ldots, u_d)$ **at a point** $(u_1, \ldots, u_d)$ of X if the partial mapping $u_i \mapsto T(u_1, \ldots, u_d)$ is differentiable at u_i and $\mathrm{d}_i T(u_1, \ldots, u_d)$ is its differential. Then:*

$$\mathrm{d}_i T(u_1, \ldots, u_d) \in \mathcal{L}(E_i; F).$$

(b) *T has **partial differentials at** $(u_1, \ldots, u_d)$ if it has for all $i \in [\![1, d]\!]$.*

(c) *T has **partial differentials** if it has any at any point of X.* ▮

Clarification. "The partial mapping $u_i \mapsto T(u_1, \ldots, u_d)$" designates, more exactly, the mapping $T_{u,i}$, from the open set $X_{u,i} = \{v_i \in E_i : (u_1, \ldots, v_i, \ldots, u_d) \in X\}$ of E_i into F, defined by

$$T_{u,i}(v_i) \stackrel{\text{def}}{=} T(u_1, \ldots, v_i, \ldots, u_d).$$

Furthermore, $\mathrm{d}_i T(u_1, \ldots, u_d) \stackrel{\text{def}}{=} \mathrm{d}T_{u,i}(u_i)$. □

Let us show that differentiability implies the existence of partial differentials.

THEOREM 20.2.– *Let E_1, $E_2 \ldots$, E_d and F be separated semi-normed spaces, X an open subset of $E_1 \times \cdots \times E_d$, T a mapping from X into F and $(u_1, \ldots, u_d) \in X$ such that:*

$$T \text{ is differentiable at } (u_1, \ldots, u_d).$$

Then, for all $(z_1, \ldots, z_d) \in E_1 \times \cdots \times E_d$ *and* $i \in [\![1, d]\!]$*:*

(a) *T has unique partial differentials at* $(u_1, \ldots, u_d)$.

(b) $$\mathrm{d}T(u_1, \ldots, u_d)(z_1, \ldots, z_d) = \sum_{i=1}^{d} \mathrm{d}_i T(u_1, \ldots, u_d) z_i.$$

(c) $$\mathrm{d}_i T(u_1, \ldots, u_d) z_i = \mathrm{d}T(u_1, \ldots, u_d)(0_{E_1}, \ldots, z_i, \ldots, 0_{E_d}).$$ ▮

Proof. (a) *and* (c). By taking $z = (0_{E_1}, \ldots, z_i, \ldots, 0_{E_d})$ in Definition 19.1 of the differentiability of T, we obtain the differentiability at u_i of the partial mapping $u_i \mapsto T(u_1, \ldots, u_d)$ as well as the stated expression of its differential, which is (Definition 20.1 (a)) the partial differential $\mathrm{d}_i T(u_1, \ldots, u_d) z_i$. The latter is unique since the differential is unique (Theorem 19.2).

(b) Since $(z_1, \ldots, z_d) = \sum_{i=1}^{d} (0_{E_1}, \ldots, z_i, \ldots, 0_{E_d})$, the linearity of the differential gives

$$\mathrm{d}T(u_1, \ldots, u_d)(z_1, \ldots, z_d) = \sum_{1 \leq i \leq d} \mathrm{d}T(u_1, \ldots, u_d)(0_{E_1}, \ldots, z_i, \ldots, 0_{E_d}) =$$
$$= \sum_{1 \leq i \leq d} \mathrm{d}_i T(u_1, \ldots, u_d) z_i. \quad \square$$

Let us show conversely that, for a mapping defined on normed spaces, the existence of continuous partial differentials leads to differentiability.

THEOREM 20.3.– *Let F be a separated semi-normed space,*

E_1, $E_2 \ldots$, E_d *normed spaces,*

X an open subset of $E_1 \times \cdots \times E_d$*, T a mapping from X into F and* $(u_1, \ldots, u_d) \in X$ *such that:*

T has partial differentials $\mathrm{d}_i T$ *on X which are continuous at* $(u_1, \ldots, u_d)$.

Then:

T is differentiable at $(u_1, \ldots, u_d)$. ▮

Proof. Introductory computation. Let $(z_1, \ldots, z_d) \in E_1 \times \cdots \times E_d$, $i \in [\![1, d]\!]$ and $t \in \mathbb{R}$. By denoting

$$v_i(t) = (u_1 + z_1, \ldots, u_{i-1} + z_{i-1}, u_i + t z_i, u_{i+1}, \ldots, u_d),$$

we decompose

$$T(u+z) - T(u) = \sum_{i=1}^{d} T(v_i(1)) - T(v_i(0)).$$

By denoting

$$S_i(t) = T(v_i(t)) - t\mathrm{d}_i T(u) z_i,$$

it yields that $T(v_i(1)) - T(v_i(0)) - \mathrm{d}_i T(u) z_i = S_i(1) - S_i(0)$ and thus

$$T(u+z) - T(u) - \sum_{i=1}^{d} \mathrm{d}_i T(u) z_i = \sum_{i=1}^{d} S_i(1) - S_i(0). \tag{20.1}$$

Differentiation. Since T has partial differentials, according to Definitions 20.1 of partial differentials and 19.1 of differentiability, each mapping S_i is differentiable and

$$\mathrm{d}_t S_i(t) 1 = \mathrm{d}_i T(v_i(t)) z_i - \mathrm{d}_i T(u) z_i.$$

The mean value theorem 19.13 (c) therefore yields, for every semi-norm of F,

$$\|S_i(1) - S_i(0)\|_{F;\mu} \leq \sup_{0 \leq t \leq 1} \|\mathrm{d}_t S_i(t) 1\|_{F;\mu} = \\ = \sup_{0 \leq t \leq 1} \|\mathrm{d}_i T(v_i(t)) z_i - \mathrm{d}_i T(u) z_i\|_{F;\mu}.$$

Since E_i is normed, the topology of $\mathcal{L}(E_i; F)$ is generated (Theorem 12.9 (a)) by the semi-norms $|||L|||_{\mathcal{L}(E_i;F);\mu} = \sup_{x_i \in E_i,\, x_i \neq 0} \|L x_i\|_{F;\mu} / \|x_i\|_{E_i}$ indexed by $\mu \in \mathcal{N}_F$ and the inequality of Theorem 12.9 (c) gives

$$\|S_i(1) - S_i(0)\|_{F;\mu} \leq \sup_{0 \leq t \leq 1} |||\mathrm{d}_i T(v_i(t)) - \mathrm{d}_i T(u)|||_{\mathcal{L}(E_i;F);\mu} \|z_i\|_{E_i}.$$

Let $\epsilon > 0$. By assumption, $\mathrm{d}_i T$ is continuous at u and therefore there exists $\eta_i > 0$ such that $\|v - u\|_{E_1 \times \cdots \times E_d} \leq \eta_i$ implies $|||\mathrm{d}_i T(v) - \mathrm{d}_i T(u)|||_{\mathcal{L}(E_i;F);\mu} \leq \epsilon / d$.

In addition, $v_i(t) - u = (z_1, \ldots, z_{i-1}, t z_i, 0_{E_{i+1}}, \ldots, 0_{E_d})$; thus, for all $t \in [0, 1]$, it follows that $\|v_i(t) - u\|_{E_1 \times \cdots \times E_d} \leq \|z\|_{E_1 \times \cdots \times E_d}$.

Therefore, $\|z\|_{E_1 \times \cdots \times E_d} \leq \inf_{1 \leq i \leq d} \eta_i$ leads to $\|S_i(1) - S_i(0)\|_{F;\mu} \leq \epsilon \|z_i\|_{E_i} / d$ from which, with (20.1),

$$\left\| T(u+z) - T(u) - \sum_{i=1}^{d} \mathrm{d}_i T(u) z_i \right\|_{F;\mu} \leq \sum_{i=1}^{d} \|S_i(1) - S_i(0)\|_{F;\mu} \leq \\ \leq \epsilon \|z\|_{E_1 \times \cdots \times E_d}.$$

It thereof follows (Definition 19.1) that T is differentiable at u and that its differential is given by $\mathrm{d}T(u) z = \sum_i \mathrm{d}_i T(u) z_i$. In effect, this mapping $\mathrm{d}T$ is continuous linear from $E_1 \times \cdots \times E_d$ into F since each mapping $z \mapsto \mathrm{d}_i T(u) z_i$ also is (being the composite of continuous mappings $z \mapsto z_i$ and $\mathrm{d}_i T(u)$). □

Generalization of Theorem 20.3 to non-normed spaces E_i. The above proof can be extended to separated semi-normed spaces E_i, assuming that the $\mathrm{d}_i T$ are "continuous" in the following sense, in which $\{\| \, \|_{E_i;\nu_i} : \nu_i \in \mathcal{N}_{E_i}\}$ is the family of semi-norms of E_i: for every $\mu \in \mathcal{N}_F$, there exist finite sets $N_j \subset \mathcal{N}_{E_j}$ such that, for any $\epsilon > 0$, there exists $\eta > 0$ such that:

$$\sup_{1 \le j \le d,\ \nu_j \in N_j} \|y_j\|_{E_j;\nu_j} \le \eta \quad \Rightarrow \quad \|\mathrm{d}_i T(u+y) z_i - \mathrm{d}_i T(u) z_i\|_{F;\mu} \le \epsilon \sup_{\nu_i \in N_i} \|z_i\|_{E_i;\nu_i}.$$

This property is stronger than continuity which is written thus: for every $\mu \in \mathcal{N}_F$, every bounded set B_i of E_i and every $\epsilon > 0$, there are finite sets $N_j \subset \mathcal{N}_{E_j}$ and $\eta > 0$ such that:

$$\sup_{1 \le j \le d,\ \nu_j \in N_j} \|y_j\|_{E_j;\nu_j} \le \eta \quad \Rightarrow \quad \sup_{z_i \in B_i} \|\mathrm{d}_i T(u+y) z_i - \mathrm{d}_i T(u) z_i\|_{F;\mu} \le \epsilon. \ \square$$

Usefulness of the continuity of partial differentials. In Theorem 20.3, it is not possible to break away from the assumption of continuity of the $\mathrm{d}_i T$. For example, the function defined on $\mathbb{R}^2$ by

$$f(x_1, x_2) = \begin{cases} \dfrac{x_1 x_2}{x_1^2 + x_2^2} & \text{if } x \neq (0,0), \\ 0 & \text{otherwise,} \end{cases}$$

has partial differentials at any point but it is not differentiable nor even continuous at $(0,0)$. □

20.2. Differentiation of a multilinear or multi-component mapping

Let us show that any continuous linear mapping is continuously differentiable and let us calculate its differential.

THEOREM 20.4.– *Let E_1, $E_2 \ldots$, E_d and F be separated semi-normed spaces and*

$$T \in \mathcal{L}^d(E_1 \times \cdots \times E_d; F).$$

Then, for all $(u_1, \ldots, u_d)$ and $(z_1, \ldots, z_d)$ in $E_1 \times \cdots \times E_d$:

(a) *T is continuously differentiable.*

(b) $$\mathrm{d}T(u_1, \ldots, u_d)(z_1, \ldots, z_d) = \sum_{i=1}^{d} T(u_1, \ldots, z_i, \ldots, u_d).$$

(c) $$\mathrm{d}_i T(u_1, \ldots, u_d) z_i = T(u_1, \ldots, z_i, \ldots, u_d). \ \blacksquare$$

Proof. (b) *and differentiability of T.* The point is to evaluate

$$\Delta = T(u_1 + z_1, \ldots, u_d + z_d) - T(u_1, \ldots, u_d) - \sum_{i=1}^{d} T(u_1, \ldots, z_i, \ldots, u_d).$$

The multilinearity implies that $T(u_1 + z_1, \ldots, u_d + z_d)$ is the sum of 2^d terms of the form $T(w_1, \ldots, w_d)$ where w_i is equal to u_i or z_i (this is proved by induction).

For $n \in [\![1, d]\!]$, let us denote S_n the sum of the terms in which n of the w_i are equal to z_i. Then, $S_0 = T(u_1, \ldots, u_d)$ and $S_1 = \sum_{i=1}^{d} T(u_1, \ldots, z_i, \ldots, u_d)$. Thus, $\Delta = \sum_{n=2}^{d} S_n$.

According to the characterization of continuous multilinear mappings of Theorem 7.20 (a), for every semi-norm $\| \ \|_{F;\mu}$ of F, there exist $c \geq 0$ and finite sets $N_1 \ldots, N_d$ respectively indexing the semi-norms of $E_1 \ldots, E_d$ such that:

$$\|T(v_1, \ldots, v_d)\|_{F;\mu} \leq c \sup_{\nu_1 \in N_1} \|v_1\|_{E_1;\nu_1} \cdots \sup_{\nu_d \in N_d} \|v_d\|_{E_d;\nu_d}.$$

Hence

$$\|S_n\|_{F;\mu} \leq 2^d c \sup_{\nu_1 \in N_1, \ldots, \nu_d \in N_d} (\|z_1\|_{E_1;\nu_1} + \cdots + \|z_d\|_{E_d;\nu_d})^n \times \\ \times (\|u_1\|_{E_1;\nu_1} + \cdots + \|u_d\|_{E_d;\nu_d})^{d-n}.$$

For all $\epsilon > 0$, if $\sup_{\nu_1 \in N_1, \ldots, \nu_d \in N_d} \|z_1\|_{E_1;\nu_1} + \cdots + \|z_d\|_{E_d;\nu_d} \leq \inf\{\epsilon, 1\}/m$, where $m = d2^d c \sup_{\nu_1 \in N_1, \ldots, \nu_d \in N_d} (1 + \|u_1\|_{E_1;\nu_1} + \cdots + \|u_d\|_{E_d;\nu_d})^d$, it follows that

$$\|\Delta\|_{F;\mu} \leq \sum_{n=2}^{d} \|S_n\|_{F;\mu} \leq \epsilon \sup_{\nu_1 \in N_1, \ldots, \nu_d \in N_d} \|z_1\|_{E_1;\nu_1} + \cdots + \|z_d\|_{E_d;\nu_d} = \\ = \epsilon \sup_{\nu_1 \in N_1, \ldots, \nu_d \in N_d} |\!|\!|(z_1, \ldots, z_d)|\!|\!|_{E_1 \times \cdots \times E_d;\nu_1, \ldots, \nu_d}$$

when $E_1 \times \cdots \times E_d$ is endowed with the semi-norms defined in Theorem 6.9 (a). Consequently (Definition 19.1), T is differentiable and its differential is given by (b).

(a) It remains to verify that the mapping $\mathrm{d}T$ is continuous from $E_1 \times \cdots \times E_d$ into $\mathcal{L}(E_1 \times \cdots \times E_d; F)$.

Theorem 12.21 (a) on the separation of variables in a multilinear mapping shows that the mapping $u \mapsto \mathrm{d}_i T(u)$ given by (c) is continuous from $E_1 \times \cdots \times E_d$ into $\mathcal{L}(E_i; F)$ (it does not depend on u_i and this theorem gives the continuity with respect to the $d - 1$ other variables).

Furthermore, the projection P_i defined by $P_i(z) = z_i$ is continuous linear (Theorem 7.28) from $E_1 \times \cdots \times E_d$ into E_i, and thus (Theorem 12.12 (b)) the mapping $\widehat{P}_i$, which to M associates $M \circ P_i$, is continuous linear from $\mathcal{L}(E_i; F)$ into $\mathcal{L}(E_1 \times \cdots \times E_d; F)$.

The composite mapping $\widehat{P}_i \circ d_i T$, that is to say $u \mapsto \mathrm{d}_i T(u) \circ P_i$, is thus (Theorem 7.9 (a)) continuous from $E_1 \times \cdots \times E_d$ into $\mathcal{L}(E_1 \times \cdots \times E_d; F)$. Being the sum in i of these mappings according to (b), the mapping $u \mapsto dT(u)$ is also continuous.

(c) Each partial mapping $u_i \mapsto T(u_1, \ldots, u_d)$ is continuous linear, and therefore (Theorem 19.8) it is differentiable and its differential $\mathrm{d}_i T$ has the value stated. □

Let us observe that a mapping that has several components is differentiable if its components are.

THEOREM 20.5.– *Let E and $F_1 \ldots, F_d$ be separated semi-normed spaces, X an open subset of E, T a mapping from X into $F_1 \times \cdots \times F_d$, $u \in X$ and denote*

$$T(v) = (T_1(v), \ldots, T_d(v))$$

that is $T_i(v) = (T(v))_i$ for all $i \in [\![1, d]\!]$. Then:

(a) $\quad T$ *is differentiable at* $u \quad \Leftrightarrow \quad$ *all* T_i *are differentiable at* u.

(b) $\quad T$ *is differentiable at* $u \quad \Rightarrow \quad \mathrm{d}T(u) = (\mathrm{d}T_1(u), \ldots, \mathrm{d}T_d(u))$. ∎

Proof. There is $T_i = P_i \circ T$, where P_i is the projection defined by $P_i(w) = w_i$. Since $P_i \in \mathcal{L}(F_1 \times \cdots \times F_d; F_i)$ (Theorem 7.28), the differentiability of T at u implies (Theorem 19.19 (a) and (b)) that of the composite $P_i \circ T$, that is T_i, and

$$\mathrm{d}T_i(u) = \mathrm{d}(P_i \circ T)(u) = P_i \circ \mathrm{d}T(u) = (\mathrm{d}T(u))_i.$$

This proves (b) and the direct part of (a).

Conversely, $T = \sum_i R_i \circ T_i$, where $R_i \in \mathcal{L}(F_i; F_1 \times \cdots \times F_d)$ is defined by $R_i(w_i) = (0_{E_1}, \ldots, w_i, \ldots, 0_{E_d})$. If the T_i are differentiable at u, the $R_i \circ T_i$ also are, again according to Theorem 19.19 (a), and therefore their sum T also is. This shows the converse part of (a). □

Other formulation of Theorem 20.5 (b). If T is differentiable at u, then $(\mathrm{d}T(u))_i = \mathrm{d}T_i(u)$, that is $(\mathrm{d}T(u)z)_i = \mathrm{d}T_i(u)z$ for all $z \in E$. □

20.3. Differentiation of a composite multilinear mapping

Let us differentiate the composite with a mapping of several variables.

THEOREM 20.6.– *Let E, $F_1 \ldots, F_d$ and G be separated semi-normed spaces, two open subsets X of E and Y of $F_1 \times \cdots \times F_d$, two mappings $T = (T_1, \ldots, T_d)$ from X into Y and S from Y into G, and $u \in X$ such that:*

each T_i is differentiable at u,

S is differentiable at $T(u)$.

Then:

(a) *The mapping* $v \mapsto S(T_1(v), \ldots, T_d(v))$, *that is* $S \circ T$, *is differentiable at* u.

(b) $$\mathrm{d}(S(T_1, \ldots, T_d))(u) = \sum_{1 \le i \le d} \mathrm{d}_i S(T_1(u), \ldots, T_d(u)) \circ \mathrm{d}T_i(u). \blacksquare$$

Complement. In (b), $\mathrm{d}(S(T_1, \ldots, T_d))(u) \in \mathcal{L}(E; G)$ and, for every i, $\mathrm{d}T_i(u) \in \mathcal{L}(E; F_i)$ and $\mathrm{d}_i S(T_1(u), \ldots, T_d(u)) \in \mathcal{L}(F_i; G)$. □

Proof of Theorem 20.6. (a) This is a particular case of the chain rule theorem 19.17 (a) because T is differentiable at u (Theorem 20.5 (a)) since its components T_i are.

(b) The expressions of the differential of a composite mapping (Theorem 19.17 (b)), a multivariable mapping (Theorem 20.2 (b)) and a mapping having several components (Theorem 20.5 (b)) give, for every $z \in E$,

$$\mathrm{d}(S \circ T)(u)z = \mathrm{d}S(T(u))(\mathrm{d}T(u)z) = \sum_{1 \le i \le d} \mathrm{d}_i S(T(u))((\mathrm{d}T(u)z)_i) =$$
$$= \sum_{1 \le i \le d} \big(\mathrm{d}_i S(T(u)) \circ \mathrm{d}T_i(u)\big) z.$$

Hence, the equality stated. □

Let us show that if the intermediate spaces F_i are normed, $S(T_1, \ldots, T_d)$ is continuously differentiable when S and the T_i are.

THEOREM 20.7.– *Suppose that the assumptions of Theorem 20.6 are verified and:*

the F_i *are normed;*

S *and the* T_i *are continuously differentiable.*

Then:

$S(T_1, \ldots, T_d)$ *is continuously differentiable.* ■

Proof. This is a particular case of Theorem 19.18 with $T = (T_1, \ldots, T_d)$ and $F = F_1 \times \cdots \times F_d$, because T is continuously differentiable with values in F which is normed according to Theorem 6.6 (c). □

QUESTION: The case where the F_i are not normed. Can Theorem 20.7 be extended to any separated semi-normed spaces F_i? The author of these lines does not know! We are going to see (Theorem 20.8 (c)) that it can be extended to in the particular case in which S is multilinear. □

Let us differentiate the composite with a multilinear mapping.

THEOREM 20.8.– *Let E, $F_1 \dots$, F_d and G be separated semi-normed spaces, an open set X of E, for each $i \in [\![1, d]\!]$ a mapping T_i from X into F_i, and $u \in X$ such that:*

$$\textit{each } T_i \textit{ is differentiable at } u$$

and let

$$S \in \mathcal{L}^d(F_1 \times \cdots \times F_d; G).$$

Then, for all $z \in E$:

(a) $$S(T_1, \dots, T_d) \textit{ is differentiable at } u.$$

(b) $$\mathrm{d}(S(T_1, \dots, T_d))(u)z = \sum_{1 \le i \le d} S(T_1(u), \dots, \mathrm{d}T_i(u)z, \dots, T_d(u)).$$

(c) *The T_i are continuously differentiable* $\Rightarrow$
$$\Rightarrow \quad S(T_1, \dots, T_d) \textit{ is continuously differentiable.}$$ ▮

Proof. (a) and (b). This follows from Theorem 20.6 since, according to parts (a) and (c) of Theorem 20.4, any multilinear mapping S is differentiable and has partial differentials $\mathrm{d}_i S(v_1, \dots, v_d) y_i = S(v_1, \dots, y_i, \dots, v_d)$.

(c) If the dT_i are continuous, then $\mathrm{d}(S(T_1, \dots, T_d))$ also is because, following (b),

$$\mathrm{d}(S(T_1, \dots, T_d)) = \sum_{1 \le i \le d} S^i(T_1, \dots, \mathrm{d}T_i, \dots, T_d)$$

where $S^i(v_1, \dots, L_i, \dots, v_d)z = S(v_1, \dots, L_i z, \dots, v_d)$, and because every composite mapping $S^i(T_1, \dots, \mathrm{d}T_i, \dots, T_d)$ is continuous according to Theorem 7.9 (a), since $(T_1, \dots, \mathrm{d}T_i, \dots, T_d)$ is continuous from X into $F_1 \times \cdots \times \mathcal{L}(E; F_i) \times \cdots \times F_d$ and $S^i \in \mathcal{L}^d(F_1 \times \cdots \times \mathcal{L}(E; F_i) \times \cdots \times F_d;\ \mathcal{L}(E; G))$ according to Theorem 12.22. □

Other proof of Theorem 20.8 (c) when the F_i are normed. This is then a special case of Theorem 20.7 because S is continuously differentiable (Theorem 20.4 (a)). □

Chapter 21

Successive Differentiations

21.1. Successive differentiations

Let us define m times differentiable mappings.

DEFINITION 21.1.– *Let E and F be two separated semi-normed spaces, X an open subset of E, T a mapping from X into F and $m \in \mathbb{N}$.*

(a) *T is m* ***times differentiable at a point*** *u of X if it has $m-1$ successive differentials $\mathrm{d}T$, d^2T ..., $\mathrm{d}^{m-1}T$ in an open set containing u and if $\mathrm{d}^{m-1}T$ has a differential $\mathrm{d}^mT(u)$ at u.*

(b) *T is m* ***times differentiable*** *if it has m successive differentials $\mathrm{d}T$, $\mathrm{d}^2T\dots$, d^mT (these are mappings from X into $\mathcal{L}(E;F)$, $\mathcal{L}(E;\mathcal{L}(E;F))\dots$, respectively).*

(c) *T is m times* ***continuously differentiable*** *if it is m times differentiable and if it is continuous as well as its m differentials.*

(d) *T is* ***indefinitely differentiable*** *if it is m times differentiable for every $m \in \mathbb{N}$.*

(e) *T is* ***indefinitely continuously differentiable*** *if it is m times continuously differentiable for every $m \in \mathbb{N}$.* ■

The case $m = 0$. According to Definition 21.1, any mapping (from X into F) is 0 times differentiable and 0 time continuously differentiable if it is continuous. Moreover, we denote

$$\mathrm{d}^0T \stackrel{\mathrm{def}}{=} T.$$

This case is included in Definition 21.1 in order to be able to include in the same statement the case of a continuous mapping and that of a continuously differentiable mapping, which is used, for example, in the proof of Theorem 21.6 (b). □

Let us observe that a mapping is m times continuously differentiable as soon as it has an m-th continuous differential.

THEOREM 21.2.– *For every mapping T from an open subset of a separated semi-normed space into a separated semi-normed space and $1 \leq m < \infty$:*

(a) *T is m times continuously differentiable* $\Leftrightarrow$ $\begin{cases} T \text{ is } m \text{ times differentiable and} \\ \mathrm{d}^m T \text{ is continuous.} \end{cases}$

(b) *T is indefinitely continuously differentiable*

$\Leftrightarrow$ *T is indefinitely differentiable.* ▮

Proof. (a) If T is m times differentiable, its differentials $\mathrm{d}T\ldots$, $\mathrm{d}^{m-1}T$ are differentiable and therefore continuous (Theorem 19.6 (b)). If in addition $d^m T$ is continuous, it is therefore m times continuously differentiable, and conversely.

(b) If T is indefinitely differentiable, all its differentials are differentiable and therefore continuous, thus T is indefinitely continuously differentiable, and conversely. □

Let us observe that continuous linear mappings are indefinitely differentiable.

THEOREM 21.3.– *Let E and F be separated semi-normed spaces and*

$$L \in \mathcal{L}(E; F).$$

Then:

(a) $$L \text{ is indefinitely differentiable.}$$

(b) $$\mathrm{d}L(u) = L, \quad \forall u \in E.$$

(c) $$\mathrm{d}^m L = 0, \quad \forall m \geq 2. \ ▮$$

Proof. (b) It has already been seen in Theorem 19.8 that L is differentiable and that $\mathrm{d}L(u) = L$.

(c) *and* (a). The mapping $\mathrm{d}L$ being constant according to (b), its differential is zero (Theorem 19.7), and more specifically $\mathrm{d}^2 L(u) = 0_{\mathcal{L}(E;\mathcal{L}(E;F))}$. The following differentials are therefore, also all equal to zero. □

Let us show that multilinear mappings are indefinitely differentiable.

THEOREM 21.4.– *Let E_1, $E_2\ldots$, E_d and F be separated semi-normed spaces and*

$$T \in \mathcal{L}^d(E_1 \times \cdots \times E_d; F).$$

Then:

(a) T *is indefinitely differentiable.*

(b) $$\mathrm{d}^m T = 0, \quad \forall m \geq d+1. \blacksquare$$

Proof. Based on Theorem 20.4, T is differentiable and $\mathrm{d}T = \sum_{1 \leq i \leq d} S_i$, where

$$S_i \in \mathcal{L}^{d-1}(E_1 \times \cdots \times E_{i-1} \times E_{i+1} \times \cdots \times E_d; \mathcal{L}(E_i; F))$$

and $S_i(u_1, \ldots, u_{i-1}, u_{i+1}, \ldots, u_d)(z_i) = T(u_1, \ldots, u_{i-1}, z_i, u_{i+1}, \ldots, u_d)$.

By utilizing m times this result, $m \leq d$, it results in T being m times differentiable and $\mathrm{d}^m T$ being a sum of multilinear mappings of $d-m$ variables. In particular, $\mathrm{d}^{d-1}T$ is a sum of mappings with 1 variable, that is of linear mappings. Therefore (Theorem 21.3), $\mathrm{d}^{d-1}T$ is indefinitely differentiable and its differentials of second and more order are equal to zero, which implies (a) and (b). □

21.2. Schwarz's symmetry principle

Let us show that the second differential depends symmetrically on the first and second variations, which is **Schwarz's principle**[1].

THEOREM 21.5.– *Let E and F be two separated semi-normed spaces, X an open subset of E and T a mapping from X into F twice differentiable at a point u of X. Then, for all $y \in E$ and $z \in E$,*

$$(\mathrm{d}^2 T(u)y)z = (\mathrm{d}^2 T(u)z)y. \blacksquare$$

Proof. Let $\{\| \ \|_{E;\nu} : \nu \in \mathcal{N}_E\}$ and $\{\| \ \|_{F;\mu} : \mu \in \mathcal{N}_F\}$ be the families of semi-norms of E and F, and let V be an open set containing u in which T is once differentiable (it has been assumed as twice differentiable only at u).

An estimation. Let $p \in E$ and $q \in E$. We will conclude based on an estimation of

$$S(p) = T(u+p+q) - T(u+p) - T(u+q) + T(u) - (\mathrm{d}^2 T(u)q)p.$$

1 **History of Schwarz's principle (Theorem 21.5).** Karl WEIERSTRASS proved in his unpublished teachings of 1861 that the second partial derivatives of a function f may be exchanged at points x where they are continuous, that is $\partial^2 f/\partial x_1 \partial x_2(x) = \partial^2 f/\partial x_2 \partial x_1(x)$.
Hermann Amandus SCHWARZ proved in 1873 that this result is valid as soon as one of the two members is continuous.

The generalization for a non-normed space E is new. This generalization is new, as is all the extension of differential calculus to semi-normed spaces, see note 1, p. 275.

More specifically, we shall subtract the estimations with respect to $p = sy$, $q = sz$ and to $p = sz$ and $q = sy$, for s being small enough.

Let us start by estimating its differential. The differential of the linear mapping $p \mapsto (\mathrm{d}^2T(u)q)p$ being $\mathrm{d}^2T(u)q$ (Theorem 19.8 (b)) and the differential of a constant mapping being zero (Theorem 19.7), it follows that

$$\begin{aligned} \mathrm{d}S(p) &= \mathrm{d}T(u+p+q) - \mathrm{d}T(u+p) - \mathrm{d}^2T(u)q = \\ &= \big(\mathrm{d}T(u+p+q) - \mathrm{d}T(u) - \mathrm{d}^2T(u)(p+q)\big) - \\ &\qquad - \big(\mathrm{d}T(u+p) - \mathrm{d}T(u) - \mathrm{d}^2T(u)p\big) \end{aligned} \tag{21.1}$$

for all p and q such that $u+p$, $u+q$ and $u+p+q$ are elements in V.

Remember that $\mathrm{d}T$ is a mapping from V into $\mathcal{L}(E;F)$ that is endowed with (Definition 12.1 (a)) the semi-norms $\|L\|_{\mathcal{L}(E;F);B,\mu} = \sup_{u\in B} \|Lu\|_{F;\mu}$ indexed by the bounded sets B of E and by $\mu \in \mathcal{N}_F$. Denote

$$|||L|||_\mu = \|L\|_{\mathcal{L}(E;F);\{y,z\},\mu} = \sup\{\|Ly\|_{F;\mu}, \|Lz\|_{F;\mu}\}. \tag{21.2}$$

Let $\mu \in \mathcal{N}_F$ and $\epsilon > 0$. The differentiability of $\mathrm{d}T$ at u implies (Definition 19.1) the existence of a finite set $N \subset \mathcal{N}_E$ and of $\eta > 0$ such that $\sup_{\nu\in N} \|p\|_{E;\nu} \le \eta$ leads to

$$|||\mathrm{d}T(u+p) - \mathrm{d}T(u) - \mathrm{d}^2T(u)p|||_\mu \le \epsilon \sup_{\nu\in N} \|p\|_{E;\nu}.$$

If $\sup_{\nu\in N} \|p\|_{E;\nu} \le \eta/2$ and $\sup_{\nu\in N} \|q\|_{E;\nu} \le \eta/2$, this inequality is also verified by $p+q$ instead of p, so (21.1) gives

$$|||\mathrm{d}S(p)|||_\mu \le \epsilon \sup_{\nu\in N} \|p\|_{E;\nu} + \|p+q\|_{E;\nu}.$$

Hence, with (21.2)

$$\|\mathrm{d}S(p)y\|_{F;\mu} \le 2\epsilon \sup_{\nu\in N} \|p\|_{E;\nu} + \|q\|_{E;\nu}. \tag{21.3}$$

Let us choose now $p = sy$, $q = sz$, where $0 < s < \eta/(2\sup_{\nu\in N} \|y\|_{E;\nu} + \|z\|_{E;\nu})$. Since $S(0_E) = 0_F$, the mean value theorem 19.13 (a) gives

$$\|S(p)\|_{F;\mu} \le \sup_{0\le t\le 1} \|\mathrm{d}S(tp)p\|_{F;\mu} = s \sup_{0\le t\le 1} \|\mathrm{d}S(tp)y\|_{F;\mu}.$$

With (21.3), this yields, by denoting $S_q(p)$ instead of $S(p)$,

$$\|S_q(p)\|_{F;\mu} \le 2\epsilon \sup_{\nu\in N} \|p\|_{E;\nu} + \|q\|_{E;\nu}. \tag{21.4}$$

Conclusion. We have

$$(\mathrm{d}^2T(u)q)p - (\mathrm{d}^2T(u)p)q = S_q(p) - S_p(q).$$

Now by choosing $p = sz$, $q = sy$ and s as above, the inequality (21.4) is obtained for $S_p(q)$ instead of $S_q(p)$. Thus,

$$\|(\mathrm{d}^2T(u)q)p - (\mathrm{d}^2T(u)p)q\|_{F;\mu} \leq 4\epsilon \sup_{\nu\in N} \|p\|_{E;\nu} + \|q\|_{E;\nu}.$$

Hence, by dividing both members by s^2,

$$\|(\mathrm{d}^2T(u)z)y - (\mathrm{d}^2T(u)y)z\|_{F;\mu} \leq 4\epsilon \sup_{\nu\in N} \|y\|_{E;\nu} + \|z\|_{E;\nu}.$$

This is true for all $\epsilon > 0$, and therefore the first member is zero. This is true for all $\mu \in \mathcal{N}_F$, from which $\mathrm{d}^2T(u)z)y - (\mathrm{d}^2T(u)y)z = 0_F$ since F is separated. This establishes the equality stated. □

Bilinear mapping associated with a second differential. When it exists, $\mathrm{d}^2T(u)$ is an element of the space $\mathcal{L}(E; \mathcal{L}(E; F))$, and thus a bilinear mapping $[\mathrm{d}^2T(u)]$ is defined from $E \times E$ into F by

$$[\mathrm{d}^2T(u)](y, z) \stackrel{\text{def}}{=} (\mathrm{d}^2T(u)y)(z).$$

This mapping is symmetric, that is $[\mathrm{d}^2T(u)](y, z) = [\mathrm{d}^2T(u)](z, y)$, according to Theorem 21.5.

If E is metrizable, $[\mathrm{d}^2T(u)] \in \mathcal{L}^2(E \times E; F)$ following Theorem 12.21 (c) and the mapping $\mathrm{d}^2T(u) \mapsto [\mathrm{d}^2T(u)]$ is an isomorphism from $\mathcal{L}(E; \mathcal{L}(E; F))$ onto $\mathcal{L}^2(E \times E; F)$.

If E is normed, the mapping $\mathrm{d}^2T(u) \mapsto [\mathrm{d}^2T(u)]$ is an isometry (Theorem 12.21 (d)) for the norms of $\mathcal{L}$ and $\mathcal{L}^2$ defined in Theorems 12.10 (a) and 12.20 (a).

Caution with the identification of $\mathcal{L}(E; \mathcal{L}(E; F))$ to $\mathcal{L}^2(E \times E; F)$. When E is normed, $\mathrm{d}^2T(u)$ is frequently identified to $[\mathrm{d}^2T(u)]$ by identifying $\mathcal{L}(E; \mathcal{L}(E; F))$ to $\mathcal{L}^2(E \times E; F)$. This last identification is still possible when E is metrizable according to Theorem 12.21 (c), but **is it valid for any separated semi-normed space** E? The author of these lines does not know.

Furthermore, we should proceed with caution, as with any identification, see § 14.6, *Dangerous identifications*, p. 216. □

Multilinear mapping associated with an m-order differential. More generally, to $\mathrm{d}^mT(u)$, a mapping $[\mathrm{d}^mT(u)] \in \mathcal{L}^m(E \times \cdots \times E; F)$ can be associated. □

Commutation of the values of successive partial differentials. For any mapping T defined on an open subset X of a product $E_1 \times \cdots \times E_d$ of semi-normed spaces and which is twice differentiable at a point u of X, it follows that, for all $z_i \in E_i$ and $z_j \in E_j$,

$$(\mathrm{d}_i\mathrm{d}_jT(u)z_i)z_j = (\mathrm{d}_j\mathrm{d}_iT(u)z_j)z_i.$$

Indeed, by denoting $\widehat{z_i} = (0_{E_1}, \ldots, z_i, \ldots, 0_{E_d})$, we get (Theorem 20.2 (c)) $\mathrm{d}_iT(u)z_i = \mathrm{d}T(u)\widehat{z_i}$, hence

$$(\mathrm{d}_i\mathrm{d}_jT(u)z_i)z_j = (\mathrm{d}^2T(u)\widehat{z_i})\widehat{z_j} = (\mathrm{d}^2T(u)\widehat{z_j})\widehat{z_i} = (\mathrm{d}_j\mathrm{d}_iT(u)z_j)z_i. \ \square$$

CAUTION. If $E_i \neq E_j$,

$$\mathrm{d}_i\mathrm{d}_jT(u) \neq \mathrm{d}_j\mathrm{d}_iT(u)$$

because they do not occupy the same space, since $\mathrm{d}_i\mathrm{d}_jT(u) \in \mathcal{L}(E_i; \mathcal{L}(E_j; F))$ which is not preserved when exchanging i and j. □

21.3. Successive differentiations of a composite mapping

Let us show that the composite of m times differentiable mappings is m times differentiable when the intermediate space is normed.

THEOREM 21.6.– *Let E, F and G be three separated semi-normed spaces, two open sets X of E and Y of F, two mappings T from X into Y and S from Y into G, $u \in X$ and $1 \leq m \leq \infty$. If*

F is normed

then:

(a) $\left\{\begin{array}{l} T \textit{ is } m \textit{ times differentiable at } u \textit{ and} \\ S \textit{ is } m \textit{ times differentiable at } T(u) \end{array}\right\} \Rightarrow S \circ T$ *is m times differentiable at u.*

(b) *T and S are m times continuously differentiable*

$\Rightarrow$ *$S \circ T$ is m times continuously differentiable.* ▮

Proof of Theorem 21.6. (a) *Case* $m = 1$. This is the chain rule theorem 19.17.

Case $2 \leq m < \infty$. Let us reason by induction assuming (a) as true for $m - 1$. In addition, suppose that T is m times differentiable at u and S is m times differentiable at $T(u)$.

Since $m \geq 2$, T is differentiable on an open set V containing u and S is differentiable on an open set W containing $T(V)$. By choosing V small enough so that $T(V) \subset W$, the chain rule theorem 19.17 shows that $S \circ T$ is then differentiable in V, and for all $v \in V$,

$$\mathrm{d}(S \circ T)(v) = \mathrm{d}S(T(v)) \circ \mathrm{d}T(v). \tag{21.5}$$

Now, $\mathrm{d}T$ is $m - 1$ times differentiable at u (with values in $\mathcal{L}(E; F)$). Furthermore, $\mathrm{d}S$ is $m - 1$ times differentiable at $T(u)$ and therefore, according to the induction assumption, the composite mapping $v \mapsto \mathrm{d}S(T(v))$ is $m - 1$-time differentiable at u (with values in $\mathcal{L}(F; G)$). Thus,

$$v \mapsto (\mathrm{d}T(v), \mathrm{d}S(T(v))) \text{ is } m - 1 \text{ times differentiable at } u$$

(with values in $\mathcal{L}(E; F) \times \mathcal{L}(F; G)$). On the other hand, the mapping

$$\left\{\begin{array}{l} (L, M) \mapsto M \circ L \text{ is indefinitely differentiable} \\ \text{from } \mathcal{L}(E; F) \times \mathcal{L}(F; G) \text{ into } \mathcal{L}(E; G). \end{array}\right. \tag{21.6}$$

In effect, F being normed, it is continuous bilinear (Theorem 12.12 (f)) and thus (Theorem 21.4 (a)) indefinitely differentiable.

Therefore, according to the induction assumption, the composite mapping

$$v \mapsto \mathrm{d}S(T(v)) \circ \mathrm{d}T(v) \text{ is } m-1 \text{ times differentiable at } u$$

(with values in $\mathcal{L}(E;G)$). That is to say, according to (21.5), that $\mathrm{d}(S \circ T)$ is $m-1$-time differentiable at u and therefore $S \circ T$ is therein m times differentiable. Thereby, (a) is true for m. By induction, it is therefore true for all $m \in \mathbb{N}$.

Case $m = \infty$. It results from the case in which m is an integer, according to Definition 21.1 (d) of an indefinitely differentiable mapping.

(b) *Case* $m = 0$. This is Theorem 7.9 (a) of the continuity of a composite mapping.

Case $1 \leq m < \infty$. Let us reason by induction by assuming (b) as true for $m-1$. In addition, suppose that T and S are m times continuously differentiable. The equality (21.5) and the property (21.6) imply then, by proceeding in the same way as (a), that $\mathrm{d}(S \circ T)$ is $m-1$ times continuously differentiable. Therefore, $S \circ T$ is m times continuously differentiable and (b) is true for m. By induction, it is therefore true for all $m \in \mathbb{N}$.

Case $m = \infty$. It results, here again, from the case in which m is an integer. □

QUESTION: The case where F is not normed. Is the assumption "F is normed" necessary in Theorem 21.6, that is for the composite of m times differentiable mappings to be m times differentiable? The author of these lines does not know.

We utilize this assumption to establish Property (21.6), and more precisely the continuity of the composition of linear mappings (Theorem 12.12 (f)) from which it derives. It would thus be interesting that this latter be generalized. We know how to avoid it only when T is linear (Theorem 21.7, below) or when S is linear or multilinear (Theorem 21.8, hereafter). Could we also avoid it when $E = \mathbb{R}^d$? □

Let us observe that the composite of an m times differentiable mapping with a continuous linear mapping is m times differentiable, even when the intermediate space is not normed.

THEOREM 21.7.– *Let E, F and G be three separated semi-normed spaces, X be an open subset of E, T a mapping from X into F, $u \in X$ and $1 \leq m \leq \infty$ such that:*

$$T \text{ is } m \text{ times differentiable at } u$$

and let

$$L \in \mathcal{L}(F;G).$$

Then:

(a) $\qquad L \circ T$ *is m times differentiable at u.*

(b) $\qquad \mathrm{d}^m(L \circ T)(u) = L \circ \mathrm{d}^m T(u).$

(c) T *is m times continuously differentiable*

$\qquad \Rightarrow \quad L \circ T$ *is m times continuously differentiable.* ▮

Proof. (a) *and* (b). *Case* $m = 1$. This is Theorem 19.19.

Case $2 \leq m < \infty$. Let us reason by induction on m assuming (a) and (b) as true for $m - 1$. If T is m times differentiable, then $\mathrm{d}T$ is $m - 1$ times differentiable, and therefore, according to the induction assumption, $L \circ \mathrm{d}T$ is $m - 1$-time differentiable; this latter being equal to $\mathrm{d}(L \circ T)$, it follows thereof that $L \circ T$ is m times differentiable and, with the formula with respect to $m - 1$,

$$\mathrm{d}^m(L \circ T) = \mathrm{d}^{m-1}(\mathrm{d}(L \circ T)) = \mathrm{d}^{m-1}(L \circ \mathrm{d}T) = L \circ \mathrm{d}^{m-1}\mathrm{d}T = L \circ \mathrm{d}^m T.$$

Then, (a) and (b) are true for m. By induction, they are thus true for all m.

Case $m = \infty$. It follows from the case in which m is an integer, according to Definition 21.1 (d) of an indefinitely differentiable mapping.

(c) If $\mathrm{d}^k T$ is continuous for all $k \leq m$, the composite mapping $L \circ \mathrm{d}^k T$, namely $\mathrm{d}^k(L \circ T)$ according to (b), is also from Theorem 7.9 (a). □

Let us more generally show that the composite of m times differentiable mappings with a continuous multilinear mapping is m times differentiable, even when the intermediate space is not normed.

THEOREM 21.8.– *Let E, $F_1 \ldots$, F_d and G be separated semi-normed spaces, X an open subset of E, for each $i \in [\![1, d]\!]$ a mapping T_i from X into F_i, $1 \leq m \leq \infty$ and*

$$S \in \mathcal{L}^d(F_1 \times \cdots \times F_d; G).$$

Then:

(a) *The T_i are m times differentiable* $\Rightarrow$ *$S(T_1, \ldots, T_d)$ is m times differentiable.*

(b) *The T_i are m times continuously differentiable*
$\Rightarrow$ *$S(T_1, \ldots, T_d)$ is m times continuously differentiable.* ■

Proof of Theorem 21.8. (a) *Case* $m = 1$. This is Theorem 20.8 (a)

Case $2 \leq m < \infty$. Let us reason by induction on m assuming that the property is true for $m - 1$. Let us observe that (Theorem 20.8 (b))

$$\mathrm{d}(S(T_1, \ldots, T_d)) = \sum_{1 \leq i \leq d} S^i(T_1, \ldots, \mathrm{d}T_i, \ldots, T_d)$$

where $S^i(v_1, \ldots, L_i, \ldots, v_d)z = S(v_1, \ldots, L_i z, \ldots, v_d)$ and thus (Theorem 12.22)

$$S^i \in \mathcal{L}^d(F_1 \times \cdots \times \mathcal{L}(E; F_i) \times \cdots \times F_d; \mathcal{L}(E; G)).$$

If the T_i are m times differentiable from X into F_i, then the $\mathrm{d}T_i$ are $m-1$ times differentiable from X into $\mathcal{L}(E;F_i)$, and thus the induction assumption implies that each $S^i(T_1,\ldots,\mathrm{d}T_i,\ldots,T_d)$, and thus $\mathrm{d}(S(T_1,\ldots,T_d))$, is $m-1$ times differentiable from X into $\mathcal{L}(E;G)$. That is to say, $S(T_1,\ldots,T_d)$ is m times differentiable from X into G. The property then holds for m. By induction, it is consequently true for all $m\in\mathbb{N}$.

Case $m=\infty$. It results from the case in which m is an integer, according to Definition 21.1 (d) of an indefinitely differentiable mapping.

(b) *Case* $m=1$. This is Theorem 20.8 (c).

Case $2\le m\le\infty$. We revisit the proof by induction of (a) replacing *differentiable* by *continuously differentiable.* □

Other proof of Theorems 21.7 and 21.8 when the intermediate spaces F or F_i are normed. These are then special cases of Theorem 21.6 because any linear or multilinear mapping is indefinitely differentiable (Theorems 21.3 and 21.4). □

Chapter 22

Derivation of Functions of One Real Variable

22.1. Derivative of a function of one real variable

We reserve the term **function** to a mapping defined on a subset of $\mathbb{R}$ or of $\mathbb{R}^d$.

Let us define the derivative[1] of a function defined on an open subset of $\mathbb{R}$.

DEFINITION 22.1.– *Let f be a function from an open set I of $\mathbb{R}$ into a separated semi-normed space E. If f is differentiable at a point s, its* **derivative** *at s is the element of E, denoted $f'(s)$, defined by*

$$f'(s) \stackrel{\text{def}}{=} \mathrm{d}f(s)1. \blacksquare$$

Remember that $\mathrm{d}f(s)$ is a linear mapping, here from $\mathbb{R}$ into E, and that $\mathrm{d}f(s)1$ refers to its value at the point 1. Conversely, the differential is deduced from the derivative by: for all $t \in \mathbb{R}$,

$$\mathrm{d}f(s)t = tf'(s). \tag{22.1}$$

In effect, $\mathrm{d}f(s)t = \mathrm{d}f(s)(t1) = t(\mathrm{d}f(s)1) = tf'(s)$.

The derivative is often denoted $\mathrm{d}f(s)/\mathrm{d}s$ instead of $f'(s)$, especially when it is desirable to specify the variable with respect to the one being differentiated.

Advantage of the derivative. The derivative f' is a function from I into E, as is f, which is more pleasing than the differential $\mathrm{d}f$ (Definition 19.1), which is a function from I into $\mathcal{L}(\mathbb{R}; E)$. □

---

1 **History of the notion of derivative of a real function.** See note 1, p. 275.

History of the notations. The notation $f'x$ was introduced by Joseph Louis LAGRANGE in 1772 [65]. The notation $\mathrm{d}y/\mathrm{d}x$ was introduced by Gottfried VON LEIBNIZ in 1675 [67].

Let us define the convergence to a limit of a function of a real variable that will be used to characterize the derivative.

DEFINITION 22.2.– *Let $I \subset \mathbb{R}$, E a separated semi-normed space, f a function from I into E and $a \in \overline{I}$.*

*We say that $f(t)$ **has a limit** (or tends to) $\ell \in E$ when t tends to a, and we denote*

$$f(t) \to \ell \text{ when } t \to a, t \neq a,$$

if, for every semi-norm $\| \ \|_{E;\nu}$ of E and every $\epsilon > 0$, there exists $\eta > 0$ such that:

$$t \in I,\ 0 < |t-a| \leq \eta \quad \Rightarrow \quad \|f(t) - \ell\|_{E;\nu} \leq \epsilon. \ ∎$$

Equivalent formulation of the convergence to a limit. Let $\widetilde{f}$ be the function defined on $I \cup \{a\}$ by

$$\widetilde{f}(t) = f(t) \text{ if } t \in I, t \neq a, \quad \widetilde{f}(a) = \ell.$$

Then:

$$f(t) \to \ell \text{ when } t \to a, t \neq a \quad \Leftrightarrow \quad \widetilde{f} \text{ is continuous at the point } a.$$

(If $a \notin I$, then $\widetilde{f}$ is an extension of f; if $a \in I$ and $\ell = f(t)$, then $\widetilde{f} = f$; finally, if $a \in I$ and $\ell \neq f(t)$, then $\widetilde{f}$ is a modification of f.) □

Let us characterize differentiability.

THEOREM 22.3.– *Let I be an open subset of $\mathbb{R}$, E a separated semi-normed space, $\{\| \ \|_{E;\nu} : \nu \in \mathcal{N}_E\}$ its family of semi-norms, f a function from I into E and $s \in I$. Then:*

f is differentiable at s

$$\Leftrightarrow \quad \begin{cases} \text{there exists an element } f'(s) \in E \text{ such that, for every } \nu \in \mathcal{N}_E \text{ and } \epsilon > 0 \\ \text{there exists } \eta > 0 \text{ such that: } t \in \mathbb{R} \text{ and } |t| \leq r \text{ imply } s+t \in I \text{ and} \\ \qquad\qquad \|f(s+t) - f(s) - tf'(s)\|_{E;\nu} \leq \epsilon |t| \end{cases}$$

$$\Leftrightarrow \quad \frac{f(s+t) - f(s)}{t} \text{ has a limit } f'(s) \text{ in } E \text{ when } t \neq 0, s+t \in I, t \to 0. \ ∎$$

Proof. The first characterization is a special case of Definition 19.1 of differentiability since the mapping $t \mapsto tf'(s)$ is an element of $\mathcal{L}(\mathbb{R}; E)$ and conversely, since any mapping $\mathrm{d}f(s)$ of $\mathcal{L}(\mathbb{R}; E)$ is under this form, with $f'(s) = \mathrm{d}f(s)1$.

The second characterization is equivalent to the first following Definition 22.2 of a limit. □

Let us give the expression for the derivative of the composite of two functions of a real variable, which is the **chain rule theorem**.

THEOREM 22.4.– *Let I and J be open subsets of $\mathbb{R}$, E a separated semi-normed space, f a function from I into J differentiable at a point s, and g a function from J into E differentiable at $f(s)$. Then:*

$$(g \circ f)'(s) = g'(f(s))\, f'(s). \blacksquare$$

Clarification. Here, $g'(f(s))\, f'(s) \in E$ is the product of $g'(f(s)) \in E$ by $f'(s) \in \mathbb{R}$. □

Proof. Definition 22.1 of the derivative, the chain rule theorem 19.17 (b) and the expression (22.1), p. 313, of the differential give

$$(g \circ f)'(s) = \mathrm{d}(g \circ f)(s)1 = \mathrm{d}g(f(s))(\mathrm{d}f(s)1) = \mathrm{d}g(f(s))(f'(s)) = \\ = g'(f(s))\, f'(s). \square$$

Let us observe that the derivative commutes with linear mappings.

THEOREM 22.5.– *Let I be an open subset of $\mathbb{R}$, E and F two separated semi-normed spaces, f a function from I into E differentiable at a point s, and $L \in \mathcal{L}(E; F)$. Then:*

$$(Lf)'(s) = L(f'(s)). \blacksquare$$

Proof. Definition 22.1 of the derivative, the chain rule when one of the mappings is linear (Theorem 19.19 (b)) and again Definition 22.1 successively give

$$(Lf)'(s) = \mathrm{d}(Lf)(s)1 = L(\mathrm{d}f(s))1 = L(\mathrm{d}f(s)1) = L(f'(s)). \square$$

22.2. Derivative of a real function of one real variable

Let us show that if a real differentiable function takes the same value at the ends of an interval, its derivative vanishes in at least one intermediate point, which is **Rolle's theorem**[2].

THEOREM 22.6.– *Let f be a function from $[a, b]$ into $\mathbb{R}$, where $-\infty < a < b < \infty$, such that:*

$$f \textit{ is continuous on } [a, b] \textit{ and differentiable on } (a, b),$$

$$f(a) = f(b).$$

Then, there exists $c \in (a, b)$ such that:

$$f'(c) = 0. \blacksquare$$

2 **History of Rolle's theorem 22.6.** Michel ROLLE demonstrated this property for polynomials in 1691, without resorting to differentiability (notion then in gestation). Pierre Ossian BONNET extended it to differentiable functions in 1860.

Proof. The interval $[a, b]$ being compact in $\mathbb{R}$ (Borel–Lebesgue theorem 4.16 (a)), the continuous function f reaches its maximum at a point c of $[a, b]$ (Theorem 8.12). If $\max f > f(a)$, then $c \in (a, b)$ from which (Theorem 19.16) $\mathrm{d}f(c) = 0$. Furthermore, $f'(c) = \mathrm{d}f(c)1 = 0$.

Similarly, if $\min f < f(a)$, this minimum is reached at a point c, where $f'(c) = 0$. Otherwise, $\max f = \min f = f(a)$, and therefore f is constant and $f' = 0$. □

Let us give the **mean value theorem** for a real function defined on an interval.

THEOREM 22.7.– *Let f be a function from $[a, b]$ into $\mathbb{R}$, where $-\infty < a < b < \infty$, such that:*

$$f \text{ is continuous on } [a, b] \text{ and is differentiable on } (a, b).$$

Then:

(a) *There exists $c \in (a, b)$ such that* $\quad f(b) - f(a) = (b - a)f'(c)$.

(b) $$(b - a) \inf_{a<s<b} f'(s) \leq f(b) - f(a) \leq (b - a) \sup_{a<s<b} f'(s).$$

(c) $$|f(b) - f(a)| \leq (b - a) \sup_{a<s<b} |f'(s)|.$$

(d) *If f is defined and differentiable on an open subset of $\mathbb{R}$ containing $[a, b]$, then, for all $t \in [a, b]$,*

$$|f(b) - f(a) - (b - a)f'(t)| \leq (b - a) \sup_{a<s<b} |f'(s) - f'(t)|. \blacksquare$$

Proof. (a) This follows from Rolle's theorem 22.6 with respect to the function

$$g(s) = f(s) - \frac{f(b) - f(a)}{b - a}(s - a).$$

In effect, $g(a) = g(b)$ $(= f(a))$ and thus there exists $c \in (a, b)$ such that:

$$g'(c) = f'(c) - \frac{f(b) - f(a)}{b - a} = 0.$$

(b) These inequalities follow from (a). (They are also given by the mean value theorem 19.15 with $u = a$ and $z = b - a$ since $\mathrm{d}f(s)(b - a) = (b - a)f'(s)$.)

(c) This inequality results from (b). (It is also given by Theorem 19.13 (a).)

(d) This inequality is obtained by subtracting $(b-a)f'(t)$ from the inequality (b). (It is also given by Theorem 19.13 (b).) □

A real function f defined on a subset I of $\mathbb{R}$ is said to be **positive** and it is denoted $f \geq 0$ if $f(t) \geq 0$ for all $t \in I$. It is denoted $f \geq g$ if $f - g \geq 0$ and $f \leq g$ if $g - f \geq 0$. It is said to be **strictly positive** and it is denoted $f > 0$ if $f(t) > 0$ for all $t \in I$.

A real function f defined on an interval of $\mathbb{R}$ is said to be **increasing** if $s \geq t$ implies $f(s) \geq f(t)$, and **decreasing** if $s \geq t$ implies $f(s) \leq f(t)$. It is **strictly increasing** if $s > t$ implies $f(s) > f(t)$, and **strictly decreasing** if $s > t$ implies $f(s) < f(t)$.

A function is **monotonous** if it is increasing or decreasing. It is referred to as **strictly monotonous** if it is strictly increasing or strictly decreasing.

Let us relate the increasing or decreasing behavior of a differentiable real function to the sign of its derivative.

THEOREM 22.8.– *Let f be a differentiable function from (a,b) into $\mathbb{R}$, where $-\infty \leq a < b \leq \infty$. Then:*

(a) $\quad f' \geq 0 \quad \Leftrightarrow \quad$ *f is increasing.*

(b) $\quad f' > 0 \quad \Rightarrow \quad$ *f is strictly increasing.*

(c) $\quad f' \leq 0 \quad \Leftrightarrow \quad$ *f is decreasing.*

(d) $\quad f' < 0 \quad \Rightarrow \quad$ *f is strictly decreasing.*

(e) $\quad f' = 0 \quad \Leftrightarrow \quad$ *f is constant.* ▮

Proof. (a) If $f' \geq 0$, the mean value theorem 22.7 (b) gives $f(s) - f(t) \geq 0$ for all $s \geq t$, namely f is increasing. Conversely, if f is increasing, $(f(s+t) - f(s))/t \geq 0$ when this quotient is defined; thus, when $t \to 0$, its limit $f'(s)$ (Theorem 22.3) is ≥ 0.

(b) If $f' > 0$, then f' is strictly increasing, because otherwise there would be $s > t$ such that $f(s) \leq f(t)$ so according to the mean value theorem 22.7 (a), there would exist $c \in (a,b)$ such that $(s-t)f'(c) = f(s) - f(t)$ and thus $f'(c) \leq 0$.

(c) *and* (d). These properties result from (a) and (b) with respect to the function $-f$.

(e) This property follows from (a) and (c) because $f' = 0$ is equivalent to $f' \geq 0$ and $f' \leq 0$, and f being constant is equivalent to f being increasing and decreasing. □

Let us differentiate the inverse of a strictly increasing function having a non-zero derivative.

THEOREM 22.9.– *Let f be a function from $[a, b]$ into $\mathbb{R}$, where $-\infty < a < b < \infty$, and $s \in (a, b)$ such that:*

$$f \text{ is continuous and strictly monotonous,}$$

$$f \text{ is differentiable at } s \text{ and } f'(s) \neq 0.$$

Then:

$$f \text{ is invertible, } f^{-1} \text{ is continuous and strictly monotonous,}$$

$$f^{-1} \text{ is differentiable at } f(s) \text{ and its derivative is } \frac{1}{f'(s)}. \blacksquare$$

Proof. Assume that f is strictly increasing; then (Theorem 22.8 (b)), $f'(s) > 0$. Since f is continuous, the intermediate value theorem 8.13 shows that, for every $v \in [f(a), f(b)]$, there exists $t \in [a, b]$ such that $f(t) = v$. Since f is strictly increasing, t is unique. Therefore, f is invertible.

Consider $t \in (a, b)$, $t \neq s$, and denote $u = f(s)$. The inverse function f^{-1} is continuous because $t \to s$ when $v \to u$. In effect, for every $\epsilon > 0$, there is $|t - s| \leq \epsilon$ if $|v - u| \leq \eta$ where $\eta = \inf\{f(s + \epsilon) - f(s), f(s) - f(s - \epsilon)\}$ which is > 0 since f is strictly increasing. In addition,

$$\frac{f^{-1}(v) - f^{-1}(u)}{v - u} = \frac{t - s}{f(t) - f(s)} = \frac{1}{\dfrac{f(t) - f(s)}{t - s}} \tag{22.2}$$

and (Theorem 22.3) $(f(t) - f(s))/(t - s) \to f'(s)$. The left-hand side of (22.2) therefore tends to $1/f'(s)$ since $x \mapsto 1/x$ is continuous (Theorem 7.26 (b)) at any point > 0. According to the characterization of the differentiability of Theorem 22.3, this shows that the derivative of f^{-1} at $f(s)$ is $1/f'(s)$.

If f is decreasing, we conclude by considering the function $-f$. $\square$

Comparison to Theorem 19.22. The invertibility of a real function, contrarily to that of a mapping in Banach spaces, is obvious since it is equivalent to the strict monotony. The differentiability of the inverse function at $f(s)$ is also easier as soon as f is differentiable at s, without resorting to continuous differentiability of f in a neighborhood of s, as in Theorem 19.22.

Let us complete Theorem 19.22 by continuous differentiability.

THEOREM 22.10.– *If a real function f is continuously differentiable in an interval I of $\mathbb{R}$ and if its derivative does not cancel, then f^{-1} is continuously differentiable in the interval $f(I)$.* $\square$

Proof. If f' is continuous and does not cancel, its sign is constant according to the intermediate value theorem 8.13, and thus (Theorem 22.9) f is invertible, f^{-1} is differentiable in $f(I)$ and $(f^{-1})' = 1/(f' \circ f^{-1})$. Since f' is continuous as well as f^{-1} (because it is differentiable, see Theorem 19.6 (b)), their composite $f' \circ f^{-1}$ is continuous (Theorem 7.9 (a)) as well as (Theorem 7.26 (b)) its inverse $1/(f' \circ f^{-1})$, that is $(f^{-1})'$. This proves that f^{-1} is also continuously differentiable. $\square$

22.3. Leibniz formula

Let us calculate the derivative of the composite of functions with a mapping of several variables.

THEOREM 22.11.– *Let I be an open subset of $\mathbb{R}$, $F_1 \ldots, F_d$ and G be separated semi-normed spaces, Y an open subset of $F_1 \times \cdots \times F_d$, $f = (f_1, \ldots, f_d)$ a function from I into Y, and S a mapping from Y into G.*

If all the f_i are differentiable at a point s of I and if S is differentiable at $f(s)$, then $S(f_1, \ldots, f_d)$ is differentiable at s and

$$(S(f_1, \ldots, f_d))'(s) = \sum_{1 \leq i \leq d} \mathrm{d}_i S(f_1(s), \ldots, f_d(s))(f_i'(s)). \blacksquare$$

Clarification. In the above equality, $(S(f_1, \ldots, f_d))'(s) \in G$, $\mathrm{d}_i S(f_1(s), \ldots, f_d(s)) \in \mathcal{L}(F_i; G)$ and $f_i'(s) \in F_i$. □

Proof. Theorem 20.6 of the differentiation of the composite with a mapping of several variables gives the differentiability and, with Definition 22.1 of the derivative,

$$(S(f_1, \ldots, f_d))'(s) = \mathrm{d}(S(f_1, \ldots, f_d))(s)1 =$$

$$= \sum_{1 \leq i \leq d} \mathrm{d}_i S(f_1(s), \ldots, f_d(s))(\mathrm{d}f_i(s)1)$$

hence the stated equality since $\mathrm{d}f_i(s)1 = f_i'(s)$. □

Let us calculate the derivative of the composite of functions with a multilinear mapping, which is a generalization of the **Leibniz formula**.

THEOREM 22.12.– *Let I be an open subset of $\mathbb{R}$, $F_1 \ldots, F_d$ and G be separated semi-normed spaces, $f_1 \ldots, f_d$ functions from I respectively into $F_1 \ldots, F_d$, and*

$$S \in \mathcal{L}^d(F_1 \times \cdots \times F_d; G).$$

If all the f_i are differentiable at a point s of I, then $S(f_1, \ldots, f_d)$ is differentiable at s and

$$(S(f_1, \ldots, f_d))'(s) = \sum_{1 \leq i \leq d} S(f_1(s), \ldots, f_i'(s), \ldots, f_d(s)). \blacksquare$$

Proof. This is a special case of Theorem 22.11 since (Theorem 20.4 (a) and (c)) every continuous multilinear mapping S is differentiable and its partial differentials are equal to $\mathrm{d}_i S(v_1, \ldots, v_d) y_i = S(v_1, \ldots, y_i, \ldots, v_d)$. □

In particular, let us calculate the derivative of a product of real functions which is the **Leibniz formula**[3].

THEOREM 22.13.– *Let I be an open subset of $\mathbb{R}$ and f_1 and f_2 differentiable functions from I into $\mathbb{R}$. Then, $f_1 f_2$ is differentiable and*

$$(f_1 f_2)' = f_1' f_2 + f_1 f_2'. \blacksquare$$

Proof. This is a special case of Theorem 22.12 since the product is continuous bilinear from $\mathbb{R}^2$ into $\mathbb{R}$ (Theorem 7.24 (b)). □

22.4. Derivatives of the power, logarithm and exponential functions

Motivation. The following volumes calling on logarithms, exponential and their derivatives, it has seemed to us useful to recall a self-contained construction of them and to establish their properties that will be used. We have included them here so as to not have to come back to some properties of differentiability that their demonstrations make use of.

Derivatives of the power function. Let us show that the function $x \mapsto x^s$ is differentiable.

THEOREM 22.14.– *Let $s \in \mathbb{R}$. The function $x \mapsto x^s$ is indefinitely differentiable from $(0, \infty)$ into $\mathbb{R}$ and*

$$\frac{\mathrm{d}x^s}{\mathrm{d}x} = sx^{s-1}. \ \square$$

Proof. (b) We shall proceed by steps.

1st *The case where s is rational and positive.* For $m \in \mathbb{N}$, Leibnitz's formula (Theorem 22.12 with $f_i(x) = x$ and $S(v_1, \ldots, v_m) = v_1 \cdots v_m$) shows that the function $g_m\colon x \mapsto x^m$ is differentiable from $(0, \infty)$ into $\mathbb{R}$ and $\mathrm{d}x^m/\mathrm{d}x = \sum_{i=1}^m x^{m-1} = mx^{m-1}$.

For $n \in \mathbb{N}^*$, the inverse function $x \mapsto x^{1/n}$ is thus differentiable (Theorem 22.9) from $(0, \infty)$ into $\mathbb{R}$ and $\mathrm{d}x^{1/n}/\mathrm{d}x = 1/(\mathrm{d}x^n/\mathrm{d}x)(x^{1/n}) = 1/(n(x^{1/n})^{n-1})) = (1/n)x^{1/n-1}$.

For $s = m/n$, the composite function $x \mapsto x^s$ is differentiable (Theorem 22.4) from $(0, \infty)$ into $\mathbb{R}$ and $\mathrm{d}x^s/\mathrm{d}x = (\mathrm{d}x^{1/n}/\mathrm{d}x)(x^m)\,\mathrm{d}x^m/\mathrm{d}x = (1/n)(x^m)^{1/n-1} m x^{m-1} = sx^{s-1}$. Then, the mean value theorem 22.7 (c) yields

$$|y^s - x^s| \leq s\,|y - x| \sup\{x^{s-1}, y^{s-1}\}. \tag{22.3}$$

This inequality is true for all $s \in \mathbb{Q}$ because, if $s < 0$, it then results from $y^s - x^s = (y^{-s} - x^{-s})(xy)^s$. Theorem 22.7 (d) then gives, for any positive rational s, if $x > 0$ and $|t| \leq x/2$,

$$\begin{aligned} |(x+t)^s - x^s - sx^{s-1}t| &\leq |t| \sup_{\tau \leq |t|} |s(x+\tau)^{s-1} - sx^{s-1}| \leq \\ &\leq t^2 |s(s-1)|\, x^{s-2} \sup\{2^{s-2}, 2^{2-s}\}. \end{aligned} \tag{22.4}$$

On the other hand, (22.3) results in $|x^s - 1| = |x^s - 1^s| \leq |s|\,|x - 1| \sup\{x^{s-1}, 1\}$ whence

$$|x^s - x^r| = |x^r(x^{s-r} - 1)| \leq |s - r| \sup\{x^s, x^r\} \sup\{x, x^{-1}\} \tag{22.5}$$

---

3 **History of the Leibniz formula.** Gottfried VON LEIBNIZ gave in 1675 the expression of the derivative of a product, under the form $\mathrm{d}xv = x\,\mathrm{d}v + v\,\mathrm{d}x$ [67, p. 467].

if $r \leq s$. By interchanging r and s else, this inequality is obtained for all r and s in $\mathbb{Q}$.

2nd *The case where s is real and positive.* For all $s \in \mathbb{R}$, there exist (Theorem 1.5) rational numbers s_n and s'_n such that $s_n \leq s \leq s'_n$ and $|s'_n - s_n| \leq 1/n$. Since x^s increases with x if $x \geq 1$ and decreases if $0 < x \leq 1$, (22.5) yields

$$|x^s - x^{s_n}| \leq |x^{s'_n} - x^{s_n}| \leq \frac{c_n}{n} \sup\{x, x^{-1}\}$$

where $c_n = \sup\{x^{s'_n}, x^{s_n}\}$. If n is large enough, $s/2 \leq s_n \leq s'_n \leq 2s$, hence $c_n \leq \sup\{x^{2s}, x^{s/2}\}$. Thus, $x^{s_n} \to x^s$ as $n \to \infty$ and the inequality (22.4) with respect to s_n results in the one with respect to any positive real number s. For such an s, and the function $x \mapsto x^s$ is then differentiable and $\mathrm{d}x^s/\mathrm{d}x = sx^{s-1}$.

3rd *The case where s is negative.* There is $1/(x+t) - 1/x + t/x^2 = t^2/(x^2(x+t))$ if $x + t \neq 0$. If $|t| \leq x/2$, the right-hand side is majored by $2x^{-3}t^2$, and thus the function $x \mapsto x^{-1}$ is differentiable and $\mathrm{d}x^{-1}/\mathrm{d}x = -x^{-2}$. By composing with the function $s \mapsto x^{-s}$ which is differentiable since now $-s > 0$, we get (Theorem 22.4) the differentiability of the function $x \mapsto x^s$ and

$$\frac{\mathrm{d}x^s}{\mathrm{d}x} = \Big(\frac{\mathrm{d}x^{-1}}{\mathrm{d}x}\Big)(x^{-s})\,\frac{\mathrm{d}x^{-s}}{\mathrm{d}x} = (-(x^{-s})^{-2})(-sx^{-s-1}) = sx^{s-1}.$$

4th *Infinite differentiability.* By iterating (or by induction), it results that, for all $m \in \mathbb{N}$, the function $x \mapsto x^s$ is m times differentiable and $\mathrm{d}^m x^s/\mathrm{d}x^m = s(s-1)\dots(s-m)x^{s-m}$. □

Let us supplement this result by a majoration of the difference of powers of two numbers.

THEOREM 22.15.– *For all real numbers $x > 0$, $y > 0$, r and s:*

(a) $$|x^s - y^s| \leq |s||x-y| \sup\{x^{s-1}, y^{s-1}\}.$$

(b) $$|x^s - x^r| \leq |s-r| \sup\{x^s, x^r\} \sup\{x, x^{-1}\}.\ \square$$

Proof. It suffices to pass to the limit on inequalities (22.3) and (22.5) with respect to rational approximations of r and s since then $x^{s_n} \to x^s$ as seen in the 2nd part of the proof of Theorem 22.14 for $s > 0$. For $s < 0$, this convergence also holds since the function $x \mapsto 1/x$ is continuous (Theorem 7.26 (b). □

Construction of logarithms. Let us first give an inequality that will be used to define logarithms.

LEMMA 22.16.– *For all real numbers $x > 0$, $t_1 \neq 0$ and $t_2 \neq 0$:*

$$t_1 \leq t_2 \quad \Rightarrow \quad \frac{x^{t_1} - 1}{t_1} \leq \frac{x^{t_2} - 1}{t_2}.\ \square$$

Outlines of the proof. From Bernoulli's inequality[4] $(1+y)^n \geq 1 + ny$ for all $y \in \mathbb{R}$, $y > -1$ and $n \in \mathbb{N}$, we deduce that $n(z^{1/n} - 1) \leq (n-1)(z^{1/(n-1)} - 1)$ for all $z \in \mathbb{R}$, $z > 0$ and therefore that $n(z^{1/n} - 1) \leq m(z^{1/m} - 1)$ for any integer $m \leq n$. This results in the expected inequality if $t_1/t_2 = n/m$, and thus for all rational numbers such that $0 < t_1 \leq t_2$. We deduce herefrom the inequality for all rational numbers t_1 and t_2 whatever be their sign and then, by continuity, for real numbers t_1 and t_2. Detailed proofs may be found in [PISOT–ZAMANSKY, 88, p. 308–310] or [SIMON, 107]. □

The **logarithm**[5] of a real number $x > 0$ is defined by

$$\log x \stackrel{\text{def}}{=} \lim_{n \in \mathbb{N},\, n \to \infty} n(x^{1/n} - 1). \tag{22.6}$$

4 **History of Bernoulli's inequality.** The inequality $(1+y)^n \geq 1 + ny$ is due to Jacob BERNOULLI.

5 **History of logarithms.** Logarithms were independently defined by John NEPER in 1614 [75] and Jost BÜRGI in 1620 (work achieved in the early years of the 17th Century). The use of the formula $\log x = \int_1^x \mathrm{d}y/y$ to define them is due to Nicolas MERCATOR [72], in 1668. This one, or more exactly the equivalent formula $\int_1^{x^n} \mathrm{d}y/y = n \int_1^x \mathrm{d}y/y$, was established in 1644 by Father GRÉGOIRE DE SAINT-VINCENT [46].

This limit exists (Theorem 1.3) since the sequence $(n(x^{1/n} - 1))_{n\in\mathbb{N}}$ is decreasing (Lemma 22.16) and minored by $1 - x^{-1}$ (Lemma 22.16 with $t_1 = -1$).

Let us give approximations of the logarithms.

THEOREM 22.17.– *For all real numbers $x > 0$ and $t > 0$:*

(a)
$$\frac{1 - x^{-t}}{t} \leq \log x \leq \frac{x^t - 1}{t}.$$

(b)
$$0 < x \leq 1 \quad \Rightarrow \quad |\log x| \leq \frac{x^{-t}}{t}.$$

(c)
$$t \to 0 \quad \Rightarrow \quad \frac{x^t - 1}{t} \to \log x. \ \square$$

Proof. (a) Let $n \geq 1/t$. Then, $(1 - x^{-t})/t \leq n(x^{1/n} - 1) \leq (x^t - 1)/t$ (by Lemma 22.16 with $-t \leq 1/n \leq t$), from which the stated inequality is deduced by passing to the limit when $n \to \infty$.

(b) If $0 < x \leq 1$, then $x^t \leq 1$ and thus (a) yields $-x^{-t}/t \leq \log x \leq 0$, hence the inequality stated.

(c) Following (a), $|\log x - (x^t - 1)/t| \leq (x^t - 1)/t - (1 - x^{-t})/t = (x^t - 1)(1 - x^{-t})/t$. Hence, the stated convergence because $x^t \to 1$ when $t \to 0$ (Theorem 22.15 (b)) and $(1 - x^{-t})/t$ remains bounded, and more precisely $1 - x^{-1} \leq (x^{-t} - 1)/(-t) \leq x - 1$ (Lemma 22.16 with $-1 \leq t \leq 1$) when $t \leq 1$. $\square$

Now, we recall specific properties of logarithms.

THEOREM 22.18.– *For all real numbers $x > 0$, $y > 0$ and s:*

$$\log(xy) = \log x + \log y, \quad \log x^s = s \log x, \quad x^{\log y} = y^{\log x}. \ \square$$

Proof. *First equality.* It is obtained by passing to the limit when $t \to 0$, $t > 0$, in the easy equality $((xy)^t - 1)/t = (x^t - 1)/t + (y^t - 1)/t + t((x^t - 1)/t)((y^t - 1/t))$ since (Theorem 22.17 (c)) $(x^t - 1)/t \to \log x$.

Second equality. If $s \neq 0$, it is obtained by passing to the limit when $t \to 0$, $t \neq 0$ and thus $st \to 0$, in the equality $((x^s)^t - 1)/t = s(x^{st} - 1)/st$. Otherwise, $\log x^0 = \log 1 = 0$ by Definition (22.6).

Third equality. The second equality implies $\log x^{\log y} = \log x \log y = \log y^{\log x}$. It follows thereof that $x^{\log y} = y^{\log x}$ since the log function is a strictly increasing. $\square$

Derivative of the logarithm function. Let us show that the logarithm function is differentiable.

THEOREM 22.19.– *The* log *function is indefinitely differentiable and strictly increasing from* $(0, \infty)$ *into* $\mathbb{R}$ *and*

$$\frac{\mathrm{d} \log x}{\mathrm{d}x} = \frac{1}{x}. \ \square$$

Proof. If $0 < s < 1$, $x > 0$ and $x + t > 0$, the first inequality of (22.4) yields, once divided by $|st|$,

$$\left| \frac{1}{t} \left(\frac{(x+t)^s - 1}{s} - \frac{x^s - 1}{s} \right) - x^{s-1} \right| \leq |(x+t)^{s-1} - x^{s-1}|.$$

When $s \to 0$, it results, at the limit (Theorems 22.17 (c) and 22.15 (b)),

$$\left| \frac{\log(x+t) - \log x}{t} - \frac{1}{x} \right| \leq |(x+t)^{-1} - x^{-1}|.$$

When $t \to 0$, the right-hand side tends to 0 (Theorem 22.15 (a) with $s = -1$); therefore (Theorem 22.3), the log function is differentiable at x and $\mathrm{d}\log x/\mathrm{d}x = 1/x$. It is indefinitely differentiable since its derivative $x \mapsto x^{-1}$ is (Theorem 22.14), and it is strictly increasing since its derivative is strictly positive (Theorem 22.8 (b)). $\square$

Construction of the exponential. There exists a unique real number e such that[6]:

$$\log \mathrm{e} \stackrel{\text{def}}{=} 1. \tag{22.7}$$

This results from the intermediate value theorem 8.13 because the log function is strictly increasing and differentiable (Theorem 22.19) and thus continuous (Theorem 19.6 (b)) and because $\log 1 = 0$ (by Definition (22.6)) and $\log 4 \geq 1$ (Theorem 22.17 (a) with $t = 1/2$).

The exponential is the inverse function of the logarithm since, according to the second and third equalities of Theorem 22.18,

$$\log \mathrm{e}^s = s, \quad \mathrm{e}^{\log x} = x. \tag{22.8}$$

Note that, for all real numbers $x > 0$ and $s > 0$:

$$0 < \mathrm{e}^{-x} \leq s^s x^{-s}. \tag{22.9}$$

Indeed, Theorem 22.17 (b) with $t = 1/s$ gives $|\log y| \leq s y^{-1/s}$ if $0 < y \leq 1$. By dividing by s and raising to the power $-s$, it follows that $y \leq s^s |\log y|^{-s}$. For $y = \mathrm{e}^{-x}$, this yields $\mathrm{e}^{-x} \leq s^s x^{-s}$ according to (22.8).

Derivative of the exponential function. Let us show that the function $x \mapsto x^s$ is differentiable.

THEOREM 22.20.– *Let $x \in \mathbb{R}$, $x > 0$. The function $s \mapsto x^s$ is indefinitely differentiable from $\mathbb{R}$ into itself and*

$$\frac{\mathrm{d}x^s}{\mathrm{d}s} = x^s \log x, \quad \frac{\mathrm{d}\mathrm{e}^s}{\mathrm{d}s} = \mathrm{e}^s. \ \square$$

Proof. According to Theorem 22.17 (c), $(x^{s+t} - x^s)/t = x^s(x^t - 1)/t \to x^s \log x$ when $t \to 0$, $t > 0$. The function $s \mapsto x^s$ is thus (Theorem 22.3) differentiable and $\mathrm{d}x^s/\mathrm{d}s = x^s \log x$. The equality $\mathrm{d}\mathrm{e}^s/\mathrm{d}s = \mathrm{e}^s$ follows since $\log \mathrm{e} = 1$ by Definition (22.7). By iterating this result, we get $\mathrm{d}^m \mathrm{e}^s/\mathrm{d}s^m = \mathrm{e}^s$ for all $m \in \mathbb{N}$. The function $s \mapsto x^s$ being its composite by the linear function $s \mapsto s \log x$ since $x^s = (\mathrm{e}^{\log x})^s = \mathrm{e}^{s \log x}$ according to (22.8) and (1.9), it is also m times differentiable (Theorem 21.6 (a)). $\square$

6 **History of the number** e. Gottfried VON LEIBNIZ identified in a letter to Christiaan HUYGENS the number, which he denoted b, such that $\log \mathrm{e} = 1$. Leonhard EULER introduced the notation e to designate it in a letter from 1731 to Christian GOLDBACH.

Bibliography

[1] ACKERMANN, W., *Die Widerspruchsfreiheit der allgemeinen Mengenlehre*, Springer, 1936.

[2] ALAOGLU, L., Weak topologies of normed linear spaces, *Ann. of Math.*, (2) 41 (1940), 252–267.

[3] ALEKSANDROV, P. S. and URYSOHN, P. S., Zur Theorie der topologischen Raümen, *Math. Ann.*, 92 (1924), 258–266.

[4] ARZELÀ, C., Un' osservazione intorno alle serie di funzioni, *Rend. Accad. R. Sci. Ist. Bologna* (1882–1883), 142–159.

[5] ASCOLI, G., Le curve limiti di una varietà data di curve. *Atti R. Accad. Lincei, Memorie della Cl. Sci. Fis. Mat. Nat.* (3) 18 (1883–1884), 521–586.

[6] BAIRE, R., Sur les fonctions de variables réelles, *Annali Mat. pura* e *Appl.* (3) 3 (1899), 1–122.

[7] BANACH, S., Sur les opérations dans les ensembles abstraits et leur application aux équations intégrales, *Fund. Mat.* 3 (1922), 133–181.

[8] BANACH, S., Sur les fonctionnelles linéaires, *Studia Mat.* 1 (1929), 211–216 et 223–239.

[9] BANACH, S., Über metrische Gruppen, *Studia Mat.* 3 (1931), 101–113.

[10] BANACH, S., *Théorie des opérations linéaires*, Monografje Matematyczne, Varsovie, 1932.

[11] BANACH, S. and STEINHAUS, H., Sur le principe de condensation de singularités, *Fund. Math.* 9 (1927), 50–61.

[12] BERGH, J. and LÖFSTRÖM, J., *Interpolation spaces: an introduction*, Springer, 1976.

[13] BOLZANO, B., Rein Analytischer Beweis der Lehrsatzes, daß zwischen je zwei Werthen, die ein entgegengesetzes Resultat gewähren, wenigstens eine reelle Wurzel der Gleichung liege, *Ostwald's Klassiker*, 153, Leipzig, 1905.

[14] BOURBAKI, N., *Théorie des ensembles, fascicule de résultats*, Hermann, 1939.

[15] BOURBAKI, N., Sur certains espaces vectoriels topologiques, *Ann. Inst. Fourier*, 2 (1950), 5–16.

[16] BOURBAKI, N., *Théorie des ensembles*, Hermann, 1954.

[17] BOURBAKI, N., *Espaces vectoriels topologiques*, Hermann, 1967.

[18] BOURBAKI, N., *Algèbre*, Hermann, 1969.

[19] BOURBAKI, N., *Topologie générale*, Hermann, 1971.

[20] BOURBAKI, N., *Éléments d'histoire des mathématiques*, Hermann, 1974.

[21] BUNIAKOWSKY, V., Sur quelques inégalités concernant les intégrales ordinaires et les intégrales aux différences finies. *Mém. Acad. Saint-Pétersbourg*, (7) 1, no. 9 (1859).

[22] CAJORI, F., *A History of Mathematical Notations*, 2 vol., Open Court (Chicago), 1974.

[23] CANTOR, G., Über eine elementare Frage der Mannigfaltigskeitslehre, *Jahresericht der Deutsch. Math. Vereing.*, vol. I (1890-1891), 75–78

[24] CANTOR, G., *Gesammelte Abhandlungen*, Springer, 1932.

[25] CAUCHY, A., *Cours d'Analyse de l'École Royale Polytechnique*, Debure (Paris), 1821 (= Œuvres *complètes*, vol. III, Gauthiers-Villars, 1897).

[26] CAUCHY, A., *Résumé des leçons données à l'École Royale Polytechnique sur le calcul infinitésimal*, Debure (Paris), 1823 (= *Œuvres complètes*, vol. IV, Gauthiers-Villars, 1899).

[27] CHUQUET, N., Le Triparty en la science des nombres, A. Marre, *Bull. Bibl. Storia Math.*, 13 (1880), 555–659 et 693–814.

[28] CLARKSON, J. A., Uniformly convex spaces, *Trans. Amer. Math. Soc.*, 40 (1936), 396–414.

[29] CURTZE, M., Über die Handschrift "Algorismes proportionum magistri Nicolay Orem ". *Zeistchr. für Math. und Phys.*, 13, Supplem. (1868), 65–79 et 101–104.

[30] DAY, M. M., Reflexive Banach spaces not isomorphic to uniformly convex spaces, *Bull. Amer. Math. Soc.*, 47 (1941), 313–317.

[31] DEDEKIND, R., *Was sind und was sollen die Zahlen*, opuscule, 1888 (included in [32]).

[32] DEDEKIND, R., *Gesammelte mathematische Werke*, 3 vol., Braunschweig, 1932.

[33] DESCARTES, R., *Discours de la méthode*, Leyde, 1637.

[34] DIEUDONNÉ, J., La dualité dans les espaces vectoriels topologiques. *Ann. Sci. École Norm. Sup.*, 59, 1942, 107–139.

[35] DIEUDONNÉ, J. and SCHWARTZ, L., La dualité dans les espaces $(\mathcal{F})$ et $(\mathcal{LF})$. *Ann. Inst. Fourier*, 1, 1949, 61–101.

[36] DONOGHUE, W. F. and SMITH, K. T., On the symmetry and bounded closure of locally convex spaces, *Trans. Amer. Math. Soc.* 73 (1952), 321–344.

[37] DUNFORD, N., Uniformity in linear spaces, *Trans. Amer. Math. Soc.* 44 (1938), 305–356.

[38] EDWARDS, R., *Functional Analysis*, Winston, 1965.

[39] ENDERTON, H. B., *Elements of Set Theory*, Elsevier, 1977.

[40] EUCLID, *Elementorum Libri* XV, Apud Hieronymum de Marnef & Gulilmum Canellat, Paris, 1573 (= *Les œuvres d'Euclide*, trad. F. Peyrard (1819), A. Blanchard, Paris, 1966).

[41] EULER, L., Principes généraux du mouvement des fluides, *Mémoires de l'Académie des Sciences de Berlin*, 11, 1755 (= *Opera omnia, Series 2*, XII, Teubner et O. Füssli, 1954, 59–91).

[42] FISCHER, E., Applications d'un théorème sur la convergence en moyenne, *C. R. Acad. Sci. Paris*, 144 (1907), 1148–1150.

[43] FRÉCHET, M., Sur quelques points du calcul fonctionnel (thèse), *Rend. Palermo*, 22 (1906), 1–74.

[44] FRÉCHET, M., Sur les ensembles de fonctions et les opérations linéaires, *C. R. Acad. Sci. Paris*, 144 (1907), 1414–1416.

[45] FRÉCHET, M., Les ensembles abstraits et le calcul fonctionnel, *Rend. Palermo*, 30 (1910), 1–26.

[46] GRÉGOIRE DE SAINT-VINCENT, P. (P. GERGORII A SANCTO VICENTIO), *Opus geometricum quadraturæ circuli et sectionum coni*, 2 vol., Antverpiæ, 1647.

[47] GOLDSTINE, H. H., Weakly complete Banach spaces, *Duke Math. J.*, 4 (1938), 125–131.

[48] GRASSMANN, H., *Die Ausdehnungslehre*, 1844.

[49] HADAMARD, J., Sur les opérations fonctionnelles, *C. R. Acad. Sci. Paris*, 136 (1903), 351–354.

[50] HAHN, H., Über Folgen linearer Operationen, *Monatsh. für Math. und Physik*, 32 (1922), 3–88.

[51] HAHN, H., Über lineare Gleichungen in linearen Räumen, *J. Reine Angew. Math. (J. de Crelle)*, 157 (1927), 214–229.

[52] HARRIOT, T., *Artis Analyticae Praxis ad Aequationes Algebraicas Resolvendas*, 1631.

[53] HAUSDORFF, F., *Grunfzüge der Mengenlehre*, Leipzig, 1st ed., 1914.

[54] HAUSDORFF, F., Zur Theorie der linearen metrischen Raüme, *J. Reine Angew. Math. (J. de Crelle)*, 167 (1932), 294–311.

[55] HEINE, E., Über trigonometrische Reihen, *J. Reine Angew. Math. (J. de Crelle)*, 71 (1870), 353–365.

[56] HEINE, E., Die Elemente der Functionenlehre, *J. Reine Angew. Math. (J. de Crelle)*, 74 (1872), 172–188.

[57] HILBERT, D., *Grundzüge einer allgemeinen Theorie des linearen Integralgleichungen*, Teubner (1912) (= *Gött. Nachr.*, 1904, 1905, 1906, 1910).

[58] HILDEBRANDT, T. H., On uniform limitedness of sets of functional operations. *Bull. Amer. Math. Soc.*, 29 (1923), 309–315.

[59] HORVÁTH, J., *Topological Vector Spaces and Distributions*, Addison-Wesley, 1966.

[60] JAMES, R. C., A non reflexive Banach space isometric with its second conjugate space, *Proc. Nat. Acad. Sc. USA*, 37 (1951), 147–177.

[61] JORDAN, P. and NEUMANN, J. VON, On inner products in linear metric spaces, *Ann. of Math.*, 36 (1935), 719–723.

[62] KAKUTANI, S., Weak topology and regularity of Banach spaces, *Proc. Imper. Acad. Tokyo*, 15 (1939), 63–67.

[63] KOLMOGOROV, A. N., Zur Normierbarkeit eines allgemeinen topologischen linearen Raumes, *Studia Math.*, 5 (1934), 29–33 (en anglais : *Selected works*, vol. 1, Dordrecht, Kluwer (1991), 183–186.

[64] KÖTHE, G., *Topological Vector Spaces I*, Springer–Verlag, 1969.

[65] LAGRANGE, J., *Sur une nouvelle espèce de calcul relatif à la différentiation et à l'intégration des quantités variables*, 1772.

[66] LANG, S., *Real Analysis*, Addison-Wesley, 1969.

[67] LEIBNIZ, W. G., Nova metodus pro maximis et minimis, itemque tangentibus, quae nec fractas nec irrationales quantitates moratur, et singulare pro illis calculi genus, *Acta Eruditorum*, 1884, 467–473 (= *Mathematische Schriften*, vol I, 220–226, Olms (Hildesheim), 1960.

[68] LIONS, J.-L. and PEETRE, J., Sur une classe d'espaces d'interpolation, *Publications Mathématiques de l'IHÉS*, 19 (1964), 5–68.

[69] MACKEY, G. W., On convex topological linear spaces, *Trans. Amer. Math. Soc.*, 59, 1946, 519–537.

[70] MAZUR, S., Über konvexe Mengen in linearen normierte Räumen, *Studia Math.*, 4 (1933), 70–84.

[71] MAZUR, S. and ORLICZ, W., Über Folgen linearer Operationen, *Studia Math.*, 4 (1933), 152–157.

[72] MERCATOR, N., *Logarithmotechnia... cui nunc accedit vera quadratura hyperbolæ...*, Londini, 1668.

[73] MILMAN, D. P., On some criteria for regularity of spaces of the type (B), *Doklady Akad. Nauk SSSR (N. S.)*, 20 (1938), 234.

[74] MINKOWSKI, H., *Geometrie der Zahlen, I*, 1896 (2nd ed., Teubner, 1910).

[75] NEPER, J., *Mirifici logarithmorum canonis descriptio*, Lyon, 1619.

[76] NEUMANN, J. VON, Allgemeine Eigenwerttheorie Hermitescher Funktionaloperationen, *Math. Ann.*, 102 (1929), 49–131.

[77] NEUMANN, J. VON, Zur Algebra der Funktionaloperationen und Theorie der Normalen Operatoren, *Math. Ann.*, 102 (1930), 370–427.

[78] NEUMANN, J. VON, On complete topological linear spaces, *Trans. Amer. Math. Soc.*, 37 (1935), 1–20.

[79] NEWTON, I., *The mathematical papers, vol. III, 1670–1673*, Cambridge University Press, 1969.

[80] OSGOOD, W. F., Non uniform convergence and the integration of series term by term, *Amer. J. of Math.*, 19 (1897), 155–190.

[81] OUGHTRED, W., *Clavis Mathematicae*, Smith, London, 1631.

[82] PEANO, G., Intégration par séries des équations différentielles linéaires, *Math. Ann.*, 32 (1888), 450–456.

[83] PEANO, G., *Calcolo geometrico secondo l'Ausdehnungslehre di Grassmann, proceduto dalle operazioni della logica deduttiva*, Turin, 1889.

[84] PEANO, G., *Arithmetices principia, nova methodo exposita*, Turin, 1889.

[85] PEANO, G., *Formulario mathematico*, vol. 5, Turin, 1895–1905.

[86] PETTIS, B. J., A proof that every uniformly convex space is reflexive, *Duke Math. J.*, 5 (1939), 249–253.

[87] PICARD, E., Mémoire sur la théorie des équations aux dérivées partielles et la méthode des approximations successives, *J. Math.*, (4) 6 (1890), 145–210.

[88] PISOT, G. AND ZAMANSKY, M., *Mathématiques générales. Algèbre–Analyse*, Dunod, Paris, 1959.

[89] RECORDE, R., *The Whetstone of Witte, which is the second part of Arithmetike, containing the Extraction of Rootes, the Cossike Practice, with the Rules of Equation, and the Woorkes of Surde Numbers*, Londres, 1557.

[90] RIESZ, F., Sur les systèmes orthogonaux de fonctions, *C. R. Acad. Sci. Paris, sér. I*, 144 (1907), 615–619.

[91] RIESZ, F., Untersuchungen über Systeme integrierbarer Funktionen, *Math. Ann.*, 69 (1910), 449–497.

[92] RIESZ, F., Sur certains systèmes singuliers d'équations intégrales, *Ann. Éc. Norm. Sup.*, (3), 28 (1911), 33–62.

[93] RIESZ, F., Über lineare Funktionalgleichungen, *Acta Mat.*, 41 (1918), 71–98.

[94] RIESZ, F., Zur Theorie des Hilbertschen Raumes, *Acta Sci. Math. Szeged*, 7 (1934), 34–38.

[95] RUSSEL, B., *Principles of Mathematics*, 1903

[96] SCHAEFER, H. H., *Topological vector spaces*, Macmillan, 1966.

[97] SCHAUDER, J., Zur Theorie stetiger Abbildungen in Funktionalräumen, *Math. Z.*, 26 (1927), 47–65 and 417–431.

[98] SCHMIDT, E., Zur Theorie der linearen und nichtlinearen Integralgleichungen. I. Teil : Entwickelung willkürlicher Funktionen nach Systemen vorgeschriebener, *Mat. Ann.*, 63 (1907), 433–476.

[99] SCHWARTZ, L., *Analyse, Topologie générale et analyse fonctionnelle*, Hermann, 1970.

[100] SCHWARTZ, L., *Analyse I. Théorie des ensembles et topologie*, Hermann, 1991.

[101] SCHWARZ, H. A., *Gesammelte Mathematische Abhandlungen*, Band I. J. Springer, 1890.

[102] SIMON, J., *Analysis for PDEs, II — Real or Neumann space-valued distributions*, ISTE Ltd, London and John Wiley & Sons, New York, forthcoming.

[103] SIMON, J., *Analysis for PDEs, III — Real or Neumann space-valued Lebesgue and Sobolev spaces*, ISTE Ltd, London and John Wiley & Sons, New York, forthcoming.

[104] SIMON, J., *Analysis for PDEs, IV — Traces on a differentiable or Lipschitz boundary*, ISTE Ltd, London and John Wiley & Sons, New York, forthcoming.

[105] SIMON, J., *Analysis for PDEs, V — Partial differential equations*, ISTE Ltd, London and John Wiley & Sons, New York, forthcoming.

[106] SIMON, J., On the identification $H = H'$ in the Lions theorem and a related error. *Ricerche Mat.*, 59 (2) (2010), 245–255.

[107] SIMON, J., Construction de $\mathbb{R}$ par coupures, `jsimon@unice.fr`, (2017).

[108] ŠMULIAN, V. L., Über lineare topologische Raüme. *Mat. Sbornik NS*, 7 (49) (1940), 425-448.

[109] STIFEL, M., *Arithmetica integra*, Nüremberg, 1544.

[110] TYCHONOFF, A. N., Über die topologische Erweiterung von Raümen, *Math. Ann.*, 102 (1930), 544–561.

[111] TYCHONOFF, A. N., Ein Fixpunktsatz, *Math. Ann.*, 111 (1935), 767–776.

[112] VAN DANTZIG, D., Zur topologischen Algebra I, *Math. Ann.*, 107 (1932), 587–626.

[113] VIÈTE, F., FRANCISCI VIETÆ, *Opera matematica, in unum Volumen congesta ac recognita, Opera atque studio Francisci a Schooten Leydensis*, Lugduni Batavorum, 1646.

[114] WALLIS, J., *Tractatus de sectionibus conicis*, 1655.

[115] WEHAUSEN, J. V., Transformations in linear topological spaces, *Duke Math. J.*, 4 (1938), 157–169.

[116] WEIL, A., *Sur les espaces à structure uniforme et sur la topologie générale*, Hermann, 1937.

[117] WIDMANN, J., *Mercantile Arithmetic* (or *Behende und hubsche Rechenung auff allen Kauffmanschafft*), Leipzig, 1489.

[118] ZORN, M., A remark on method in transfinite algebra, *Bull. Amer. Mat. Soc.*, 41 (1935), 667–670.

Cited Authors

1 For detailed bibliographic records: `http://www-history.mcs.st-and.ac.uk`

Index

U

V, W, Z

Other titles from

iSTE

in

Mathematics and Statistics

2017

HARLAMOV Boris
Stochastic Analysis of Risk and Management (Stochastic Models in Survival Analysis and Reliability Set – Volume 2)

2016

CELANT Giorgio, BRONIATOWSKI Michel
Interpolation and Extrapolation Optimal Designs 1

CHIASSERINI Carla Fabiana, GRIBAUDO Marco, MANINI Daniele
Analytical Modeling of Wireless Communication Systems (Stochastic Models in Computer Science and Telecommunication Networks Set – Volume 1)

GOUDON Thierry
Mathematics for Modeling and Scientific Computing

KAHLE Waltraud, MERCIER Sophie, PAROISSIN Christian
Degradation Processes in Reliability
(Mathematial Models and Methods in Reliability Set – Volume 3)

KERN Michel
Numerical Methods for Inverse Problems

RYKOV Vladimir
Reliability of Engineering Systems and Technological Risks
(Stochastic Models in Survival Analysis and Reliability Set – Volume 1)

2015

DE SAPORTA Benoîte, DUFOUR François, ZHANG Huilong
Numerical Methods for Simulation and Optimization of Piecewise Deterministic Markov Processes

DEVOLDER Pierre, JANSSEN Jacques, MANCA Raimondo
Basic Stochastic Processes

LE GAT Yves
Recurrent Event Modeling Based on the Yule Process
(Mathematical Models and Methods in Reliability Set – Volume 2)

2014

COOKE Roger M., NIEBOER Daan, MISIEWICZ Jolanta
Fat-tailed Distributions: Data, Diagnostics and Dependence
(Mathematical Models and Methods in Reliability Set – Volume 1)

MACKEVIČIUS Vigirdas
Integral and Measure: From Rather Simple to Rather Complex

PASCHOS Vangelis Th
Combinatorial Optimization – 3-volume series – 2nd edition
Concepts of Combinatorial Optimization / Concepts and Fundamentals – volume 1
Paradigms of Combinatorial Optimization – volume 2
Applications of Combinatorial Optimization – volume 3

2013

COUALLIER Vincent, GERVILLE-RÉACHE Léo, HUBER Catherine, LIMNIOS Nikolaos, MESBAH Mounir
Statistical Models and Methods for Reliability and Survival Analysis

JANSSEN Jacques, MANCA Oronzio, MANCA Raimondo
Applied Diffusion Processes from Engineering to Finance

SERICOLA Bruno
Markov Chains: Theory, Algorithms and Applications

2012

BOSQ Denis
Mathematical Statistics and Stochastic Processes

CHRISTENSEN Karl Bang, KREINER Svend, MESBAH Mounir
Rasch Models in Health

DEVOLDER Pierre, JANSSEN Jacques, MANCA Raimondo
Stochastic Methods for Pension Funds

2011

MACKEVIČIUS Vigirdas
Introduction to Stochastic Analysis: Integrals and Differential Equations

MAHJOUB Ridha
Recent Progress in Combinatorial Optimization – ISCO2010

RAYNAUD Hervé, ARROW Kenneth
Managerial Logic

2010

BAGDONAVIČIUS Vilijandas, KRUOPIS Julius, NIKULIN Mikhail
Nonparametric Tests for Censored Data

BAGDONAVIČIUS Vilijandas, KRUOPIS Julius, NIKULIN Mikhail
Nonparametric Tests for Complete Data

IOSIFESCU Marius *et al.*
Introduction to Stochastic Models

VASSILIOU PCG
Discrete-time Asset Pricing Models in Applied Stochastic Finance

2008

ANISIMOV Vladimir
Switching Processes in Queuing Models

FICHE Georges, HÉBUTERNE Gérard
Mathematics for Engineers

HUBER Catherine, LIMNIOS Nikolaos *et al.*
Mathematical Methods in Survival Analysis, Reliability and Quality of Life

JANSSEN Jacques, MANCA Raimondo, VOLPE Ernesto
Mathematical Finance

2007

HARLAMOV Boris
Continuous Semi-Markov Processes

2006

CLERC Maurice
Particle Swarm Optimization

Printed and bound by CPI Group (UK) Ltd, Croydon, CR0 4YY

16/09/2026

14963078-0007